Electrospinning

Electrospinning

Fundamentals, Methods, and Applications

Edited by Liming Wang and Xiaohong Qin

WILEY-VCH

Editors

Prof. Liming Wang
Donghua University
College of Textiles
Songjiang District
200051 Shanghai
China

Prof. Xiaohong Qin
Donghua University
College of Textiles
Songjiang District
200051 Shanghai
China

Cover: © Sander Van den Berg/500px/ Getty Images

Library of Congress Card No.: applied for

British Library Cataloguing-in-Publication Data
A catalogue record for this book is available from the British Library.

Bibliographic information published by the Deutsche Nationalbibliothek
The Deutsche Nationalbibliothek lists this publication in the Deutsche Nationalbibliografie; detailed bibliographic data are available on the Internet at <http://dnb.d-nb.de>.

Print ISBN: 978-3-527-35197-8
ePDF ISBN: 978-3-527-84145-5
ePub ISBN: 978-3-527-84146-2
oBook ISBN: 978-3-527-84147-9

Typesetting Straive, Chennai, India
Printing CPI Group (UK) Ltd, Croydon CR0 4YY

C9783527351978_260124

Contents

Preface

Fiber materials have played an indispensable role in the process of human civilization. About ten thousand years ago, people could use natural fibers such as wool and hemp fibers to cover their bodies and keep warm. Along with the fast development of advanced science and technology, all kinds of man-made and synthetic fibers occurred and have been widely used in our daily lives for the past two hundred years. The diameter of these fibers ranges generally from a few microns to dozens of microns. If the fiber diameters were refined to the micro-/nanoscale (<1000 nm), their application performances would be significantly improved.

Electrospinning, maybe the most effective and facile method to continuously produce micro-/nanofibers from various materials, has drawn wide attention for its powerful capacity. The electrospun fibers demonstrated attractive performances such as ultrahigh surface area and length–diameter ratio. Therefore, electrospinning is a rapidly growing frontier field, and thousands of papers on the topic of electrospinning are published every year. Remarkable progress has been made with regard to the development of electrospinning methods and engineering of electrospun nanofibers to suit or enable various applications. In this up-to-date book, we aim to provide a comprehensive overview of electrospinning, including the principle, newly developed methods, and emerging applications. This book will provide researchers opportunities and new directions for the future development of electrospinning. We believe this book will provide a broad readership to scientists, researchers, and academicians working on the design of electrospinning equipment, production, and application of nanofibers.

From Chapter 1 to Chapter 4, we make an introduction to the history and development of the electrospinning theories and the state-of-the-art methods for fiber structure regulation, mass production of electrospun fibers, and manufacturing of electrospun fiber yarns in depth. Afterward, we highlight the most relevant and recent applications associated with the remarkable features of electrospun fibers, including air filtration, oil–water separation, interfacial solar-driven steam generation, waterproof and breathable fabrics, heat insulation, sound absorption, triboelectric nanogenerators, thermoelectric conversion, water-induced electric generators, flexible sensors, liquid-metal-based stretchable electronics, photocatalysts, smart actuators, and biomedical applications.

This book was co-edited by Prof. Xiaohong Qin, who has been devoted to the research field of electrospinning fundamentals, methods, and applications for more than 20 years. I sincerely appreciate all the authors for their contributions to this book. I am also grateful to the editors of Wiley for their kind invitation and warm support.

March 1, 2023
Shanghai

Liming Wang

1

Electrospinning Theory

Liang Wei

Xi'an Polytechnic University, School of Textile Science and Engineering, Key Laboratory of Functional Textile Material and Product Ministry of Education, Xi'an, Shaanxi, 710048, China

1.1 Nanotechnology and Nanofibers

Nanotechnology is a science and technology that studies the properties and applications of materials with structure sizes ranging from 1 to 100 nm. After the invention of scanning tunneling microscope in 1981, a molecular world with a length of 1 to 100 nm was born. The ultimate goal is to directly construct products with specific functions with atoms or molecules. Therefore, nanotechnology is a kind of technology that uses a single atom or molecule to manufacture material.

Nanotechnology is an interdisciplinary and comprehensive subject. The research content involves the broad field of modern science and technology. Nanoscience and technology mainly include seven relatively independent and mutually penetrating disciplines (nano-system physics, nanochemical, nanomaterials, nanobiology, nanoelectronics, nanoprocessing, and nanomechanics) and three research fields (nanomaterials, nanodevices, and nanoscale detection and characterization). The preparation and research of nanomaterials are the basis of the whole nanotechnology. Among them, nanophysics and nanochemistry are the theoretical basis of nanotechnology, and nanoelectronics is the most important content of nanotechnology.

1.1.1 Development History of Nanotechnology

In 1993, the first International Nanotechnology Conference (INTC) was held in the United States, which divided nanotechnology into six branches: nanophysics, nanobiology, nanochemistry, nanoelectronics, nano-processing technology, and nanometrology, which promoted the development of nanotechnology. Because of the particularity, magic, and universality of this technology, many outstanding scientists from all over the world have been working hard for it.

Nanotechnology generally refers to the technology of materials, design, manufacturing, measurement, control, and products at the nanometer level (0.1–100 nm).

Electrospinning: Fundamentals, Methods, and Applications, First Edition.
Edited by Liming Wang and Xiaohong Qin.

Nanotechnology mainly includes nano-level measurement technology, nano-level surface physical and mechanical properties detection technology, nano-level processing technology, preparation technology of nanoparticles, nanometer material, nanobiological technology, and nanoassembly technology.

In the 1970s, scientists began to put forward ideas about nanotechnology from different perspectives. In 1974, scientist Tony Gucci first used the term nanotechnology to describe precision machining.

In 1981, scientists invented the scanning tunneling microscope, an important tool for nanoresearch, which revealed a visible atomic and molecular world for us and played a positive role in promoting the development of nanotechnology.

In 1990, scientists at IBM's Almaden Research Center successfully rearranged individual atoms, making a key breakthrough in nanotechnology.

In July 1990, the first International Conference on nanoscience and technology was held in Baltimore, the United States, marking the official birth of nanoscience and technology.

In 1997, American scientists successfully used single electrons to move single electrons for the first time. Using this technology, it is expected that in 20 years, a quantum computer with a speed and a storage capacity that is thousands of times higher than that of today will be successfully developed.

In 1999, scientists from Brazil and the United States invented the world's smallest "scale" when conducting carbon nanotube experiments. It can weigh an object one billionth of a gram, which is equivalent to the weight of a virus. Soon after that, German scientists developed a scale capable of weighing individual atoms, breaking the record set jointly by American and Brazilian scientists.

By 1999, nanotechnology gradually entered the market, and the annual turnover based on nanotechnology products reached US$50 billion. Some countries have formulated relevant strategies or plans, and invested huge sums to seize the strategic heights of nanotechnology.

At present, the research and application of nanotechnology are mainly in materials and preparation, microelectronics and computer technology, medicine and health, aerospace and aviation, environment and energy, biotechnology, and agricultural products. The equipment made of nanomaterials has lighter weight, stronger hardness, longer service life, lower maintenance cost, and more convenient design. Nanomaterials can also be used to make materials with specific properties or materials that do not exist in nature, and to make biomaterials and biomimetic materials.

1.1.2 Introduction to Nanofibers

Nanofiber: refers to a linear material with a certain length-to-diameter ratio with a diameter of nanoscale and a large length. In addition, the fiber modified by filling nanoparticles into ordinary fibers is also called nanofiber. In a narrow sense, the diameter of nanofibers is between 1 and 100 nm, but in a broad sense, fibers with a fiber diameter of less than 1000 nm are called nanofibers.

At present, the thinnest nanofibers are single-carbon atom chains. Chinese scientists have been able to produce carbon nanotubes with a diameter of less than 0.4 nm, which is the leading level of the world. This kind of carbon nanotube is known as the king of nanomaterials. The reason is that this material, which is so thin that it is difficult to be observed by ordinary instruments, has magical abilities: super high strength, super flexibility, and strange magnetism. Due to the short spacing of carbon atoms and small diameter in carbon nanotubes, the fiber structure is not prone to defects. Its strength is 100 times that of steel, and its density is only 1/6 of that of steel, 200 times that of ordinary fibers. The rope made of carbon nanotubes can be pulled from the earth to the moon without being broken by its own weight; it has strange conductivity. Carbon nanotubes have both metal and semiconductor conductivity. Even different parts of a nanotube can show different conductivity due to structural changes. Using it to make rectifying tubes can replace silicon chips, which will cause great changes in electronics and make computers very small; nanodevices made of carbon nanotubes can be used to assemble nanorobots, mosquito planes, and tanks, etc. Carbon nanotubes can be used as hydrogen storage materials to develop hydrogen into clean energy for human services. In addition, carbon nanotubes can also be used as stealth materials, catalyst carriers, and electrode materials. Nanofibers can support the arrangement of "nanomachines" and connect the integrated and arranged "nanomachines" into a large-scale system.

1.1.3 Main Characteristics of Nanofibers

1. The smaller the size of surface effect particles, the larger the surface area. Because the surface particles lack the coordination of adjacent atoms, the surface energy becomes extremely unstable, and it is easy to combine with other atoms, showing strong activity.
2. Small size effect: when the size of particles is small enough to be close to or smaller than the wavelength of light wave, the de Broglie wavelength of conducting electrons, and the coherence length transmission depth of superconducting state, the periodic boundary conditions will be destroyed, and the acoustic, optical, electromagnetic, and thermodynamic properties of particles will be changed, such as melting point reduction, color separation and discoloration, ultraviolet absorption, and shielding electromagnetic wave.
3. Quantum size effect: when the particle size is small enough, the electronic energy level near the Fermi level changes from quasi-continuous to discrete. At this time, the substance that was originally a conductor may become an insulator, and conversely, the insulator may become a superconductor.
4. The macro-quantum Yang tunnel effect tunnel effect refers to small particles that can pass through objects under certain conditions, just like there is a tunnel inside.

1.2 Research History of Electrospinning

Electrospinning method is to apply electric field force to polymer solution or melt. Under the action of electrostatic field force, polymer jet is ejected and stretched into nanofiber in the process of movement. Electrospinning equipment mainly includes

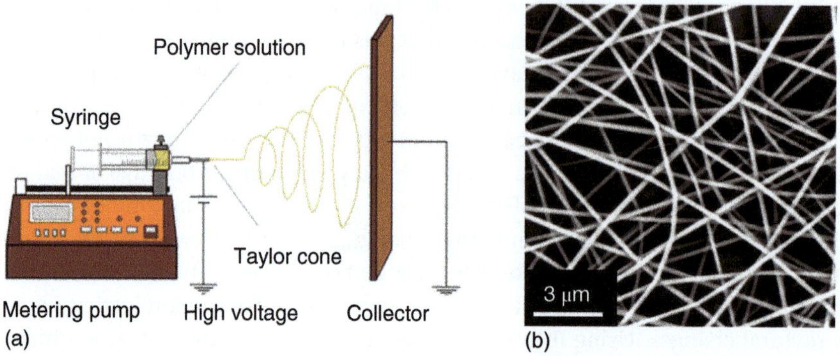

Figure 1.1 (a) Schematic of the electrospinning device. Source: Riazi et al. [1]. Reproduced with permission of Elsevier. (b) SEM image of nanofibers. Source: Xian Wen.

spinneret and solution supply device, high-voltage generator, and fiber collection device (Figure 1.1a). Apply high-voltage static electricity to the polymer solution or melt, and the charged droplets will deform under the action of electric field force and form Taylor cone at the top end of the spinneret. When the applied electric field force exceeds the surface tension of the solution, the polymer jet is ejected from the top of the Taylor cone. Under the action of electric field force and viscoelastic force of solution, the charged jet first moves along a straight line and then moves to a certain position. When it is disturbed by the outside world, it bends and deforms in a spiral curve. The jet is stretched and refined in the process of movement, and finally falls on the receiving device to form polymer nanofibers [2] (Figure 1.1b).

A brief history of the development of electrospinning and the number of publications of electrospinning in recent 2 decades are summarized in Figure 1.2. The predecessors invented the electrospinning technology on the basis of studying the water droplet morphology under the action of electric field force.

As early as 1600, William Gilbert [4] found that for a spherical water drop on a dry surface, a friction amber can be used to stretch it at an appropriate place to make the spherical water drop taper.

In 1882, Raleigh [5] theoretically studied the stability of independent droplets under the action of electrostatic field force. If the electric field force continues to increase, these droplets will become unstable; especially when the stable effect of charge and surface tension is broken, the droplets will split.

In 1899, Rayleigh [6] again studied the stability of liquid jet with and without electric field force, and proposed a linear theory of the stability of charged droplets. The theory shows that the spherical particle with uniform charge will keep its stable shape until the total charge on its surface exceeds the following critical value.

$$Q = (64\pi^2 \varepsilon_0 \gamma R)^{1/2}$$

where

Q is the total charge number of spherical droplets,
ε_0 is the vacuum dielectric constant,
γ is the surface tension of liquid, and
R is the radius of the droplet.

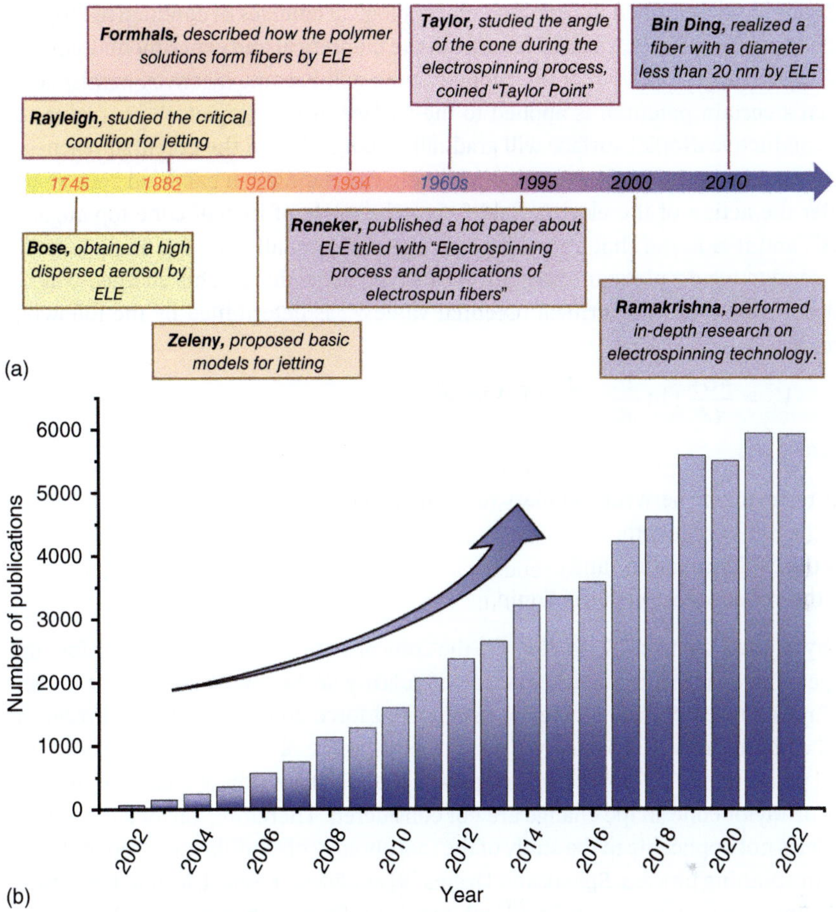

Figure 1.2 (a) Brief history and the development of electrospinning. Source: Shi et al. [3]. Reproduced with permission of John Wiley and Sons. (b) Number of publications in recent 2 decades (Data obtained from Web of Science by searching the keyword of "electrospinning"). Source: Xian Wen.

In 1917, Zeleny [7] studied experimentally the splitting phenomenon of the droplet at the end of the capillary and observed several different jet patterns. He also pointed out that the decomposition of the droplet was due to the instability of the droplet, not the formation of the surface charge of the droplet, and the critical voltage for the formation of the instability was:

$$V^2/r_0T = C$$

where

V is the applied voltage,
r_0 is the radius of liquid drop,
T is the surface tension of solution, and
C is the constant.

Taylor [4, 8] described in detail the morphological changes of conductive droplets under the action of electric field force from the theoretical and experimental aspects. The liquid drop at the top of the capillary tube will become a convex hemisphere. When a certain potential is applied to the surface of the droplet, the curvature of the conductive droplet surface will gradually change. When the potential reaches a certain critical value V_c, the conductive liquid can maintain a balanced cone shape under the action of the electric field force. The angle of its half cone top angle is 49.3°, and it is found that only when the angle of the half cone top angle is 49.3°, the conductive droplet can maintain a balanced cone shape. This charged cone is called Taylor cone. The critical potential value V_c is determined by the following formula:

$$V_c^2 = \frac{4H^2}{L^2}\left(\ln\frac{2L}{R} - \frac{3}{2}\right)(0.117\pi\gamma R)$$

where

H is the distance between capillary tube and receiving device,
L is the capillary length,
R is the capillary radius, and
γ is the surface tension of the liquid.

Taylor used the formula to describe the conical state of the liquid drop before the jet generated. There is a mathematical singularity at the top of the cone of the jet. In Taylor's research, only the electrostatic field force and surface tension received by the liquid drop are considered, while the influence of external electric field and fluid pressure on the Taylor cone shape and the redistribution of charge in the process of Taylor cone shape change are not considered. Therefore, Taylor's theoretical model is not applicable to the study of the morphology of equilibrium droplets in the electrospinning process. Spivak and Dzenis [9] analyzed the electric field force, fluid pressure, surface tension, and coulomb repulsion force on polymer droplets during electrospinning. It is assumed that the top of Taylor cone is a parabola shape, and the conditions required for the meniscus droplet to maintain its equilibrium state are studied, namely

$$\sigma \geq 0.447378 * \delta\emptyset\varepsilon_0 E$$

among them

σ is the surface tension coefficient of solution,
$\emptyset$ is the electrostatic field energy,
δ is the charge density carried by the droplet,
ε_0 is the medium permeability of vacuum, and
E is the applied voltage.

This formula also determines a maximum voltage to maintain the equilibrium state of the meniscus droplet, that is, the state of the meniscus droplet is stable only when the applied voltage is less than a certain value, which is determined by the medium permeability, surface tension, and the charge density carried by the jet. This assumption can avoid the mathematical singularity of the cone top.

Yarin et al. [10] found that for a specific self-similar solution, it will form a cone shape similar to the Taylor cone under the electric field force, while for a nonself similar solution, it will form droplets of other shapes under the electric field force. Yarin et al. assumed that the shape of Taylor cone top of nonself similar solution is hyperbolic and calculated the equipotential line with high curvature that is close to the cone top. The cone top angle calculated by hyperboloid shape is very small, and the semi-cone top angle deduced theoretically is 33.5°. This theory applies to non-viscous Newtonian fluid and viscoelastic fluid after stress relaxation. For non-viscous Newtonian fluids, this specific half-cone apex angle does not depend on the performance of the fluid, but on the viscoelasticity and surface tension of the solution.

Liu et al. [11, 12] established the mechanism of bubble electrospinning under the action of electric field force by using Maxwell's current mechanics equation, charge conservation equation, and fluid constitutive equation. In the process of bubble electrospinning, the bubble is subject to the applied electric field force, the surface tension of the solution, and the internal and external pressures of the bubble. The bubble gradually increases and changes from a circle to a cone. When the applied voltage exceeds the critical voltage, multiple jets are ejected from the top of the bubble cone. Liu approximates the bubble in two-dimensional state as a parabola and theoretically analyzes that the critical voltage for jet formation is:

$$E_c = \frac{k\gamma\sqrt{1+(\tan\overline{\alpha})^2}}{q\alpha_0\sqrt{1+4(\tan\overline{\alpha})^2}\tan\overline{\alpha}}$$

where

E_c is the critical electric field strength,
q is the electric quantity per unit area,
γ is the surface tension of solution, and
α is the angle between the tangent line of any point on the bubble surface and the x-axis.

Therefore, in bubble electrospinning, the critical voltage of jet formation is related to the charge density of the bubble surface, the surface tension of the solution, the height of the bubble, and the radius of the bottom. When the height of the bubble is very low, that is, close to 0, the radius of curvature of the bubble is infinite, the liquid surface is approximately flat, and an infinite voltage is required for jet ejection, so electrostatic spinning cannot be carried out. In addition, Liu et al. [13, 14] also measured by experimental method that the critical spinning half cone apex angle formed by jet is 23.4°.

In the process of electrospinning, the change of Taylor cone shape and the formation of jet are the basis of electrospun nanofibers, and the surface tension of solution has a great influence on the change of Taylor cone shape and the formation of jet. Therefore, researchers, based on the change of Taylor cone shape under different surface tension conditions, further studied the critical voltage of jet formation under different surface tension conditions, and established a theoretical model between the surface tension of solution and the critical voltage between jet formation.

During electrospinning, after leaving the top of Taylor cone, the jet is accelerated due to the coulomb repulsion force. The coulomb repulsion force generated by the charge carried by the front part of the jet pushes the jet to move toward the receiver. Even though there is radial disturbance in practice, the liquid jet still maintains a straight line motion, which is called the stable section jet. Spivak et al. [15] proposed a model of steady-state jet in the electrospinning process. They assumed that the jet flow was axisymmetric along the direction of the external magnetic field, simplified the equation into a one-dimensional problem, and selected the dimensionless Weber number (Weber number) N_W, Reynolds number N_R, and Euler number (Euler number) N_E. The reciprocal of E takes the potential at a point in the jet as the sum of the potential generated by surface charges and external charges. The differential equation of jet radius R with respect to coordinate axis Z is obtained by solving one-dimensional differential equation:

$$\frac{d}{dZ}\left[R^{-4} + (N_W R)^{-1} - N_E{}^{-1}R^2 - N_R{}^{-1}\left(\frac{dR^{-2}}{dZ}\right)^m\right] = 1$$

where

$N_W = \dfrac{\rho Q^2}{2\pi^2 R_0{}^3 \sigma}$, which is the ratio function of inertial force and solution surface tension,

$N_E = \dfrac{4\varepsilon_0 \rho Q^4}{\pi^2 J^2 R_0{}^6}$, which is the ratio function of inertial force and pressure generated by electric field force,

$N_R = \dfrac{Q^2\rho}{2\pi^2 R_0{}^4 \mu}\left[\dfrac{6\pi EJR_0{}^2}{Q^2\rho}\right]^{-m}$, which is the ratio function of inertial force and viscous resistance of solution.

Holman et al. [16, 17], using the concept of leakage medium model, considered the free charge when establishing the jet dynamics model, established a slender body theory involving electric field force, charge transport and jet stretching, and further elaborated the influence of surrounding electric fields on the jet surface charge through the model. Taking the solution as Newtonian viscous fluid, the stretching model of electrospun jet was established and the stability mechanism of the jet in the stable section was studied. The stability of a charged jet in a tangential electric field with all fluid parameters is comprehensively analyzed by coupling the stretching of the jet, the transfer of polarized charges, and the external electric field. The axisymmetric instability and whip instability of the charged jet are further analyzed. However, he used the Newtonian fluid constitutive equation, so he could not predict the effect of solution viscoelasticity on the stability of the jet.

Feng [18] improved and extended the slender body model of Hohman, and adopted Giesekus viscoelasticity model, the solution of the model is extended to non-newtonian fluid, and a one-dimensional non-newtonian fluid stretching model is established. He also studied the changes of charge density and applied electric field intensity on the jet surface with the axial distance of the jet. The equation of the applied electric field intensity changing with the axial distance of the jet is:

$$E(z) = E\infty(z) - \left[\frac{1}{\varepsilon_0}(\delta R)' - \left(\frac{\varepsilon}{\varepsilon_0} - 1\right)\frac{(ER^2)''}{2}\right]\ln\left(\frac{d}{R_0}\right)$$

where

E is the electric field strength,
Δ is the charge density,
R_0 is the initial jet diameter,
D is the spinning distance, and
E_∞ is the electric field strength.

Carroll et al. [19, 20], based on Feng's slender body theory, modeled and analyzed the stretching of isothermal polymer melt jet, introduced Oldroyd-B and FENE-P models related to fluid momentum equation and Gaussian law, and established the momentum equation of jet stretching:

$$\rho vv' = \rho g + \frac{F_T'}{\pi R^2} + \frac{\gamma R'}{R^2} + \frac{\sigma\sigma'}{\varepsilon_0} + (\varepsilon - \varepsilon_0)EE' + \frac{2\sigma E}{R}$$

where

F_T is the tensile force,
E is the electric field strength,
Σ is the charge density,
R is the initial jet diameter, and
γ is the surface tension.

Wan et al. [21, 22] improved Spivak's model by adding charges due to strong electric fields. The three improved conservation equations for the influence of polarization on electrospinning are as follows:

Charge conservation equation:

$$\frac{\partial q_e}{\partial t} + \nabla \cdot j = 0$$

Force balance equation:

$$\rho\frac{du}{dt} = \nabla \cdot t + \rho f + q_e E + (\nabla E) \cdot p$$

Energy conservation equation:

$$\rho c_p \frac{dT}{dt} = Q_h + \nabla \cdot q + J \cdot E + E \cdot \frac{dp}{dt}$$

where

$\nabla \cdot t$ is the viscous resistance,
Pf is the gravity,
$q_e E$ is the electric field force, and
$(\nabla E) \cdot p$ is the polarization force.

He et al. [23] studied the critical length and critical radius of the electrospun straight section jet according to the force on the jet in the stable section during the electrospinning process, using the mass conservation equation, the charge conservation equation, the N–S momentum conservation equation, and the famous Cauchy inequality:

$$L = \frac{4kQ^3}{\pi\rho^2 I^2}\left(R_0^{-2} - r_0^{-2}\right)$$

$$R_c = \frac{r_0}{\sqrt{\beta r_0^2 L + 1}}$$

where

$$R_0 = \left(\frac{2\sigma Q}{\pi k\rho E}\right)^{1/3}$$

L is the critical length of straight section jet and
R_c is the critical radius of jet in straight section.

1.3 Development Prospect of Electrospinning

1.3.1 Application of Electrostatic Spinning Technology

With the development of nanotechnology, electrospinning, as a simple and effective new processing technology that can produce nanofibers, will play a huge role in biomedical materials, filtration and protection, catalysis, energy, photoelectric, food engineering, cosmetics, and other fields (Figure 1.3)[24].

1. In the biomedical field, the diameter of nanofibers is smaller than cells, which can simulate the structure and biological function of natural extracellular matrix; most human tissues and organs are similar to nanofibers in form and structure, which makes it possible for nanofibers to be used for tissue and organ repair; some electrospun materials have good biocompatibility and degradability, which can be used as carriers to enter the human body and be easily absorbed; in addition, electrospun nanofibers also have large specific surface area, porosity, and other excellent characteristics. Therefore, they have attracted continuous attention of researchers in the biomedical field, and have been well applied in drug-controlled release, wound repair, biological tissue engineering, and other aspects.
2. The filtration efficiency of fiber filter materials will increase with the decrease of fiber diameter. Therefore, reducing the fiber diameter becomes an effective method to improve the filtration performance of fiber filter materials. In addition to small diameter, electrospun fibers also have the advantages of small pore size, high porosity, and good fiber uniformity, which makes them show great application potential in the fields of gas filtration, liquid filtration, and personal protection.
3. Electrospun fiber can effectively control the fine structure of fiber. Combined with materials with low surface energy, it can obtain materials with super

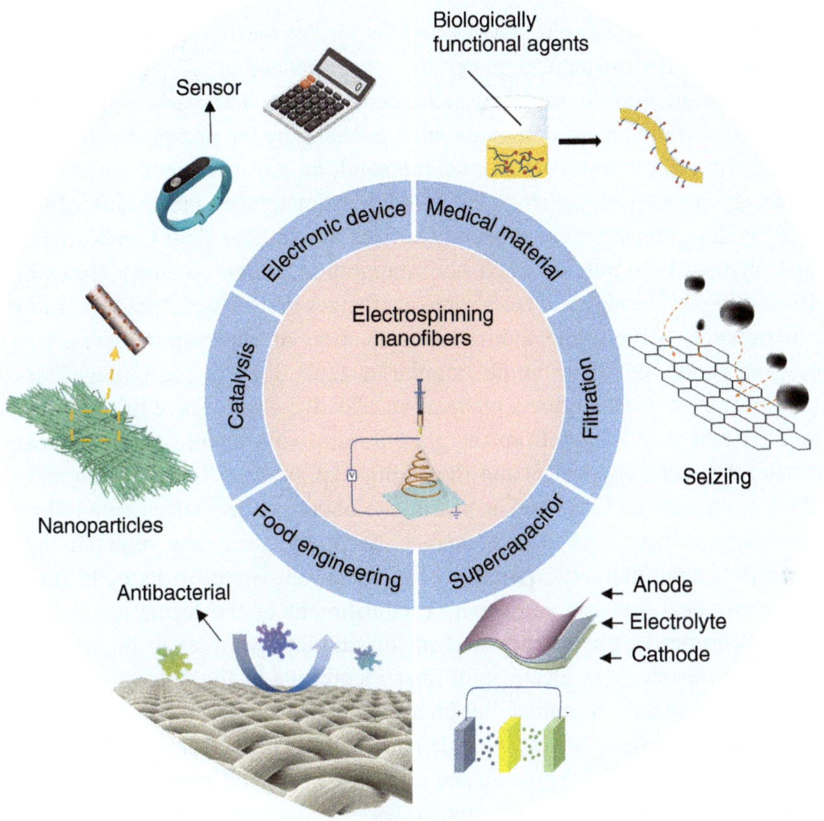

Figure 1.3 Applications of electrospinning nanofibers. Source: Xian Wen.

hydrophobic properties, and is expected to be applied to the shell of ships, the inner wall of oil pipelines, high-rise glass, automotive glass, etc. However, in order to realize the application of electrospun fiber materials in the above self-cleaning field, it is necessary to improve its strength, abrasion resistance, and the combination fastness of fiber membrane materials and matrix materials.

4. Catalyst particles with nanostructures are easy to agglomerate, thus affecting their dispersibility and utilization. Therefore, electrospun fiber materials can be used as templates to uniformly disperse, and also play a role in flexibility and operability of polymer carriers. In addition, strong synergistic effects can be generated by the composite of catalytic materials and polymer micro nanosize surfaces to improve catalytic efficiency.
5. Electrospun nanofibers have high specific surface area and porosity, which can increase the interaction area between the sensing material and the detected object, and are expected to greatly improve the sensor performance. In addition, electrospun nanofibers can also be used in energy, photoelectric, food engineering, and other fields.

1.3.2 Development Direction of Electrospinning Technology

Electrospinning technology has played a very important role in the field of building one-dimensional nanostructural materials. The application of electrospinning technology has successfully prepared nanofiber materials with various structures. Through different preparation methods, such as changing the nozzle structure and controlling the experimental conditions, the solid, hollow, core shell structure of microfiber or spider web structure of two-dimensional fiber membrane can be obtained; by designing different collecting devices, single-fiber, fiber bundle, highly oriented fiber, or randomly oriented fiber membrane can be obtained. However, electrospinning technology still faces some challenges in fiber structure regulation: first, in order to achieve the industrial application of electrospun fibers, it is necessary to obtain nanofiber bundles similar to staple fibers or continuous fibers. The preparation of oriented fibers provides an effective way to solve this problem, but there is still a lot of gap from the goal. Future work should try to improve the nozzle, the receiving device, and the addition of auxiliary electrodes to make the fibers as straight and oriented as possible to obtain an array of oriented fibers with excellent comprehensive performance. Secondly, as a new research field of electrospun nanofibers, the research of nanocobweb is still in its early stage, and the theoretical analysis and model establishment of the formation process of nanocobweb need to be further studied. In addition, in order to improve the application performance of electrospun fiber membrane in the field of ultra fine filtration, it is necessary to reduce the fiber diameter. How to reduce the average fiber diameter to less than 20 nm is a challenge for electrospun technology. In order to improve the application performance of fibers in sensor, catalyst, and other fields, it is an effective method to improve the specific surface area of fibers by preparing nanofibers with porous or hollow structures, but further research is still needed.

References

1 Riazi, K., Kubel, J., Abbasi, M. et al. (2016). Polystyrene comb architectures as model systems for the optimized solution electrospinning of branched polymers. *Polymer* 104: 240–250.

2 Reneker, D.H. and Chun, I. (1996). Nanometre diameter fibres of polymer, produced by electrospinning. *Nanotechnology* 7 (3): 216–223.

3 Shi, S., Si, Y.F., Han, Y.T. et al. (2022). Recent progress in protective membranes fabricated via electrospinning: advanced materials, biomimetic structures, and functional applications. *Advanced Materials* 34 (17): 2107938.

4 Gilbert, W. (1958). *De Magnete*. New York: Courier.

5 Rayleigh, L. (1882). On the equilibrium of liquid conducting masses charged with electricity. *Philosophical Magazine A* 14 (5): 184–186.

6 Rayleigh, L. (1899). Investigations in capillarity: the size of drops. *Philosophical Magazine A* 48: 321–337.

7 Zeleny, J. (1917). Instability of electrified liquid surfaces. *Physical Review* 10 (1): 1–6.

8 Taylor, G. (1964). Disintegration of water drops in an electric field. *Proceedings of the Royal Society of London A: Mathematical, Physical and Engineering Sciences* 280: 383–397.

9 Spivak, A.F. and Dzenis, Y.A. (1999). A condition of the existence of a conductive liquid meniscus in an external electric field. *Journal of Applied Mechanics* 66 (4): 1026–1028.

10 Yarin, A.L., Koombhongse, S., and Reneker, D.H. (2001). Taylor cone and jetting from liquid droplets in electrospinning of nanofibers. *Journal of Applied Physics* 90 (9): 4836–4846.

11 Liu, Y., Ren, Z.F., and Wang, S.Q. (2010). Theoretical analysis and experimental verification of the effect of surface tension on morphology of nanofibers in bubble electrospinning. *International Journal of Nonlinear Sciences and Numerical Simulation* 11 (8): 625–630.

12 Liu, Y., He, J.H., and Yu, J.Y. (2008). Bubble-electrospinning: a novel method for making nanofibers. *Journal of Physics: Conference Series* 96: 012001.

13 Liu, Y., He, J.H., Xu, L. et al. (2008). The principle of bubble electrospinning and its experimental verification. *Journal of Polymer Engineering* 28: 55–66.

14 Yong, L. (2008). *Research on Bubble Electrospinning Technology and Its Mechanism*. Shanghai: Donghua University.

15 Spivak, A.F., Dzenis, Y.A., and Reneker, D.H. (2000). A model of steady state jet in the electrospinning process. *Mechanics Research Communications* 27 (1): 37–42.

16 Hohman, M.M., Shin, M., Rutledge, G. et al. (2001). Electrospinning and electrically forced jets I. Stability theory. *Physics of Fluids* 13 (8): 2201–2220.

17 Hohman, M.M., Shin, M., Rutledge, G. et al. (2001). Electrospinning and electrically forced jets II. Applications. *Physics of Fluids* 13 (8): 2221–2236.

18 Feng, J.J. (2003). Stretching of a straight electrically charged viscoelastic jet. *Journal of Non-Newtonian Fluid Mechanics* 116 (1): 55–70.

19 Carroll, C.P. and Joo, Y.L. (2009). Axisymmetric instabilities in electrospinning of highly conducting, viscoelastic polymer solutions. *Physics of Fluids* 21 (10): 103101.

20 Carroll, C.P. and Joo, Y.L. (2008). Axisymmetric instabilities of electrically driven viscoelastic jets. *Journal of Non-Newtonian Fluid Mechanics* 153 (2): 130–148.

21 Wan, Y.Q., He, J.H., and Yu, J.Y. (2006). Experimental verification of scaling law between current and applied voltage in electrospinning. *Iranian Polymer Journal* 15 (3): 265–268.

22 Yuqin, W. (2006). *Research on Electrospinning Process Behavior and Vibration Electrospinning Technology*. Shanghai: Donghua University.

23 He, J.H., Wu, Y., and Zuo, W.W. (2005). Critical length of straight jet in electrospinning. *Polymer* 46 (26): 12637–12640.

24 Wen, X., Xiong, J., Lei, S.L. et al. (2021). Diameter refinement of electrospun nanofibers: from mechanism, strategies to applications. *Advanced Fiber Materials* 4 (2): 145–161.

2

Regulation of Electrospun Fiber Structure

Jiatai Gu and Liming Wang

Donghua University, Key Laboratory of Textile Science & Technology, Ministry of Education, College of Textiles, 2999 Renmin North Road, Songjiang District, Shanghai 201620, China

2.1 Introduction

Electrospinning is a promising technology for the preparation of micro-nanofibers because of its remarkable advantages regarding the simple preparation process, controllable fiber structure, and diversity of polymer raw materials. Electrospinning has made significant progress in fabricating micro-nanofibers with well-controlled structures after nearly two decades of development. Spindles, porous, hollow, core–shell, and other fiber structures have been developed through electrospinning and have shown a great application prospect in medical, textile, water treatment, air filtration, and other fields [1–5].

In electrospinning, the main influence factors over fiber structure include the properties of spinning solutions, spinning process parameters, nozzle structure, and receiving device [6–8]. The solution properties mainly include the solution viscosity, solution conductivity, and surface tension [9–11]. The spinning process parameters involve voltage, feed rate, spinning distance, and temperature and humidity environment [12–14]. The shell–core, hollow structure, and multi-component composite structure fibers can be prepared by designing the nozzle structures [15–17]. In addition, the receiving device significantly influences the volatilization of solvent and the discharge of charge loaded on fibers, which in turn affects the structure of electrospun fibers [18, 19].

In this chapter, the influence of four important components of electrospinning including spinning solutions, spinning process parameters, nozzle structure, and receiving device (collector) on fiber structure is systematically discussed.

2.2 Solution Properties

In the electrospinning process, the electric force overcomes the surface tension of the polymer solutions to create the polymer jets under the electrostatic field. This jet

Electrospinning: Fundamentals, Methods, and Applications, First Edition.
Edited by Liming Wang and Xiaohong Qin.

is split into multiple jets under electrical force and surface tension. These jets are accompanied by the volatilization of solvent in the travel process and finally solidify on the receiving substrate to form fibers [20, 21]. So, the solution viscosity, conductivity, and surface tension of the solution are the main factors in the formation of electrospun fibers. In addition, based on the phase separation effect, the volatility of the solvent affects the fiber structure [22].

2.2.1 Concentration of Polymer Solution

The solution concentration plays an essential role in the electrospinning process because the solution concentration affects the viscosity and surface tension of the solution. Generally, high concentration polymer solution brings about a high solution viscosity, inhibiting the solution flow and the drafting effect of electrostatic force on the solution jet. At low concentrations, beads were formed due to the capillary breakup of solution jets driven by the surface tension and the low solution viscosity [23]. Therefore, the proper concentration can be determined with the surface tension and viscosity of the spun solution. Deitzel et al. evaluated the effects of poly(ethylene oxide) (PEO) spun solution concentration on the morphology of the fibers [6]. The results show that the mixture of fibers and beads was formed at concentrations below 4 wt% and spun was inhibited due to the high solution viscosity.at concentrations higher than 10 wt%. They believed that if all other parameters remain unchanged, the solution viscosity and surface tension will determine the upper and lower boundaries of the processing window.

Lee et al. prepared polystyrene (PS) fibers by electrospinning and investigated the relationship between the bead morphology and polymer concentrations [24]. The bead structure was produced in obtained PS fibers at below 15 wt% concentration, and the aspect ratio of the bead increased with the increased solution concentration. The bead structure disappeared due to the high solution viscosity when the concentration was higher than 15%.

2.2.2 Molecular Weight

At the same solution concentration, the molecular weight of the polymer has an important influence on the viscosity, surface tension, conductivity, and other properties of the solution [25, 26]. Therefore, the molecular weight of the polymer would further affect the electrospun fiber structure. Koski et al. investigated the transformation of fibrous structures with molecular weight by electrospinning polyvinyl alcohol (PVA) [27]. At the same solution concentration, a bead-on-string structure could be obtained when the PVA molecular weight is 9000–13 000 g mol^{-1}. The uniform fibrous structure was formed as the molecular weight increased to 13 000–23 000 g mol^{-1}. Furthermore, the molecular weight of 31 000–50 000 g mol^{-1} shows a flat fiber structure. Meanwhile, with the increase in molecular weight, the corresponding fiber diameter also shows a significant increase.

2.2.3 Solution Conductivity

As previously described, electrospinning is the result of the interaction between electrostatic force and solution surface tension. Therefore, the electrospun fiber structure can be effectively adjusted by optimizing electrostatic force. The charge density of the electrospun jet reflects the strength of the electrostatic force [28]. Increasing the conductivity of the solution or enhancing the electric field strength can enhance the charge density on the surface of the electrospun jet. At present, much work has been done to improve the conductivity of the solution by adding organic/inorganic salts or organic/inorganic conductive materials to the solution or to optimize the electric field strength by increasing the spinning voltage or introducing an external electric field, so as to achieve the control of electrostatic force, and then adjust the electrospun fiber structure [29, 30].

Gu et al. adjusted the polyurethane (PU) fibrous structure by adding NaCl to the PU electrospinning solutions [31]. As shown in Figure 2.1, the conductivity of PU solutions increased gradually with the increase of NaCl concentrations. The diameter of PU fibers added with NaCl was significantly smaller than that of pristine PU fibers. They believed that the increase in the conductivity of the solution could stimulate the stretching effect of the jet and intensify the whipping effect of the spinning jet.

In addition, a novel two-dimensional (2D) nano-nets structure has also been developed in recent years by the electro-netting process. 2D nano-nets were first proposed by Ding et al. in 2006 [32]. They believe that the formation of such nano-nets is due to the rapid phase separation of charged droplets in the electrospinning process. On this basis, Wang et al. further optimized the electro-netting process by

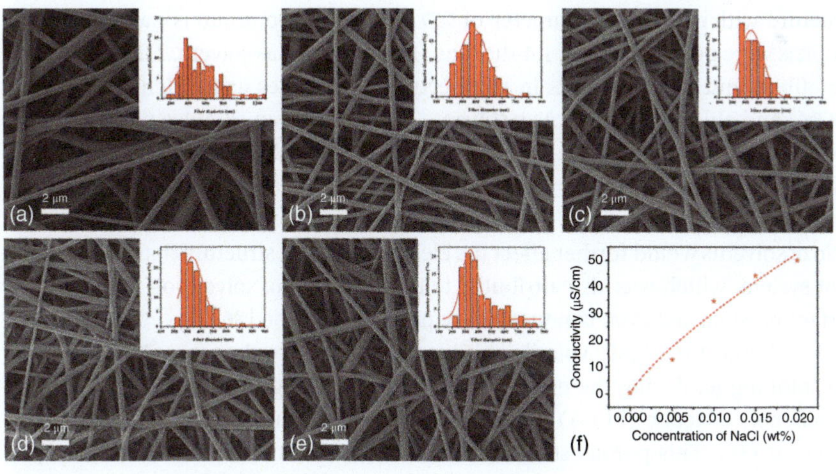

Figure 2.1 FE-SEM images of PU/NaCl fibrous membranes containing various NaCl contents: (a) 0, (b) 0.005, (c) 0.010, (d) 0.015, and (e) 0.020 wt%. (f) Conductivity of PU/NaCl solutions containing different concentrations of NaCl. The insets in (a, b, c, d, e) show the fiber diameter distribution relatively. Source: Gu et al. [31]/with permission of Elsevier.

adding a certain amount of sodium chloride (NaCl) into the spinning solution [33]. Polyacrylic acid (PAA) nano-net was formed with large quantities and uniform size after adding NaCl. The introduction of NaCl increased the conductivity of the PAA spinning solution, which led to the increase of instability of the Taylor cone during spinning.

Li et al. developed a tree-like polyvinylidene fluoride (PVDF) nanofiber by introducing tetrabutylammonium chloride (TBAC) into a PVDF solution via one-step electrospinning [34]. The dendritic fiber was composed of a coarse trunk and delicate branch fiber. TBAC, as an organic branched chain salt, was introduced into the spinning solution to improve the conductivity of the solution significantly. As a result, the straight section of PVDF became short, and the bending disturbance increased during the electrospinning process. When the charge density at the bend was higher than a certain threshold, the electrostatic force overcame the solution surface tension and caused the jet to split, and then further pulled and refined in the electric field to form branched fibers.

2.2.4 Solvent

The surface tension, solubility parameter, conductivity, boiling point, and other characteristics of solvent have an important effect on the properties of polymer solutions (surface tension, viscosity, and conductivity), so the solvent can effectively control the structure of the electrospun fiber.

Casasola et al. explored the influence of different solvents on the structure of electrospun poly lactic acid (PLA) fibers [35]. The results have shown that the sufficient quantity of nanofibers prepared with acetone as solvent demonstrated bead string structure with an average diameter of 757 nm. The discontinuous nanofibers and droplets were produced using 1,4-dioxane (DX) and THF as solvent. There were few nanofibers and many droplets in the electrospun samples using dichloromethane (DCM) and chloroform (CHL) as solvent. The electrospun samples prepared with DMF and dimethylacetamide (DMAc) as solvent have only droplet structure. These differences in fiber structure are caused by the various solvent properties such as surface tension, conductivity, and boiling point. Accordingly, the composition and ratio of solvents would further affect the electrospun fiber structure in the mixed solvent system, which was also attributed to the influence of solvent on the viscosity, surface tension, and even conductivity of polymer solution [36, 37].

In addition, the highly volatile solvent can also cause the phase separation of the spinning jet during the electrospinning process. Bognitzki et al. prepared the porous poly-L-lactide (PLLA) fibers by electrospinning using dichloromethane as a solvent [38]. This porous structure on the PLLA fiber was caused by the rapid volatilization of solvent and rapid curing of the polymer during the electrospinning (phase separation). In addition, the hole elongated along the fiber axis due to the drafting effect of the jet.

Based on the same phase separation mechanism, Han et al. reported an ultrafine porous cellulose triacetate (CTA) fiber via electrospinning with methylene chloride (MC) [39]. There are many isolated circular holes distributed on the

CTA fibers, which is the result of phase separation induced by rapid evaporation of MC.

Hybrid solvent systems containing highly volatile solvents and low volatile non-solvents are also often used to prepare porous fibers. In the electrospinning process, with the difference between solvent and non-solvent volatilization, the polymer solution appears as phase separation of the polymer-rich and -lean phases. Porous fibers are formed with the further volatilization of solvent and non-solvent. Qi et al. prepared porous poly(L-lactic acid) PLLA fiber by electrospinning using highly volatile DCM as a solvent and low volatile *n*-butanol (BuOH) as a non-solvent [40]. PLLA, dichloromethane, and n-butanol ternary phase diagram were used to analyze the forming process of porous fiber. Because DCM has a lower volatilization temperature than BuOH, the composition of the polymer solution in the electrospinning process entered the two-phase region (PLLA-rich phase and PLLA-poor phase) through the phase boundary in a short time. The further evaporation of DCM and BuOH would cause the polymer to solidify and form porous fibers. In addition, with the decrease in the ratio of non-solvent/solvent, the porous structure on the fiber surface gradually disappeared. The lower the ratio of nonsolvent/solvent, the longer the distance between the initial concentration point and the bimodal deinking curve, which meant that under the same conditions, it is difficult for polymer solution to enter the two-phase zone, thus reducing the probability of phase separation, so the porous structure on the fiber gradually disappeared.

2.3 Spinning Parameters

Spinning voltage, propulsion pump flow rate, needle to receiving plate distance, and temperature and humidity environment are essential electrospinning parameters. The stable spinning of the polymer solution can be achieved by adjusting these parameters. Therefore, it is essential to adjust the spinning voltage, flow rate, distance, and temperature and humidity environment to regulate the electrospun fiber structure.

2.3.1 Voltage

In the electrospinning process, the spinning voltage provides electrostatic force for the drafting of the jet. Much work has been done to regulate the fiber structure by adjusting the spinning voltage. However, there is no uniform conclusion about the structure adjustment of fibers. Banikazemi et al. prepared shape memory polyurethane fibers with different fiber structures by adjusting three different spinning voltages (12, 14.5, and 17 kV) at a fixed flow rate (1.5 ml h^{-1}) and spinning distance (21 cm) [41]. The results show that when the spinning voltage increases from 12 to 17 kV, the fiber diameter decreases first and then increases. Banikazemi et al. believed that there were two competing effects of high voltage on fiber diameter. A high electrostatic force generated by increasing the spinning voltage was more accessible to overcome the surface energy tension of the solution, resulting in a reduction in fiber diameter. In addition, a high electrostatic force may also reduce the movement time of the jet in the air, resulting in insufficient time for polymer molecular chains to be arranged, which further causes an increase in fiber diameter. Differently, Zhang et al. adjusted the diameter of electrospun polyacrylonitrile (PAN) fiber by tuning the spinning voltage [42]. The fiber diameter gradually increased from 495 nm to 760 nm with the spinning voltage improving from 10 kV to 20 kV. Zhang et al. believed that the increase in electrostatic force caused the increase in the mass throughput of the polymer solution, which led to the increase in fiber diameter.

Mo et al. found that the diameter of electrospun poly(L-lactide-*co*-ε-caprolactone) (P(LLA-CL)) fiber tended to decrease with the electrospun voltage increased from 9 kV to 15 kV [43]. They believed that the increased electrospun voltage would help to increase the electrostatic repulsion force on the jet, which would eventually be conducive to refine the fibers.

2.3.2 Spinning Distance

The spinning distance usually refers to the distance between the spinning nozzle and the receiving substrate. The spinning distance affects the electric field strength during electrospinning [44]. Therefore, the spinning distance will affect the draft, motion trajectory, and curing effect of the jet, which will further affect the structure of the spun fiber. Supaphol et al. prepared PVA nanofiber membranes by electrospinning and explored the influence of collection distance on fiber structure [45]. When the collection distance was relatively close, the insufficient jet drafting and incomplete solvent volatilization due to the increase of electric field strength led to forming beaded fibers and the adhesion structure. With the increase in collection distance, smooth and bead-free nanofibers were obtained due to the entire drawing of the jet. Zhang et al. investigated the influence of electrospun distance on the silk sericin fiber structure [46]. The beads and smooth fibers could be obtained by tuning the electrospun distance. This different fiber structure was caused by the change of electric field strength and electric field force.

2.3.3 Flow Rate

In electrospinning, the flow rate usually represents the amount of polymer solution extruded from the spinning nozzle per unit of time. When the flow rate is low, the Taylor cone is difficult to be observed at the nozzle due to an insufficient supply of solution, and the continuity of spinning will be affected. When the flow rate is high, the spinning liquid easily accumulates at the nozzle, which affects the spinning effect. Therefore, adjusting the spinning liquid flow rate for fiber morphology is significant.

Vashisth et al. prepared gellan-polyvinyl alcohol (gellan-PVA) nanofibers by electrospinning and explored the influence of flow rates on the structure of nanofibers [47]. The results indicated that the flow rate of the spinning solution has a significant impact on the fiber structure. The fiber diameter gradually increased with the increase of flow rates. In particular, some polymer solution was directly sprayed onto the collector due to the extrusion of a large amount of solution under a high flow rate. Similarly, Acik et al. prepared polyvinyl acetate (PVAc) hydrophobic microfiber by electrospinning and observed the coarse beaded fibers and bonding phenomenon between adjacent fibers at high flow rates [48]. This structure was generated due to the insufficient drafting and drying of the solution jet when a large amount of polymer solution is extruded from the nozzle at high flow rates.

2.3.4 Temperature and Humidity

Temperature and humidity are important environmental factors in the process of electrospinning. Temperature affects the viscosity of the polymer solution and the evaporation rate of the solvent. Humidity affects the curing and phase separation of the polymer solution. Therefore, temperature and humidity play an important role in the electrospinning process.

Vrieze et al. explored the influence of temperature and humidity on the formation of electrospun poly (vinyl pyrrolidone) (PVP) nanofibers [49]. When the ambient humidity was 60%, it was difficult to obtain great nanofibers because the water-soluble PVP polymer absorbed the surrounding water vapor. As the ambient humidity increased from 20% to 45%, the fiber diameter decreased gradually due to the slow curing of the fiber and the increase in the drafting time of the jet. In addition, the fiber diameter at 283 and 303 k is smaller than that at 293 k. Vrieze et al. believed that at low temperatures, the slow solidification of the jet and the extension of the drafting time of the jet might lead to fiber thinning. Correspondingly, at higher temperatures, the viscosity of the polymer solution decreased, and the jet draft effect increased, which can produce finer fibers.

At the same time, humidity affects the curing and phase separation of the polymer solution. Much work has been done to prepare porous fibers by increasing ambient humidity. Casper et al. studied the effect of environmental humidity on the structure of electrospun PS fiber [50]., The prepared PS fiber has a smooth surface when the ambient humidity is less than 25%. The porous structure appeared on the fiber surface with a further increase in humidity. Casper et al. believed that two reasons mainly caused the formation of porous structures on the surface of electrospun PS fiber. The first is breath figures. During electrospinning, the spinning jet was cooled due to solvent evaporation, and the water in the air was condensed on the fiber surface. When the fiber was dry, water droplets left a mark behind. The second is phase separation. Generally, phase separation includes nucleation and growth (NG) mechanism and/or spinodal decomposition (SD) mechanism. In this case, spinodal decomposition was considered the most probable mechanism for phase separation in the electrospinning of PS/THF fibers. Similarly, Szewczyk prepared PVDF nanofibers with internal voids and surface wrinkling by water vapor-initiated phase separation at 60% humidity [51].

2.4 Nozzles

In the electrospinning process, the spinning nozzle has an important influence on the adjustment of fiber structure. To prepare specific fiber structures, many novel spinning nozzles have been developed by scientific researchers, such as skin core structural fibers prepared with coaxial nozzles.

The principle of coaxial electrospun is the same as that of traditional electrospinning, composed of a high-voltage power supply, spinneret, and receiver. The special feature of coaxial electrospinning is the nozzle structure. Zhang et al. used coaxial electrospinning technology to prepare two-component core–shell nanofibers [52]. Compared with the traditional electrospinning nozzle, the coaxial nozzle introduced an inner capillary tube. The inner capillary tube was also used as an electrode. The two immiscible liquid solutions were injected into the concentric capillary tubes, respectively, the internal flow rate was adjusted through the propulsion pump, and the external was connected with the atmospheric environment. Under a certain applied voltage and flow rate, a composite jet was generated. After solvent evaporation, the core–shell structure bicomponent nanofibers were produced. Typical TEM images of bicomponent nanofibers exhibited the core–shell structure boundary along the fiber length.

Li et al. prepared a novel profiled PAN-PS core–shell nanofiber using eccentric-axial electrospinning technology [53]. The spinneret includes two eccentric-axial stainless-steel capillaries in the eccentric-axial electrospinning experimental device. The unique groove structure along the fiber axis could be obtained in the eccentric-axial electrospun fibers. In addition, with the gradual decrease of the voltage, the jet tended to be more stable, and the length of the groove on the fiber was extended.

Li et al. prepared hollow nanofibers with circular cross-sections and uniform size by electrospinning using coaxial, two-capillary spinneret [54]. Heavy mineral oil and ethanol solution of PVP and $Ti(OiPr)_4$ were used as core and sheath materials for electrospinning, where the ethanol solution containing PVP and $Ti(O\textit{i}Pr)_4$ was the raw materials for amorphous TiO_2 and PVP composite nanofibers. The TEM images show the hollow nanofibers with tubular structures after extracting heavy mineral oil with octane.

Zhao et al. used a novel multifluid compound-jet electrospinning technology to prepare bio-mimic multichannel microtubules [55]. The structure of the three-metal capillary nozzle was similar to that of the coaxial spinning nozzle, which was serviced as the inner fluid containers and electrodes. The ethanol solution of $Ti(O\textit{i}Pr)_4$ andPVP was used as the external liquid, while the commercially available harmless paraffin oil was selected as the internal liquid. After electrospinning, the obtained fiber membranes were further calcined to remove the organic matter. As a result, TiO_2 three-channel microtubes were obtained (Figure 2.2). Furthermore, the multifluid compound-jet electrospinning technique successfully developed the microtubes with two to five channels. Zhao et al. believed that a tube with six or more internal channels could be manufactured by tuning the experimental parameters.

Figure 2.2 Side-view SEM image of sample after the organics have been removed. Source: Zhao et al. [55]/with permission from American Chemical Society.

2.5 Collectors

In the process of electrospinning, the nature of the receiving substrate affects the dissipation of residual charge on the fiber and the evaporation of the residual solvent, which will further affect the structure of the fiber. Liu et al. explored the influence of different target materials, including copper mesh, aluminum foil, water, and paper, on electrospun cellulose acetate (CA) fiber structure [56]. The results showed that the conductivity and porosity of the target material would affect the accumulation of fibers on the target material. The accumulation of fibers obtained by using conductive collectors was denser, while the non-conductive collectors got a loose structure. The conductive collector was conducive to the discharge of electric charge and the reduction of repulsion between fibers, contributing to the close accumulation of fibers. Porous targets (paper and copper mesh) produced more loose fiber structures than solid targets (water and aluminum foil). Porous targets with a high specific surface area were more conducive to solvent volatilization than solid targets. The solvent diffused from the fiber on the solid target may pull the fiber together before the solvent slowly evaporates.

In Section 2.2.3, we mentioned the preparation of 2D nano-nets through the electronetting technique, where these nano-nets were small pieces of product formed locally with the formation of nanofibers. It lacks controllable structural characteristics and large-scale preparation prospects. To this end, Zhang et al. developed a collector inductively coupled direct electronetting technique to fabricate self-assembled 2D nano-nets (Figure 2.3) [57]. By adjusting the dielectric properties (conductive or dielectric) and configurations (flat or concave) of the

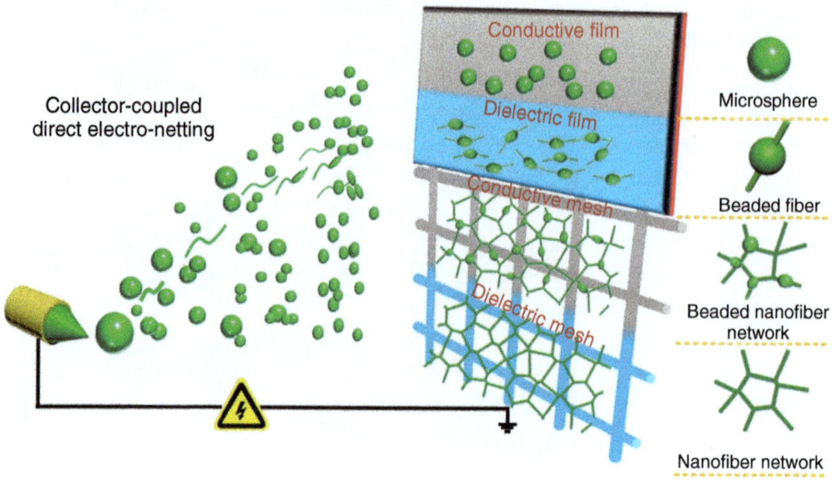

Figure 2.3 Schematic showing the collector inductively coupled direct electronetting strategy. Four designed collectors trigger the assembly of four different architectures. Source: Zhang et al. [57]/with permission of Springer Nature.

collectors, various micro-/nanoarchitectures involving microspheres, beaded fibers, beaded nanofiber nets, and nano-nets could be precisely prepared. This direct electronetting technology required extremely diluted (1–3 wt%) precursor polymer solutions, in which the polymers possessed extremely high molecular weights. Besides, the possible formation process for 2D nanofibrous networks has been proposed. In direct electronetting, the charged droplets may undergo levitating, rapid stretching deformation, and phase separation processes before the forming 2D nanofiber network architecture.

McCann et al. immersed the collector in a liquid nitrogen bath and prepared a series of porous fibers by electrospinning [58]. Poly(styrene), poly(acrylonitrile), poly(vinylidene fluoride), and poly(ε-caprolactone) porous fibers have been obtained by this method. Thermally induced phase separation was considered to be the driving force for the formation of porous fibers. The solution jet hit the liquid nitrogen bath before reaching the collector. The polymer and the remaining solvent in the jet were frozen together. In the freezing process, the polymer solution underwent phase separation to form solvent-rich and solvent-poor regions. In addition, vitrification may occur in the polymer-rich regions when the polymer was cooled to a lower than its glass transition temperature. Finally, the porous fibers could be obtained by vacuum drying.

2.6 Conclusions

In summary, tremendous work has been carried out in adjusting fiber structure. As mentioned above, spinning solutions, spinning process parameters, nozzle structure, and receiving devices (collectors) have been investigated for the regulation of electrospun fiber structure. Beaded, porous, core–shell, hollow, and other fiber structures have been developed. These unique fibrous structures show great potential in many applications, such as air filtration, distillation, catalysis, and other fields. However, there is still insufficiency in the stability of electrospinning and fiber structure and the difficulty in industrialization. Considering these problems, it is very urgent to explore the underlying reasons behind them and develop new electrospinning concepts. In the future, the control of the electric field, the stability of the Taylor cone, the adjustment of the jet path, and the design of the spinning nozzle structure may play important roles in the further development of electrospinning.

References

1 Cui, J., Li, F., Wang, Y. et al. (2020). *Separation and Purification Technology* 250: 117116.

2 Jiang, G., Luo, L., Tan, L. et al. (2018). *Materials Interfaces* 10: 28210.

3 Huang, C. and Thomas, N.L. (2020). *Polymer Reviews* 60: 595.

4 Tiwari, S.K., Tzezana, R., Zussman, E., and Venkatraman, S.S. (2010). *International Journal of Pharmaceutics* 392: 209.

5 Duan, G. and Greiner, A. (2019). *Macromolecular Materials and Engineering* 304: 1800669.

6 Deitzel, J.M., Kleinmeyer, J., Harris, D.E.A., and Tan, N.B. (2001). *Polymer* 42: 261.

7 Subbiah, T., Bhat, G.S., Tock, R.W. et al. (2005). *Journal of Applied Polymer Science* 96: 557.

8 Zheng, J., Zhang, H., Zhao, Z., and Han, C.C. (2012). *Polymer* 53: 546.

9 Haghi, A.K. and Akbari, M. (2007). *Physica Status Solidi* 204: 1830.

10 Theron, S.A., Zussman, E., and Yarin, A.L. (2004). *Polymer* 45: 2017.

11 Cengiz, F., Dao, T.A., and Jirsak, O. (2010). *Polymer Engineering and Science* 50: 936.

12 Kirecci, A., Özkoç, Ü., and İçoğlu, H.İ. (2012). *Journal of Applied Polymer Science* 124: 4961.

13 Mazoochi, T., Hamadanian, M., Ahmadi, M., and Jabbari, V. (2012). *International Journal of Industrial Chemistry* 3: 1.

14 Torres-Giner, S., Gimenez, E., and Lagaron, J.M. (2008). *Food Hydrocolloids* 22: 601.

15 Khalf, A., Singarapu, K., and Madihally, S.V. (2015). *Cellulose* 22: 1389.

16 Halaui, R., Zussman, E., Khalfin, R. et al. (2017). *Polymers for Advanced Technologies* 28: 570.

17 Magalhaes, W.L.E., Cao, X., and Lucia, L.A. (2009). *Langmuir* 25: 13250.

18 Sun, Y., Cheng, S., Lu, W. et al. (2019). *RSC Advances* 9: 25712.

19 Wang, C., Wang, J., Zeng, L. et al. (2019). *Molecules* 24: 834.

20 Reneker, D.H. and Chun, I. (1996). *Nanotechnology* 7: 216.

21 Kim, J.S. and Reneker, D.H. (1999). *Polymer Engineering and Science* 39: 849.

22 Megelski, S., Stephens, J.S., Chase, D.B., and Rabolt, J.F. (2002). *Macromolecules* 35: 8456.
23 Fong, H., Chun, I., and Reneker, D.H. (1999). *Polymer* 40: 4585.
24 Lee, K.H., Kim, H.Y., Bang, H.J. et al. (2003). *Polymer* 44: 4029.
25 Gupta, P., Elkins, C., Long, T.E., and Wilkes, G.L. (2005). *Polymer* 46: 4799.
26 Yahya, G.O., Ali, S.A., and Hamad, E.Z. (1996). *Polymer* 37: 1183.
27 Koski, A., Yim, K., and Shivkumar, S.J.M.L. (2004). *Materials Letters* 58: 493.
28 Qin, X., Wang, S., Torres, S., and Lukas, D. (2005). *Materials Letters* 59: 3102.
29 García-Mateos, F.J., Ruiz-Rosas, R., Rosas, J.M. et al. (2019). *Front Mater* 6: 114.
30 Chen, J.Y., Wu, H.C., Chiu, Y.C. et al. (2015). *Advanced Electronic Materials* 1: 1400028.
31 Gu, J., Gu, H., Cao, J. et al. (2018). *Applied Surface Science* 439: 589.
32 Ding, B., Li, C., Miyauchi, Y. et al. (2006). *Nanotechnology* 17: 3685.
33 Wang, X., Ding, B., Yu, J. et al. (2011). *Nanoscale* 3: 911.
34 Li, Z., Xu, Y., Fan, L. et al. (2016). *Materials and Design* 92: 95.
35 Casasola, R., Thomas, N.L., Trybala, A., and Georgiadou, S. (2014). *Polymer* 55: 4728.
36 Song, Z., Chiang, S.W., Chu, X. et al. (2018). *Journal of Applied Polymer Science* 135: 45787.
37 Choktaweesap, N., Arayanarakul, K., Aht-Ong, D. et al. (2007). *Polymer Journal* 39: 622.
38 Bognitzki, M., Czado, W., Frese, T. et al. (2001). *Maternité* 13: 70.
39 Han, S.O., Son, W.K., Youk, J.H. et al. (2005). *Materials Letters* 59: 2998.
40 Qi, Z., Yu, H., Chen, Y., and Zhu, M. (2009). *Materials Letters* 63: 415.
41 Banikazemi, S., Rezaei, M., Rezaei, P. et al. (2020). *Polymers for Advanced Technologies* 31: 2199.
42 Zhang, W.X., Wang, Y.Z., and Sun, C.F. (2007). *Journal of Polymer Research* 14: 467.
43 Mo, X.M., Xu, C.Y., Kotaki, M.E.A., and Ramakrishna, S. (2004). *Biomaterials* 25: 1883.
44 Robb, B. and Lennox, B. (2011). The electrospinning process, conditions and control. In: *Electrospinning for tissue regeneration* (ed. L.A. Bosworth and S. Downes), 51–66. Cambridge: Woodhead Publishing.
45 Supaphol, P. and Chuangchote, S. (2008). *Journal of Applied Polymer Science* 108: 969.
46 Zhang, X., Khan, M.M.R., Yamamoto, T. et al. (2012). *International Journal of Biological Macromolecules* 50: 337.
47 Vashisth, P., Pruthi, P.A., Singh, R.P., and Pruthi, V. (2014). *Carbohydrate Polymers* 109: 16.
48 Acik, G., Cansoy, C.E., and Kamaci, M. (2019). *Colloid and Polymer Science* 297: 77.
49 De Vrieze, S., Van Camp, T., Nelvig, A. et al. (2009). *Journal of Materials Science* 44: 1357.
50 Casper, C.L., Stephens, J.S., Tassi, N.G. et al. (2004). *Macromolecules* 37: 573.
51 Szewczyk, P.K., Gradys, A., Kim, S.K. et al. (2020). *Mater. Interfaces* 12: 13575.

52 Zhang, Y., Huang, Z.M., Xu, X. et al. (2004). *Chemistry of Materials* 16: 3406.

53 Li, X., Deng, L., Yu, X. et al. (2016). *Journal of Materials Chemistry A* 4: 14453.

54 Li, D. and Xia, Y. (2004). *Nano Letters* 4: 933.

55 Zhao, Y., Cao, X., and Jiang, L. (2007). *Journal of the American Chemical Society* 129: 764.

56 Liu, H. and Hsieh, Y.L. (2002). *Journal of Polymer Science Part B: Polymer Physics* 40: 2119.

57 Zhang, S., Liu, H., Tang, N. et al. (2019). *Nature Communications* 10: 1458.

58 McCann, J.T., Marquez, M., and Xia, Y. (2006). *Journal of the American Chemical Society* 128: 1436.

3

Mass Production of Electrospun Nanofibers

Xian Wen and Liming Wang

Donghua University, College of Textiles, Key Laboratory of Textile Science & Technology, Ministry of Education, 2999 Renmin North Road, Songjiang District, Shanghai, 201620, China

3.1 Introduction

One-dimensional nanofiber materials are characterized by large aspect ratio, high specific surface area, heterogeneous electronic structure, and are not easy to agglomerate. The fiber aggregate materials constructed by them exhibit abundant properties in flexibility, pore structure, surface effect, electric transfer, heat transfer, and mass transfer [1–4]. The research of nanofiber materials is widely involved in the fields of particle filtration [5], tissue engineering [6], energy [7], catalysis [8], and sensors [9].

Electrospinning technology has become one of the important methods to prepare continuous nanofibers due to its advantages of short spinning process, simple equipment, easy operation, wide range of processing materials, and relatively controllable process. Electrospinning was proposed and patented by Formhals in 1934 [10]. The principle of electrospinning is that the polymer solution or melt in the syringe overcomes its surface tension under the action of electric field and is ejected from the spinneret needle to form jet. Solvent evaporation or melt solidification of the jet during electrospinning eventually forms fibers on the collector. At present, a variety of fibrous materials have been successfully prepared by electrospinning, such as hollow, porous, beaded, and other structural nanofibers, as well as organic, inorganic, and organic/inorganic hybrid nanofibers of various chemical compositions [11]. With the vigorous development of electrospinning nanofiber materials and application research, the problem of low nanofiber preparation efficiency has become increasingly prominent. For example, the typical single-needle electrospinning only excites one jet at a time, with a production efficiency of 0.01–1.0 g h^{-1}, and is prone to clogging, which cannot meet the needs of industrial applications and seriously restricts the development of new materials, especially three-dimensional nanofiber aggregates (such as nanofiber aerogels).

Electrospinning: Fundamentals, Methods, and Applications, First Edition.
Edited by Liming Wang and Xiaohong Qin.

In order to realize the mass production of nanofibers without compromising fiber quality, researchers have designed various forms of spinnerets, including increasing the number of needles, using needle-free electrospinning, and assisted electrospinning. This chapter summarizes the existing electrospinning techniques with large-scale performance, focusing on the spinneret configuration and production rate of nanofibers and divides them into (i) multiple-needle electrospinning, (ii) multiple-hole electrospinning, (iii) free-surface electrospinning, (iv) melt electrospinning, and (v) multifield-assisted electrospinning.

3.2 Multiple-needle Electrospinning

Traditional single-needle electrospinning usually uses a hollow metal capillary as spinneret containing polymer solution, which excites jet to move toward a collector under high voltage. This arrangement typically produces a single jet and a low fiber production rate, which hinders its commercialization for large-scale applications [12, 13]. In order to enhance the nanofiber production rate of needle-based electrospinning, multiple needles/nozzles have been introduced in electrospinning [14–17], which involve multiple nozzles arranged in different configurations and arrays to produce multiple jets. Tomaszewksi et al. [15] assembled three types of multiple-needle electrospinning spinnerets (series, elliptic, and concentric) to study the effects of different configurations on the spinning stability and efficiency. The distance between adjacent needles was set to 3–15 mm. Their results demonstrated that the nanofiber yield of multi-needle electrospinning in a compact arrangement is much lower than expected, and there is a serious interference problem between adjacent needles. Among these three configurations, the multiple-needle spinning efficiency of concentric arrangement was the best, and whose productivity of polyvinyl alcohol (PVA) nanofibers can reach 0.06 $g\,h^{-1}$. Theron et al. [16] studied the jet–jet interaction of various multi-needle nozzles from theoretical and experimental perspectives. They designed three types of needle arrays, 3×3, 9×1, and 7×1, and the spacing between adjacent needles was set to 1–5 cm. The mathematical models were established according to the multi-jet trajectory. They found that during multiple-jet electrospinning processes, not only the self-induced Coulombic interactions and the applied electric field affect the jet path, but also mutual-Coulombic interactions between different jets devote, leading to significant interference in multi- needles. Therefore, the jets arranged on the outside were deflected greatly, resulting in poor uniformity of nanofibers. Moreover, such a spinning method only allowed nine needles arranged on a square of about 4 cm^2 in the form of a 3×3 matrix, which limits the effective increase of jets in a unit space. Salehhudin et al. [12] proposed that the arrangement of multiple needles is the key to affecting the electric field and the nanofiber morphology. They showed that the unequally spaced setup contributes to a more uniform electric field near the nozzles, improving the uniformity of nanofibers.

Multiple-needle electrospinning can not only increase yield but also produce fibers with special structures, such as core–shell fibers. The main concern of multiple-needle electrospinning is the deterioration of fiber uniformity and quality, which is attributed to the uneven distribution of the electric field near the nozzle tips and the nonuniform formation of jet due to the interference of the electric field between needles [18, 19]. Though researchers have applied extra auxiliary electrodes to control the electric field and behavior of jets in electrospinning [20–22], the introduction of auxiliary electrodes usually requires additional energy and cost. Besides, improving the complexity and easy clogging of multiple-needle electrospinning is also important for its development.

3.3 Multiple-hole Electrospinning

Multiple-hole spinneret has similar mechanism to multiple-needle spinneret, which continues the pore structure of single-needle nozzle and spinning by unique porous distribution on a smooth or curved surface. At the very beginning, a polyethylene (PE) porous-walled cylindrical tube was designed with fittings at the top to connect to a pressurized air supply and seal at the bottom [23]. The pressurized airflow of 0.4–0.8 kPa is introduced by the nozzle, which can squeeze the polymer solution in the tube to form many droplets on the outer surface. Under the action of the high-voltage electric field, the outer wall droplets form multiple jets thus significantly improving the nanofiber production rate. Finally, the hollow cylindrical porous nozzle with a diameter of 2 cm and a height of 6 cm produced nanofibers at a rate greater than $5\,g\,h^{-1}$, reaching 250 times that of single-needle electrospinning. And the average diameter of nanofibers produced by this device is similar to that of single-needle electrospinning, but with wider distribution. Based on the abovementioned work, a horizontally configured cylindrical polytetrafluoroethylene (PTFE) porous hollow tube was invented with an average pore size of 20–40 μm [24]. A wire electrode was inserted into the porous tube to maintain the same electrical potential in each hole. Compared with the previous device, this device can better control the direction of jet movement and nanofiber collection. The production rate of nanofibers can reach $0.3–0.5\,g\,h^{-1}$ for a 13 cm long porous tube with 20 holes. It can be seen from the SEM that the uniformity of nanofibers produced by the porous tube inserted with wire electrode had been improved. Furthermore, by comparing the electric field strength, jet morphology, and fiber diameter of multiple-hole and multiple-needle electrospinning, Zeng's group [25, 26] showed that multiple-hole system was able to generate a stronger and more uniform electric field during the spinning process, resulting in finer and narrow diameter distribution fiber production (Figure 3.1a). By further designing the structure of the spinneret [27, 28], this effect can be enhanced, achieving an increase in fiber production rate, and improving fiber quality. Although multiple-hole electrospinning could partly improve the quality of nanofibers, the fine control of the structural parameters of spinneret is essential for mass production.

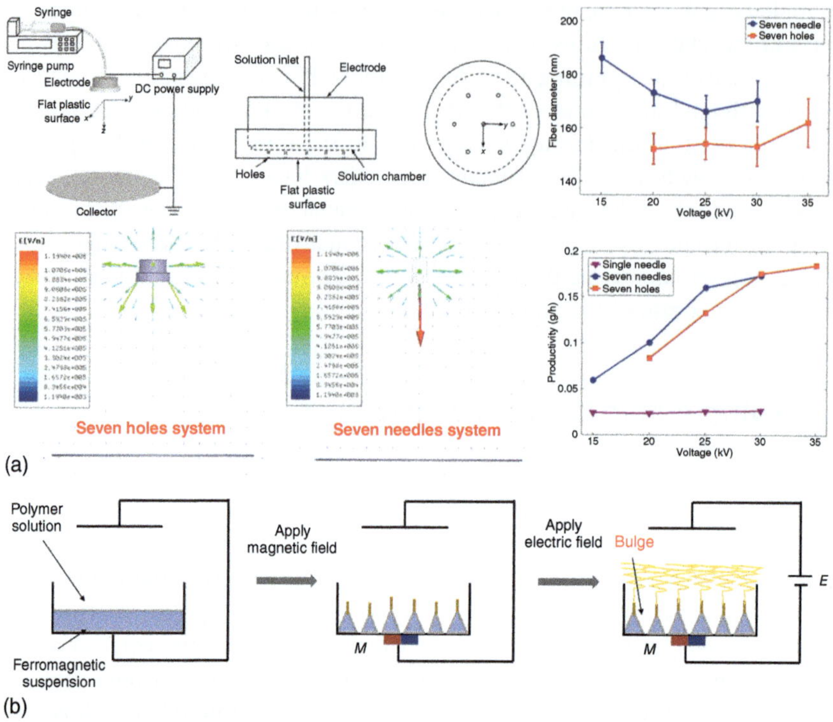

Figure 3.1 (a) Comparison of seven-hole system and seven-needle system. Source: Zheng et al. [25]. Reproduced with permission of John Wiley and Sons. (b) Magnetic fluid-induced free-surface electrospinning. Source: Xian Wen.

3.4 Free-surface Electrospinning

Free-surface electrospinning, also known as needleless electrospinning, generates jets from an open solution surface on the spinneret, and the jets are unconstrained by capillary effect occurring in a needle spinneret. To this end, researchers designed a variety of free-surface electrospinning devices. This free liquid surface spinning method is beneficial to reduce the arrangement interval of jets and greatly increasing the number of jets per unit area. The key to the formation of multiple jets is to promote and control the disturbance of the free liquid surface, for which the researchers designed a variety of free-surface electrospinning devices.

Here we divide them into (i) static electrode free-surface electrospinning, (ii) rotating electrode free-surface electrospinning, and (iii) slit-surface electrospinning.

3.4.1 Static Electrode Free-surface Electrospinning

Static electrode free-surface electrospinning could be classified into two types, fluid-induced free-surface electrospinning [29–31] and gravity-adaptive free-surface electrospinning [32–35]. Typical fluid-induced free-surface electrospinning mainly includes magnetic fluid-induced and bubble-induced electrospinning. The shape of the spinning solution is formed by the induction of the fluid medium rather than metal nozzles.

In 2003, a ferromagnetic suspension was introduced in electrospinning [29], which could generate spikes under the action of a magnetic field. The delicate design consists of a two-layer system, in which the lower layer is a ferromagnetic suspension and the upper layer is a polymer spinning solution. The two liquids are insoluble with each other, and the density of the liquid in the upper layer is smaller than that in the lower layer. As shown in Figure 3.1b, when the permanent magnet or electromagnet acts on the lower layer of suspension, lots of spikes formed at the interface of the two layers of solution. Affected by this, the upper solution surface morphology was directly changed and produced multiple spikes as well. When a high-voltage electrostatic field is further applied, multiple jets would be excited at the top of each spike, thus significantly improving the productivity of nanofibers. Although the magnetofluid electrospinning effectively avoided the clogging of needle spinneret and reduced the mutual interference of the electric field between adjacent needles, it is seriously limited by the requirements of the compatibility, density, and viscosity of the two layers of the solution. In addition, the instability of the magnetic fluid-induced spikes also leads to poor controllability of the spinning system. In 2008, a bubble electrospinning device was invented by He and Liu [30]. They introduced nitrogen gas at the bottom of the spinning solution to form a large number of bubbles on an open solution surface. When an electric field is present, the bubbles deformed and ruptured, facilitating the initiation of jets. This method constructed a protuberant bubble structure while minimizing the surface tension of the spinning solution, thus enabling the preparation of ultrafine nanofibers with a diameter of 50 nm at a low spinning voltage. However, the size and distribution of bubbles, which are extremely difficult to control, determine the number and morphology of jets. That is to say, the diameter of the prepared nanofibers is uncontrollable and the fiber diameter distribution is wide.

Gravity-adaptive free-surface electrospinning mainly includes bowl-shaped and stepped pyramid-shaped electrospinning, which relies on the gravity of spinning solution and the edge of the spinneret to form electric field-induced instabilities (i.e. jets). The edge is usually symmetrical and circular, inducing multiple jet symmetry excitations. Compared with fluid-induced free-surface electrospinning, this type of electrospinning supply solution mildly and have relatively stable morphology of spinning solution. A metal bowl spinneret was designed filled with polymer solution that could spontaneously break into a series of droplets with symmetrically spaced arrangements along the lip of the bowl [34, 35]. Under high-voltage initiation, many symmetrical jets were directly generated, and the nanofibers production rate of this device is about 40 times that of single-needle electrospinning. Jiang et al. [33] demonstrated a free-surface electrospinning setup using a stepped pyramid-shaped spinneret to expand the free surface layer given the bowl spinneret. The spinning solution is injected from the top layer of the pyramid-shaped spinneret. After the first layer is filled with the spinning solution, the second and third layers of the spinneret are filled in sequence. Dozens of jets can be excited at the edge of each free-surface layer of the spinneret at high voltage. The production rate of this system was about $4\,g\,h^{-1}$, reaching several hundred times higher than that of single-needle electrospinning. In addition, they also revealed that the strongest electric field concentrated at the edge of the spinneret was responsible for the jet excitation from the edge by electric field simulation. However, the jets excited from the multilayer free surface have different motion states, significantly decreasing the diameter uniformity of the nanofibers.

Gravity-adaptive electrospinning achieves simultaneous excitation of multiple jets at the edge of the free surface. However, this type of spinneret cannot be continuously spun for a long time because the jet excitation only occurs at the edge of the large static free surface, and the solvent of the spinning solution is easily volatilized, resulting in the rapid solidification of the edge spinning solution.

3.4.2 Rotating Electrode Free-surface Electrospinning

Rotating electrode free-surface electrospinning mainly includes rotating cylinder [36–38], rotating disc [37, 39], rotating ball [40, 41], rotating wires [42, 43], and rotating helical coil electrospinning [44, 45], which relies on curved surface to shape the jet. These types of free-surface electrospinning enable continuous liquid supply and continuous electrospinning through slow rotation with large free liquid surface. A thin layer of spinning solution with relatively stable shape adheres to the surface of the spinneret, exciting abundant jets under the action of a high-voltage electrostatic field, thus improving the production rate of nanofibers.

In 2005, the European patent WO 2005/024101 issued a patent for "a method of nanofibres production from a polymer solution using electrostatic spinning and a device for carrying out the method" [46]. The patent inventor Jirsak Oldrich used a rotating charged cylinder electrode as the spinneret and designed various surface structures of the spinneret. The cylindrical spinneret was immersed in a spinning solution, attaching a thin liquid layer of spinning solution by rotation. Under the action of the high-voltage electrostatic field, a large number of jets were excited on the cylindrical surface and formed many protrusions similar to Taylor cones in needle electrospinning. Then, the nanofibers were obtained by further stretching the protrusions into bundles of jets, which significantly increased the production rate. The Czech company Elmarco subsequently commercialized this device, and became the first company to invent commercial large-scale nanofibers production line. After that, researchers [36] found that the rotating disk spinneret requires relatively low applied voltage in electrospinning to initiate fiber formation compared to the rotating cylinder spinneret (Figure 3.2a). Furthermore, at the same spinning conditions, the rotating disk electrospinning produced finer and more uniform diameter distribution nanofibers, which could be ascribed to the effectively concentrated electric field in disk spinneret. While the cylinder had relatively low intensity and scattered distribution of electric field, the electric field on the cylinder ends was more concentrated than that on the middle surface area. Affected by the different electric field distributions, the production rate of the two free-surface electrospinning systems

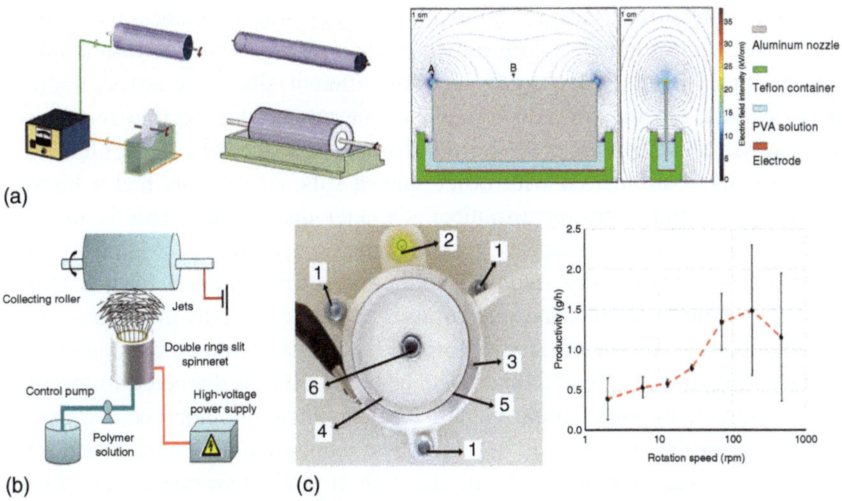

Figure 3.2 (a) Rotating cylinder and rotating disc free-surface electrospinning. Source: [36]. Reproduced with permission of John Wiley and Sons. (b) Double rings slit electrospinning. Source: Wei et al. [47]. Reproduced with permission of John Wiley and Sons. (c) Ring slit electrospinning with relative motion of two walls. 1: fixing screws, 2: bubble leveler, 3: stator, 4: rotor, 5: annular orifice, 6: cap nut that helps the disassembly. Source: Kara et al. [48]. Reproduced with permission of John Wiley and Sons.

was similar. Subsequently, they added chamfers at the ends of the rotating cylinder to reduce the electric field intensity difference between the middle and ends of the cylindrical surface [37], significantly facilitating the production rate of the cylinder electrospinning nanofibers. In 2011, Lin et al. [49] published a patent for a rotating helical coil spinneret, and the top of the helix can generate a lot of jets at the same time. Like other free-surface electrospinning, it had the ability to continuously spin, thus facilitating mass production of nanofibers. Although the axisymmetric rotating helical coil spinneret improved the stability of the spinning solution morphology, the nanofiber diameter distribution is still nonuniform and the in-depth mechanism is lacking.

3.4.3 Slit-surface Electrospinning

Slit can be used as a spinneret for electrospinning, which combines the advantages of needle electrospinning (e.g. low solvent evaporation during electrospinning) and needleless electrospinning (e.g. high fiber production rate). Slit extrusion needleless electrospinning mainly includes linear slit [50], curved slit [51], and ring slit [47, 48, 52–55] electrospinning. This type of needleless electrospinning relies on the extrusion of the spinning solution by the slit to shape the spinning solution shape. Compared with other needleless electrospinning nozzles, the free liquid surface exposed by this type of nozzle is in the shape of a long line, the free liquid surface is small, and the spinning solution is strongly restrained, which is beneficial to reduce the solvent volatilization of the spinning solution to be spun at the nozzle.

In 2015, a slit-surface electrospinning technology was developed that could spontaneously emit numerous core sheath jets at a rate of $1\,L\,h^{-1}$ [50]. Soon after, researchers [51] conducted a study on four different shapes of slit spinneret: straight, curved, rectangular, and triangular. The results showed that the shape of the slit has an important effect on the diameter, productivity, and uniformity of the nanofibers. Compared with other shaped slits, curved slits had a higher production rate and narrower nanofiber diameter distribution. This is mainly because the larger curvature could result in a higher local electric field intensity and a better electric field intensity uniformity of the curved slit spinneret. In 2018, Qin's team [47] designed a double-ring slit spinneret and systematically investigated the effect of processing parameters on the morphology, diameter, and production rate of nanofibers (Figure 3.2b). In addition, it can be found that the diameter uniformity of the slit-surface electrospinning nanofibers is significantly improved compared with the free-surface electrospinning. Finally, nanofibrous membranes with low diameter dispersion were obtained, and the highest production rate of nanofibers reached $4.5\,g\,h^{-1}$ [52]. However, with the reduction of the solution free surface, the number of jets is also significantly reduced. In 2020, studies [48, 54] found that by promoting the relative motion of the two walls of the ring slit spinneret, the solidification of the spinning solution on the spinneret surface could be prevented and the viscosity of the spinning solution could be reduced, promoting the excitation of multiple jets (Figure 3.2c). Besides, the productivity of the spinneret is 10 times higher than that of a single needle. However, the spinning solution in the shear slit is liable to cause uncontrollable deformation of the solution surface, resulting in a wide distribution of the obtained nanofibers diameter.

3.5 Melt Electrospinning

The abovementioned electrospinning refers to solution electrospinning, which is usually accompanied by the volatilization of toxic solvents. But the melt electrospinning is solvent-free and completely environmentally friendly [56]. Although the basic principles of these two methods are similar, the fiber formation process is quite different. The solution electrospinning fibers are formed by the evaporation of a large amount of solvent and the solidification of the remaining solute, while the melt electrospun fibers are formed by gradually condensing and solidifying polymer melt through heat dissipation [57].

The typical melt electrospinning device consists of feeder, heater, nozzle, high-voltage generator, and collector. The melt can be uniformly distributed around the rim by a cone-shaped and umbrella-like spinneret, which could eject numerous jets when high voltage is applied and fluid plugging is avoided [58]. Almost 30 jets were produced by this umbrella spinneret with 10 mm bottom rim diameter. And the melt viscosity could be affected by spinning temperature, which finally influences the fiber fineness and spinning efficiency. Through an unconfined melt edge electrospinning, multiple self-organized parallel jets spontaneously formed when high voltage was applied at the plate edge. And the jets sites generation

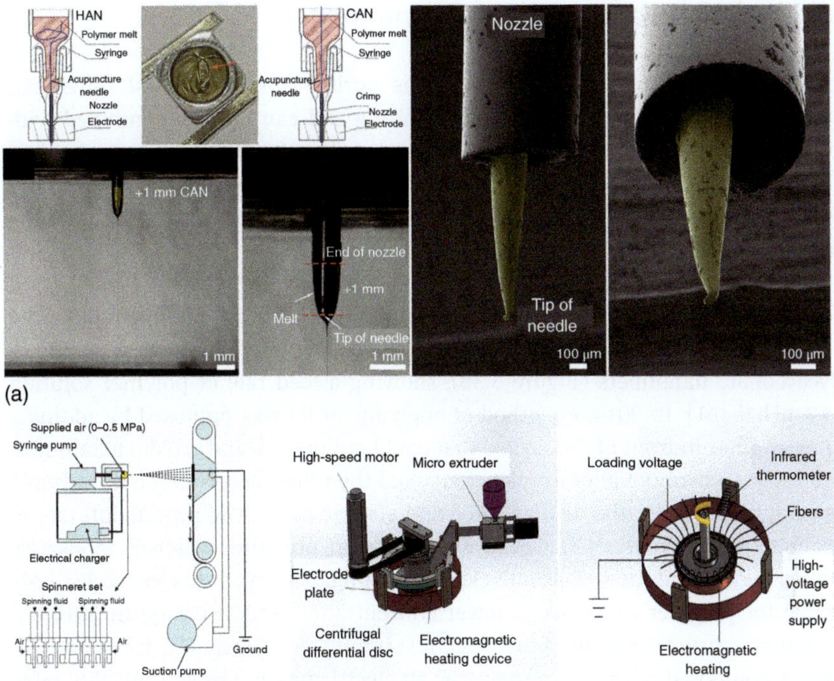

(a) (b) (c)

Figure 3.3 (a) Melt electrospinning with a modified nozzle. Source: Grosshaus et al. [60]. Reproduced with permission of John Wiley and Sons. (b) Blowing-assisted electrospinning. Source: Hsiao et al. [61]. Reproduced with permission of John Wiley and Sons. (c) Melt differential centrifugal electrospinning. Source: Liu et al. [62]. Reproduced with permission of John Wiley and Sons.

depends on the fluid perturbations induced by electric field instead of mechanical driving force. Such a device obtained ~20–25 jets with a spacing of about 5 mm and PE fibers of diameter ranging from 3 to 30 μm, achieving a productivity of up to 80 mg min^{-1}. And the resultant viscosity of jets could be effectively decreased by increasing voltage, which is beneficial to spin high-quality fibers for high-viscosity melt systems [59]. An electrospun medical-grade poly(ε-caprolactone) with a minimum mean diameter of 275 ± 86 nm could be obtained by using a modified nozzle (Figure 3.3a) [60].

Compared with solution electrospinning, melt electrospinning was relatively developed slowly and is less attractive for applications. These mainly due to three aspects: first, melt electrospinning equipment includes heating device that makes it more complicated than solution electrospinning equipment; next, the temperature required for melt electrospinning is not only higher than the melting point of the polymer, but also generally higher than the normal processing temperature which limits its industrialization; last, the diameter of nanofibers obtained by melt electrospinning is generally an order of magnitude higher than that of solution electrospinning, which is not conducive to its delicate application.

3.6 Multifield-assisted Electrospinning

Introducing extra auxiliary forces, such as airflow and centrifugal force, can achieve a high flow rate during electrospinning and can be well combined with the abovementioned spinning methods, which is beneficial for mass production. The jets move under multiple actions such as electric field force, charge repulsion, air-assisted force, or centrifugal force, which further stretch the jet and improve the quality of the nanofibers [63–65].

Air-assisted electrospinning can generally be classified into two categories: air-blowing and air-suction-assisted electrospinning. In 2011, blowing-assisted electrospinning was combined with a multiple-needle setup to prepare electrospun polycarbonate nanofibers (Figure 3.3b), showing a feed rate of polymer solution of 3.0 ml h^{-1} [61]. In 2014, a method of applying air jet was proposed by adding a 700-mesh filter in front of the nozzle that could cut the jets and crush and atomize them into massively smaller droplets [66]. And then these jets were directly drawn into nanofibers under the airflow force and electric force. The production rate of nanofibers was as high as 75.6 g h^{-1} at 0.8 MPa air pressure, which is thousands of times higher than that of needle electrospinning. However, this method is only suitable for polymer solutions of lower concentration and viscosity. In 2016, an air-suction-assisted differential melt electrospinning technology was reported [67], which is composed of three components: an electric heating ring, a metallic tube, and a metallic sleeve. Different from single-needle electrospinning, the high voltage was applied to the hollow disc electrode of air input instead of the spinneret, which forms multiple jets and is earthed. The production rate could reach 80-fold than that of single-needle system, implying its ability for mass production. Furthermore, although affected by flowing air, the stability of the jets was far superior to other air-assisted electrospinning methods.

In centrifugal electrospinning, the extra centrifugal force is not only beneficial to overcome the surface tension of polymer solution, but also to further stretch the jets to obtain lower diameter fibers. In 2013, Valipouri et al. [68] prepared high-quality and fine nanofibers using an air-sealed centrifugal electrospinning device, in which the spinneret and collector rotated at the same speed. Besides, the skillful design enables spinning of polymer solutions with highly volatile solvents. By integrating the advantages of melt differential electrospinning and centrifugal spinning, polypropylene nanofibers with a high flow rate of 124.26 g h^{-1} were successfully prepared [62]. The melt differential centrifugal electrospinning method could generate multiple jets from the rim of the centrifugal differential disk without solvent pollution (Figure 3.3c). And the coupling effect of high-voltage electrostatic and centrifugal forces can also benefit the crystal orientation of nanofibers, thereby improving their mechanical properties.

3.7 Future and Prospects

Electrospinning to effectively prepare nanofibers with unique properties (such as large aspect ratio, high specific surface area, heterogeneous electronic structure, and not easy to agglomerate) has been rapidly developed recently. Due to the advantages of nanofibers, they have been widely applied in energy, electronics, biotechnology, sensors, and so on, which undoubtedly stimulates continued demand for electrospun nanofibers in the future, opening up opportunities for further commercialization prospects of nanomaterials. Many techniques have been devoted to improving the industrial productivity of electrospun nanofibers. There is still much room for further improvement.

Although researchers have tried various strategies to improve the production rate of electrospinning, it is still difficult to balance the requirements of continuous, stable, high-yield, and controllable nanofiber diameters for industrial electrospinning. In needle-type industrial electrospinning technology, the ability to assemble multiple needles to improve the yield of nanofibers is limited and the quality of nanofibers is significantly deteriorated due to the mutual interference by the electric fields of adjacent needles. In addition, the problem of easy clogging of needles also affects the stability of multi-needle electrospinning.

Different from the single jet of single-needle electrospinning, in the process of free-surface electrospinning, multiple jets are simultaneously excited from the free liquid surface, which puts forward higher requirements for the control of multiple-jet excitation sites and the control of multiple-jet motion processes. These methods have common problems with solution utilization ratio and solvent evaporation, which resulted in the insufficient utilization of the polymer solutions. In addition, although researchers have tried various methods to construct different free surfaces, the research on the excitation mechanism of multiple jets in the free surface is still limited. How to design a surface-free electrospinning device meeting these criteria will be the key to producing high-quality nanofibers with much improved productivity.

References

1. Xue, J.J., Wu, T., Dai, Y.Q. et al. (2019). Electrospinning and electrospun nanofibers: methods, materials, and applications. *Chemical Reviews* 119 (8): 5298–5415.
2. Wen, X., Xiong, J., Lei, S.L. et al. (2021). Diameter refinement of electrospun nanofibers: from mechanism, strategies to applications. *Advanced Fiber Materials* 4 (2): 145–161.
3. Akampumuza, O., Gao, H.C., Zhang, H.N. et al. (2018). Raising nanofiber output: the progress, mechanisms, challenges, and reasons for the pursuit. *Macromolecular Materials and Engineering* 303 (1): 1700269.
4. Shi, S., Si, Y.F., Han, Y.T. et al. (2022). Recent progress in protective membranes fabricated via electrospinning: advanced materials, biomimetic structures, and functional applications. *Advanced Materials* 34 (17): 2107938.
5. Yun, K.M., Suryamas, A.B., Iskandar, F. et al. (2010). Morphology optimization of polymer nanofiber for applications in aerosol particle filtration. *Separation and Purification Technology* 75 (3): 340–345.

6 Zhou, Y.G. and Tan, G.Z. (2017). Fabrication of nanofiber mats with microstructure gradient by cone electrospinning. *Nanomaterials and Nanotechnology* 7: https://doi.org/10.1177/1847980417748478.

7 Liu, X., Wang, X., Zhao, H. et al. (2014). Myocardial cell pattern on piezoelectric nanofiber mats for energy harvesting. *Journal of Physics: Conference Series* 557: 012057.

8 Cho, E.J., Kim, N.H., Kang, J.K. et al. (2009). One-pot fabrication of a self-assembled organic-inorganic nanofiber embedded palladium ion as a catalyst. *Chemistry of Materials* 21 (1): 3–5.

9 Chen, D.J., Lei, S., Chen, Y.Q. et al. (2011). A single polyaniline nanofiber field effect transistor and its gas sensing mechanisms. *Sensors* 11 (7): 6509–6516.

10 Formhals, A. (1934). Process and apparatus for preparing artificial threads. *US Patent* 1,975: 504. RICHARD SCHREIBER GASTELL.

11 Sun, Y.J., Zhang, X., Zhang, M. et al. (2022). Rational design of electrospun nanofibers for gas purification: principles, opportunities, and challenges. *Chemical Engineering Journal* 446 (1): 137099.

12 SalehHudin, H.S., Mohamad, N., Mahadi, W.N.L. et al. (2021). Simulation and experimental study of parameters in multiple-nozzle electrospinning: effects of nozzle arrangement on jet paths and fiber formation. *Journal of Manufacturing Processes* 62: 440–449.

13 Theron, S.A., Zussman, E., Yarin, A.L. et al. (2004). Experimental investigation of the governing parameters in the electrospinning of polymer solutions. *Polymer* 45 (6): 2017–2030.

14 Ding, B., Kimura, E., Sato, T. et al. (2004). Fabrication of blend biodegradable nanofibrous nonwoven mats via multi-jet electrospinning. *Polymer* 45 (6): 1895–1902.

15 Tomaszewski, W. and Szadkowski, M. (2005). Investigation of electrospinning with the use of a multi-jet electrospinning head. *Fibres & Textiles in Eastern Europe* 13 (4): 22–26.

16 Theron, S.A., Yarin, A.L., Zussman, E. et al. (2005). Multiple jets in electrospinning: experiment and modeling. *Polymer* 46 (9): 2889–2899.

17 Jiang, J.X., Zheng, G.F., Wang, X. et al. (2019). Arced multi-nozzle electrospinning spinneret for high-throughput production of nanofibers. *Micromachines* 11 (1): 27.

18 Xie, S. and Zeng, Y.C. (2012). Effects of electric field on multineedle electrospinning: experiment and simulation study. *Industrial & Engineering Chemistry Research* 51 (14): 5336–5345.

19 Angammana, C.J. and Jayaram, S.H. (2011). The effects of electric field on the multijet electrospinning process and fiber morphology. *IEEE Transactions on Industry Applications* 47 (2): 1028–1035.

20 Kim, G., Cho, Y.S., and Kim, W.D. (2006). Stability analysis for multi-jets electrospinning process modified with a cylindrical electrode. *European Polymer Journal* 42 (9): 2031–2038.

21 Chen, R.X., Wu, Y.K., Fan, J. et al. (2019). Numerical approach to controlling a moving jet's vibration in an electrospinning system: an auxiliary electrode

and uniform electric field. *Journal of Low Frequency Noise Vibration and Active Control* 38 (3–4): 1687–1698.

22 Gupta, A., Ayithapu, P., and Singhal, R. (2021). Study of the electric field distribution of various electrospinning geometries and its effect on the resultant nanofibers using finite element simulation. *Chemical Engineering Science* 235: 116463.

23 Dosunmu, O.O., Chase, G.G., Kataphinan, W. et al. (2006). Electrospinning of polymer nanofibres from multiple jets on a porous tubular surface. *Nanotechnology* 17 (4): 1123–1127.

24 Varabhas, J.S., Chase, G.G., and Reneker, D.H. (2008). Electrospun nanofibers from a porous hollow tube. *Polymer* 49 (19): 4226–4229.

25 Zheng, Y.S., Liu, X.K., and Zeng, Y.C. (2013). Electrospun nanofibers from a multihole spinneret with uniform electric field. *Journal of Applied Polymer Science* 130 (5): 3221–3228.

26 Zheng, Y.S. and Zeng, Y.C. (2014). Jet repulsion in multi-jet electrospinning systems: from needle to needleless. *Advanced Materials Research* 852: 624–628.

27 Zheng, Y.S. and Zeng, Y.C. (2013). Electric field analysis of spinneret design for multihole electrospinning system. *Journal of Materials Science* 49 (5): 1964–1972.

28 Zheng, Y.S., Zhuang, C.M., and Gong, R.H. (2014). Electric field design for multijet electropsinning with uniform electric field. *Industrial & Engineering Chemistry Research* 53 (38): 14876–14884.

29 Yarin, A.L. and Zussman, E. (2004). Upward needleless electrospinning of multiple nanofibers. *Polymer* 45 (9): 2977–2980.

30 He, J.H., Liu, Y., Xu, L. et al. (2008). BioMimic fabrication of electrospun nanofibers with high-throughput. *Chaos, Solitons & Fractals* 37 (3): 643–651.

31 Higham, A.K., Tang, C., Landry, A.M. et al. (2014). Foam electrospinning: a multiple jet, needle-less process for nanofiber production. *AICHE Journal* 60 (4): 1355–1364.

32 Jiang, G.J. and Qin, X.H. (2014). An improved free surface electrospinning for high throughput manufacturing of core-shell nanofibers. *Materials Letters* 128: 259–262.

33 Jiang, G.J., Zhang, S., and Qin, X.H. (2013). High throughput of quality nanofibers via one stepped pyramid-shaped spinneret. *Materials Letters* 106: 56–58.

34 Thoppey, N.M., Bochinski, J.R., and Clarke, L.I. (2011). Edge electrospinning for high throughput production of quality nanofibers. *Nanotechnology* 22 (34): 345301.

35 Thoppey, N.M., Gorga, R.E., Bochinski, J.R. et al. (2012). Effect of solution parameters on spontaneous jet formation and throughput in edge electrospinning from a fluid-filled bowl. *Macromolecules* 45 (16): 6527–6537.

36 Niu, H.T., Lin, T., and Wang, X.G. (2009). Needleless electrospinning. I. A comparison of cylinder and disk nozzles. *Journal of Applied Polymer Science* 114 (6): 3524–3530.

37 Niu, H.T., Wang, X.G., and Lin, T. (2012). Needleless electrospinning: influences of fibre generator geometry. *Journal of the Textile Institute* 103 (7): 787–794.

38 Ali, U., Niu, H.T., and Khurshid, M.F. (2019). Electrospinning behavior of needleless spinneret with a popular mace shape. *Journal of the Textile Institute* 110 (3): 349–357.

39 Ahmad, A., Ali, U., and Nazir, A. (2019). Toothed wheel needleless electrospinning: a versatile way to fabricate uniform and finer nanomembrane. *Journal of Materials Science* 54 (21): 13834–13847.

40 Xiong, J., Li, A.L., Liu, Y. et al. (2012). Scalable and hierarchically designed MOF fabrics by netting MOFs into nanofiber networks for high-performance solar-driven water purification. *Journal of Materials Chemistry A* 9 (37): 21005–21012.

41 Quan, Z.Z., Wang, Y.H., and Zu, Y. (2021). A rotary spinneret for high output of electrospun fibers with bimodal distribution. *European Polymer Journal* 159: 110707.

42 Forward, K.M. and Rutledge, G.C. (2012). Free surface electrospinning from a wire electrode. *Chemical Engineering Journal* 183: 492–503.

43 Bhattacharyya, I., Molaro, M.C., and Braatz, R.D. (2016). Free surface electrospinning of aqueous polymer solutions from a wire electrode. *Chemical Engineering Journal* 289: 203–211.

44 Liu, Z., Chen, R.X., and He, J.H. (2016). Active generation of multiple jets for producing nanofibres with high quality and high throughput. *Materials & Design* 94: 496–501.

45 Zheng, G.F., Jiang, J.X., Wang, X. et al. (2018). Self-cleaning threaded rod spinneret for high-efficiency needleless electrospinning. *Applied Physics A-A-materials Science & Processing* 124 (7): 473.

46 Jirsak, O., Sanetrnik, F., Lukas, D. et al. (2009). Method of nanofibres production from a polymer solution using electrostatic spinning and a device for carrying out the method. US Patent US10570806. Technicka Univerzita v Liberc.

47 Wei, L., Qiu, Q.H., Wang, R.W. et al. (2018). Influence of the processing parameters on needleless electrospinning from double ring slits spinneret using response surface methodology. *Journal of Applied Polymer Science* 135 (27): 46407.

48 Kara, Y., He, H.J., and Molnar, K. (2020). Shear-aided high-throughput electrospinning: a needleless method with enhanced jet formation. *Journal of Applied Polymer Science* 137 (37): e49104.

49 Lin, T, Wang X.G., Wang, X. et al. (2011). Electrostatic spinning assembly. *US Patent US20110311671A1*. NEWTECH TEXTILE TECHNOLOGY DEVELOPMENT (SHANGHAI) Co Ltd.

50 Yan, X.R., Marini, J., Mulligan, R. et al. (2015). Slit-surface electrospinning: a novel process developed for high-throughput fabrication of core-sheath fibers. *PLoS One* 10 (5): e0125407.

51 Yan, G.L., Niu, H.T., Shao, H. et al. (2017). Curved convex slot: an effective needleless electrospinning spinneret. *Journal of Materials Science* 52 (19): 11749–11758.

52 Wei, L., Sun, R.J., Liu, C.K. et al. (2019). Mass production of nanofibers from needleless electrospinning by a novel annular spinneret. *Materials & Design* 179: 107885.

53 Yang, Z.H., Zhang, X.L., Qin, Z.K. et al. (2022). Airflow synergistic needleless electrospinning of instant noodle-like curly nanofibrous membranes for high-efficiency air filtration. *Small* 18 (14): 2107250.

54 Molnar, K. (2019). Shear-aided annular needleless electrospinning. *Materials Research Express* 6 (7): 075304.

55 Molnar, K. and Nagy, Z.K. (2016). Corona-electrospinning: needleless method for high-throughput continuous nanofiber production. *European Polymer Journal* 74: 279–286.

56 Zhmayev, E., Cho, D., and Joo, Y.L. (2010). Nanofibers from gas-assisted polymer melt electrospinning. *Polymer* 51 (18): 4140–4144.

57 Hutmacher, D.W. and Dalton, P.D. (2011). Melt electrospinning. *Chemistry, an Asian Journal* 6 (1): 44–56.

58 Li, H.Y., Bubakir, M.M., Xia, T. et al. (2014). Mass production of ultra-fine fibre by melt electrospinning method using umbellate spinneret. *Materials Research Innovations* 18 (sup4): 921–925.

59 Wang, Q.Q., Curtis, C.K., Thoppey, N. et al. (2014). Unconfined, melt edge electrospinning from multiple, spontaneous, self-organized polymer jets. *Materials Research Express* 1 (4): 045304.

60 Grosshaus, C., Bakirci, E., Berthel, M. et al. (2020). Melt electrospinning of nanofibers from medical-grade poly(epsilon-caprolactone) with a modified nozzle. *Small* 16 (44): 2003471.

61 Hsiao, H.Y., Huang, C.M., Liu, Y.Y. et al. (2011). Effect of air blowing on the morphology and nanofiber properties of blowing-assisted electrospun polycarbonates. *Journal of Applied Polymer Science* 124 (6): 4904–4914.

62 Liu, Y.J., Tan, J., Yu, S.Y. et al. (2019). High-efficiency preparation of polypropylene nanofiber by melt differential centrifugal electrospinning. *Journal of Applied Polymer Science* 137 (3): 48299.

63 Sinha-Ray, S., Lee, M.W., Sinha-Ray, S. et al. (2013). Supersonic nanoblowing: a new ultra-stiff phase of nylon 6 in 20–50 nm confinement. *Journal of Materials Chemistry C* 1 (21): 3491–3498.

64 Zhang, Q.Q., Wang, X.X., Fu, J. et al. (2018). Electrospinning of Ultrafine conducting polymer composite nanofibers with diameter less than 70 nm as high sensitive gas sensor. *Materials* 11 (9): 1744.

65 Müller, F., Jokisch, S., Bargel, H. et al. (2020). Centrifugal electrospinning enables the production of meshes of ultrathin polymer fibers. *ACS Applied Polymer Materials* 2 (11): 4360–4367.

66 He, J.X., Wang, L.D., Liu, R.T. et al. (2014). Fabrication of nanofibers with ultrahigh production by a facile high pressure air-jet atomized electrospinning. *Fibers and Polymers* 15 (11): 2283–2289.

67 Chen, H.B., Li, H.Y., Ma, X.L. et al. (2016). Large scaled fabrication of microfibers by air-suction assisted needleless melt electrospinning. *Fibers and Polymers* 17 (4): 576–581.

68 Valipouri, A., Ravandi, S.A.H., and Pishevar, A.R. (2014). A novel method for manufacturing nanofibers. *Fibers and Polymers* 14 (6): 941–949.

4

Manufacturing and Application of Electrospinning Nanofiber Yarn

Ailin Li, Liming Wang, and Xiaohong Qin

Donghua University, Ministry of Education, College of Textiles, Key Laboratory of Textile Science & Technology, 2999 Renmin North Road, Songjiang District, Shanghai, 201620, China

4.1 Introduction

Fibers and yarns are the most widely used forms of materials in the textile industry and have been used in various textile processing technologies for thousands of years. As far as the current textile processing technology is concerned, micron-diameter fibers are sufficient for daily wear purposes. However, more functional demands and better properties of fibers and yarns are required with the rapid development of emerging fields like smart textiles, functional textiles, and biomedical textiles. Nanofibers have been extensively attended to for the high specific surface area compared to normal micron-scale fibers, and more surfaced molecules and functional groups can be exposed, providing more possibilities for expanding the functionalization of fibers. Nanofiber yarns have greatly enhanced processability and structural expandability while maintaining the advantages of nanofibers and providing more opportunities in the practice application.

Electrospinning technology provides ideas for the processing of nanofiber yarns. From a series of US patents published by Anton Formhals in 1934, electrospinning technology was originally proposed to prepare linear nanofiber materials, similar to traditional textile yarns, applied to the textile field. Various nanofiber yarn production strategies have been reported now. Conjugate electrospinning, airflow-assisted electrospinning, and friction electrospinning methods have emerged in recent years, and the output of nanofiber yarn has been significantly increased. Moreover, the structural and functional design of the nanofiber yarns has also drawn the attention of researchers. For instance, pure nanofiber yarn, core-spun yarn, micro-/nanofiber composite yarn, and other kinds of nanofiber-based yarn with a complex structure are developed to meet the desire for practical application.

In this chapter, the development of fabrication strategies for nanofiber yarns is comprehensively reviewed. Various yarns based on nanofibers are summarized and categorized by the yarn architecture and fabricating methods, and the technological

Electrospinning: Fundamentals, Methods, and Applications, First Edition.
Edited by Liming Wang and Xiaohong Qin.

limitations of conventional strategies will be discussed while the novel yarn structure and corresponding application are highlighted. The remaining challenges and future development trends of fabrication of electrospinning nanofiber yarn strategies and their applications will also be discussed.

4.2 Electrospun Pure Nanofiber Yarns

4.2.1 Processing of Electrospun Pure Nanofiber Yarns

Pure nanofiber yarn refers to yarn composed entirely of nanofibers. The forming of pure nanofiber yarns is essentially a process of nanofiber forming, bundling, twisting, and winding. The nanofiber forming is based on electrospinning technology. The bundling, twisting, and wounding of nanofibers are controlled by the collecting units. So far, many researchers have reported different methods and devices for preparing nanofiber yarn. According to the method of nanofiber bundling and twisting, pure nanofiber yarn production methods can be mainly divided into the following six types.

4.2.1.1 Pure Nanofiber Yarn Bundling by Parallel Collector

This method was first proposed by Dolton et al. [1]. Two grounded rings were used to collect directionally aligned polycaprolactone (PCL) nanofibers, and the nanofiber bundles were twisted as the two discs were rotated in opposite directions, as shown in Figure 4.1a. The length of nanofiber yarns prepared by this method is subject to the distance between parallel receivers, and the prepared yarns are discontinuous fibrous short segments. In 2011, Yan et al. [2] improved this production method by inserting a rotating plastic tube in two rotating discs to collect the nanofiber yarn, and by controlling the speed of the rotating device to adjust the twist and diameter of the nanofiber yarn (Figure 4.1b).

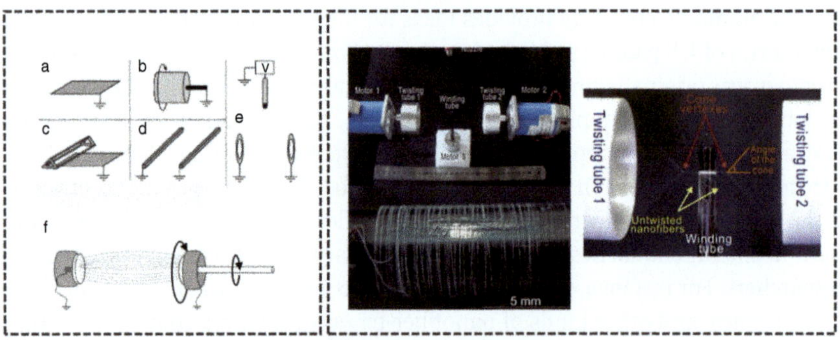

(a) (b)

Figure 4.1 (a) Typical electrospinning collectors. a. single plate, b. rotating drum, c. triangular frame placed near single plate, d. parallel dual plate, e. dual-grounded ring, f dual-rotating plates. Source: Dalton et al. [1]. Reproduced with permission of Elsevier, (b) Nanofiber collector equipped with two twisting tubes and a winding tube. Source: Yan et al. [2]. Reproduced with permission of Elsevier.

Teo et al. [3] designed a method for processing nanofiber bundles. They arranged two metal blades in a line with a certain spacing as a pair of receivers. The two metal blades are wrapped with insulating tape except for the blade part to narrow the deposition area of nanofibers, and the nanofiber bundles are collected between the two metal blades. Afterward, the nanofiber bundles are washed, dried, and twisted to obtain fixed-length nanofiber yarns. This kind of nanofiber bundling is usually done with high orientation. However, the yarn length is affected by the distance between the collectors and the continuous nanofiber yarns are hardly prepared parallel collectors.

4.2.1.2 Pure Nanofiber Yarn-Producing Methods by Rotating Collectors

Rotating collectors are the most common fiber collectors for nanofiber yarn fabrication because nanofibers can be bundled and twisted simultaneously with the collector rotation. The rotation collecting units increase the yarn-producing efficiency and make it possible for continuous nanofiber yarn manufacture.

Configuration of the Ring Collector System Lin et al. [4] reported a ring collector and two injectors system (Figure 4.2a). First, a nanofibers web was formed on the surface of the ring collector. Then the nanofiber yarn was deformed by drawing the nanofiber web from the web center. The nanofiber cone became stable with the formation of later-spun nanofibers. The stable nanofiber cone led to the formation of continuous yarns with the rotation of the ring collector. Similarly, Stylios et al. [5] prepared nanofiber yarns containing carbon nanotubes by controlling two vertically placed rotating discs. As shown in b of Figure 4.2b, the nanofibers are prepared from the downward-facing spinneret and held by the vertically set collecting discs. The rotation of disc M1 is used for twisting nanofibers into yarns, and the other is for yarn winding. The nanofiber web became very fragile by the tension in different directions because M1 and M2 rotated in different directions at the same time.

Electrospun nanofibers form a fibrous web first and then twisted into yarns with the rotation of the collecting unit and the nanofiber arrangement and orientation can be partly improved. On the other hand, the nanofibrous webs are usually thin and fragile during the yarn twisting and winding process. As a result, it is difficult to keep the nanofibrous web stable for continuous yarn fabrication.

Configuration of Rotating Drum System The method was first reported by Ko et al. [6], as shown in Figure 4.3a. This collector system is a combination of a ventilation device and a high-speed rotating drum to orient and bundle the nanofibers and then twist and collect the nanofiber yarns. The deposition area of the oriented nanofibers obtained by this method is large, which is not favorable for bundling and twisting of nanofibers. Fennessey and Farris [7] (Figures 4.3b) used an oriented arranged PAN nanofiber bundle prepared using a high-speed rotating roller, which is then twisted with an electric spinning machine. When preparing oriented nanofiber yarns with this method, the centrifugal force generated by the high-speed rotating receiver affects the collection of orientated nanofibers, and it is difficult for nanofiber bundles to deposit and twist continuously and stably. Therefore, there have been few reports of the rotating drum system in recent years.

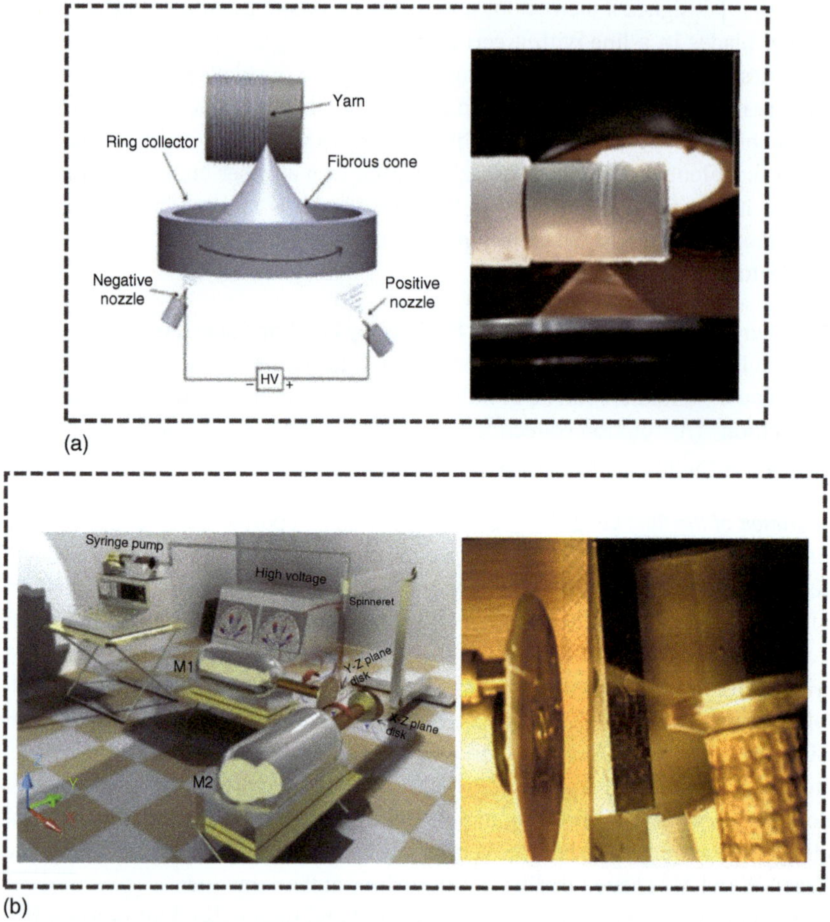

Figure 4.2 (a) Ring collector system. Source: Shuakat et al. [4]. Reproduced with permission of Wiley. (b) Ring twisting-winding system. Source: Bazbouz et al. [5]. Reproduced with permission of Elsevier.

4.2.1.3 Pure Nanofiber Yarn Producing Methods by Water Bath Collecting System

This method promotes fiber bundle orientation through the conductivity or surface tension and utilizes the liquid vortex to facilitate nanofiber bundling to twist into yarn. Both Khil et al. [8] and Smit et al. [9] have explored the methods of processing nanofiber bundles using liquid water baths, as shown in Figure 4.4a,b, where electrospun nanofibers are first deposited in a reservoir and then the nanofiber bundles are fished out and wound on a rotating drum. Unfortunately, serious adhesion between the nanofibers is produced by this method, and the polymers used are required to be insoluble in the coagulation bath, which is not suitable for all polymeric nanofibers. Teo et al. [10] improved the above device by chiseling a small hole in the middle of the reservoir, as shown in Figure 4.4c. They utilized the vortex of water to pull the nanofibers deposited in the water down out of the small hole and wind it in the form

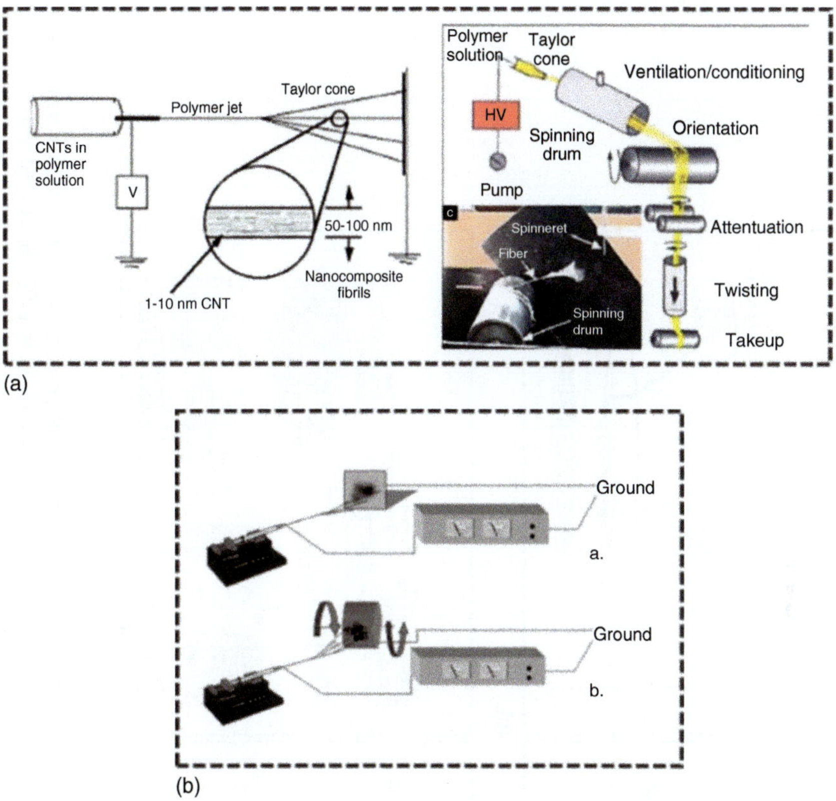

Figure 4.3 (a) Rotating drum collector. Source: Ko et al. [6]. Reproduced with permission of Wiley. (b) High-speed rotating roller. (a. stationary collector, b. rotating collector). Source: Fennessey et al. [7]. Reproduced with permission of Elsevier.

of yarn on a rotating drum. The method successfully prepared a nanofiber yarn of the copolymer of polyvinylidene fluoride and hexafluoropropylene (PVDF-co-HFP). Yousefzadeh et al. [12] added a disc-type device at the spinning needle to limit the deposition area of nanofibers in the aqueous solution to further increase the yarn twisting. The twisted nanofiber yarns were collected onto the rotating drum by a winding roller. Tian et al. [11] optimized the drafting and winding mechanism after bundling by adding a heating mechanism for drying the nanofiber yarn to further promote the alignment of nanofibers (Figure 4.4d). The orientation of nanofiber in the yarn can be improved by water bath system, but the types of polymer are limited.

4.2.1.4 Pure Nanofiber Yarn by Electric Field-Assisted System

An electric field-assisted system refers to the method of adding an auxiliary electrode in the electrospinning zone to adjust the electric field distribution, thereby enabling the agglomeration of nanofibers and the preparation of nanofiber yarns by mechanical twisting. Liu et al. [13] equipped an aluminum disc collector with an annular stainless-steel electrode, as shown in Figure 4.5a. The nanofibers are deposited on the stainless-steel electrode to obtain annular nanofiber bundles, and then the

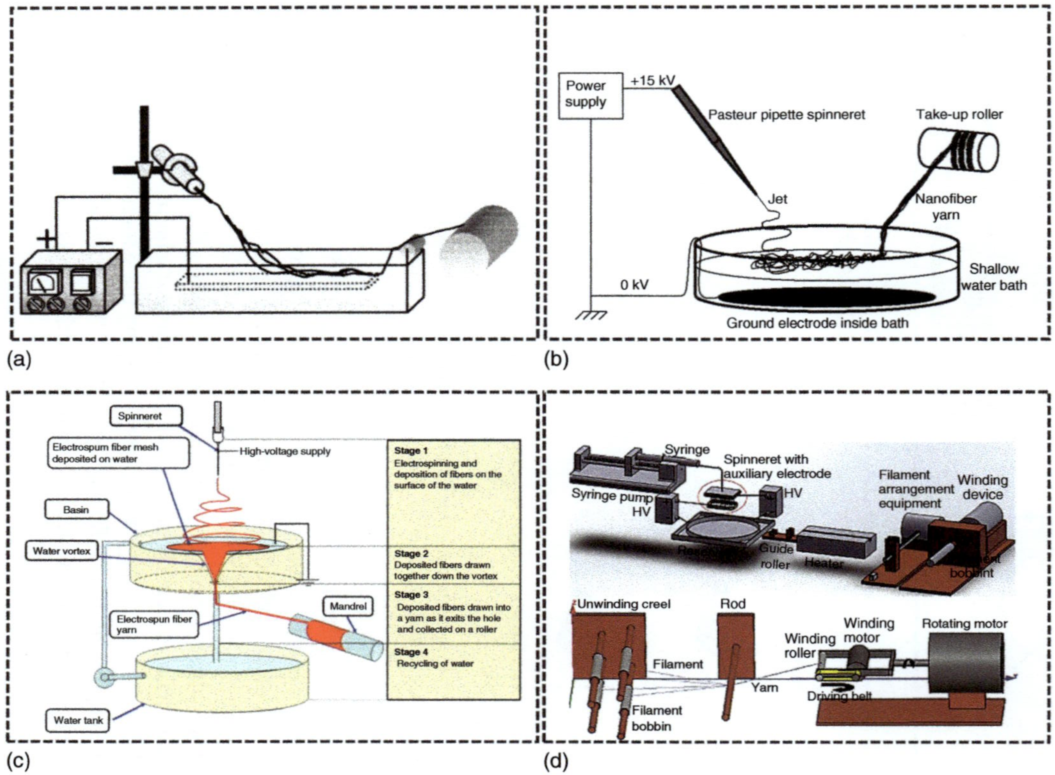

Figure 4.4 (a, b) Single-injector electrospinning nanofibers collected by water bath. Source: (a) Khil et al. [8]. Reproduced with permission of John Wiley &Sons. (b) Smit et al. [9]. Reproduced with permission of Elsevier. (c) Vortex-fluid water bath collection. Source: Teo et al. [10]. Reproduced with permission of Elsevier, (d) Multi-needle liquid bath collection system. Source: Tian et al. [11]/Springer Nature.

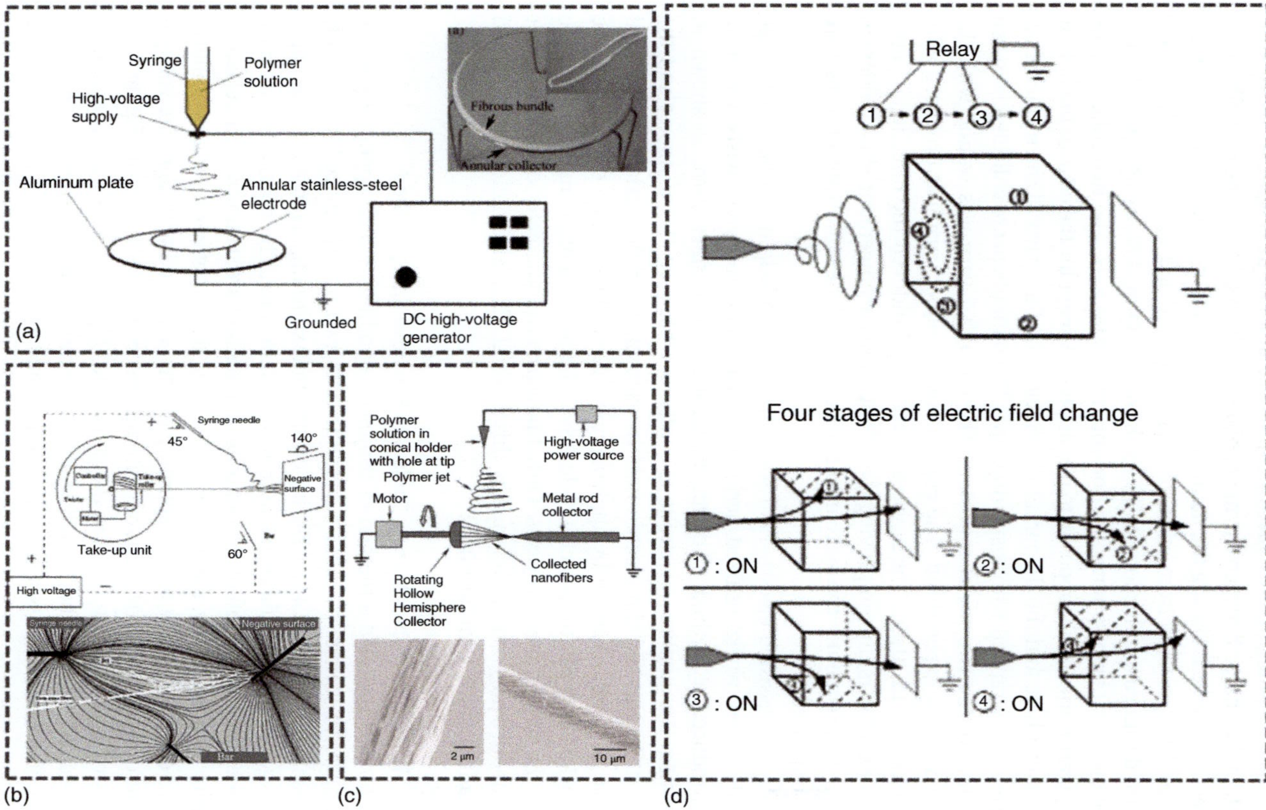

Figure 4.5 (a) Annular stainless-steel electrode. Source: Liu et al. [13]. Reproduced with permission of Elsevier. (b) Bar electrode. Source: Dabirian et al. [14]. Reproduced with permission of Wiley. (c) Spinner electrode. Source: Lotus et al. [15]. Reproduced with permission of American Institute of Physics. (d) Variable directional electric field. ①–④ are the side under an applied voltage at different moments. Source: Gu et al. [16]. Reproduced with permission of American Institute of Physics.

nanofiber bundles are drafted and twisted to form nanofiber yarns. The nanofiber yarns prepared by this method were not uniform and had hairy surfaces.

Dabirian et al. [14] fabricated nanofiber yarns using a pair of negatively charged metal rods and a metal plate collector configured at an angle (Figure 4.5b). By concentrating the charge between the metal plate and the metal rod, the nanofibers are oriented and aligned. During the nanofiber yarn twisting, the later-spun nanofibers can enter the yarn body from one side, then be twisted and wound together. Lotus et al. [15] designed a collection device, as shown in Figure 4.5c, using a pair of rotating hollow metal hemispheres and non-rotating metal rods. The two collectors are placed apart horizontally opposite to one another. Except for the edge of the hollow metal hemisphere and the tip part of the metal rod, other parts of the metal collector system are wrapped with insulating tape. The nanofibers are collected between the surface of the two metal collectors. The nanofiber yarn was wound with the movement of the tapered rod. The continuous nanofiber yarn was produced with a diameter of 5 to 30 μm. The assisted electric field induced the directional deposition of nanofibers thus improving the orientational arrangement of nanofibers in the yarns. Consequently, the mechanical properties and the uniformity of yarns were greatly improved.

Usually, the assisted electric field was introduced to help nanofibers to arrange parallelly; a few works utilized periodic electric fields to apply twists. Gu et al. [16] introduced a method of direct processing of nanofiber yarn, as shown in Figure 4.5d. An auxiliary electrode is configured between the spinneret and the collection device, and the electric field in each plane of the variable directional electrode is rapidly changed by applying one side of the auxiliary electrode at one time clockwise. A rotating electric field drives the rotation of the nanofiber bundle to twist the nanofibers into yarns. Only fine yarns (the diameter of the nanofiber yarn is about 5 μm) can be obtained by this method. Excessive nanofibers will lead to the failure of the yarn forming.

4.2.1.5 Twisted Nanofiber Yarn by Conjugate Electrospinning Method

Pan et al. [17] invented a method of conjugate bundling, which applies a positive and a negative voltage of equal magnitude onto two needles placed horizontally facing each other (Figure 4.6a). During electrostatic spinning, nanofibers with opposite charges on their surfaces are spun from two needles. The nanofibers are bundled under the interaction of the opposite charges and are wound and collected by the drum. Unfortunately, the device cannot twist the nanofiber bundles. Similarly, Dabirian et al. [18] use a pair of spinning injectors applied with opposite voltages and ungrounded collectors to produce nanofiber yarns, as shown in Figure 4.6b, but the orientation of the fibers in the yarn is not high. Conjugate electrospinning achieves effective nanofiber bundling due to the attraction of positive and negative charges on the nanofibers, while the orientational arrangement of nanofibers and yarn twists cannot be reached. Many works have been done to combine conjugate electrospinning and rotation collecting units for better nanofiber orientation and yarn twist. For instance, Ali et al. [19] used an ungrounded funnel-shaped metal target as a nanofiber collector, using two oppositely charged electrospinning nozzles to

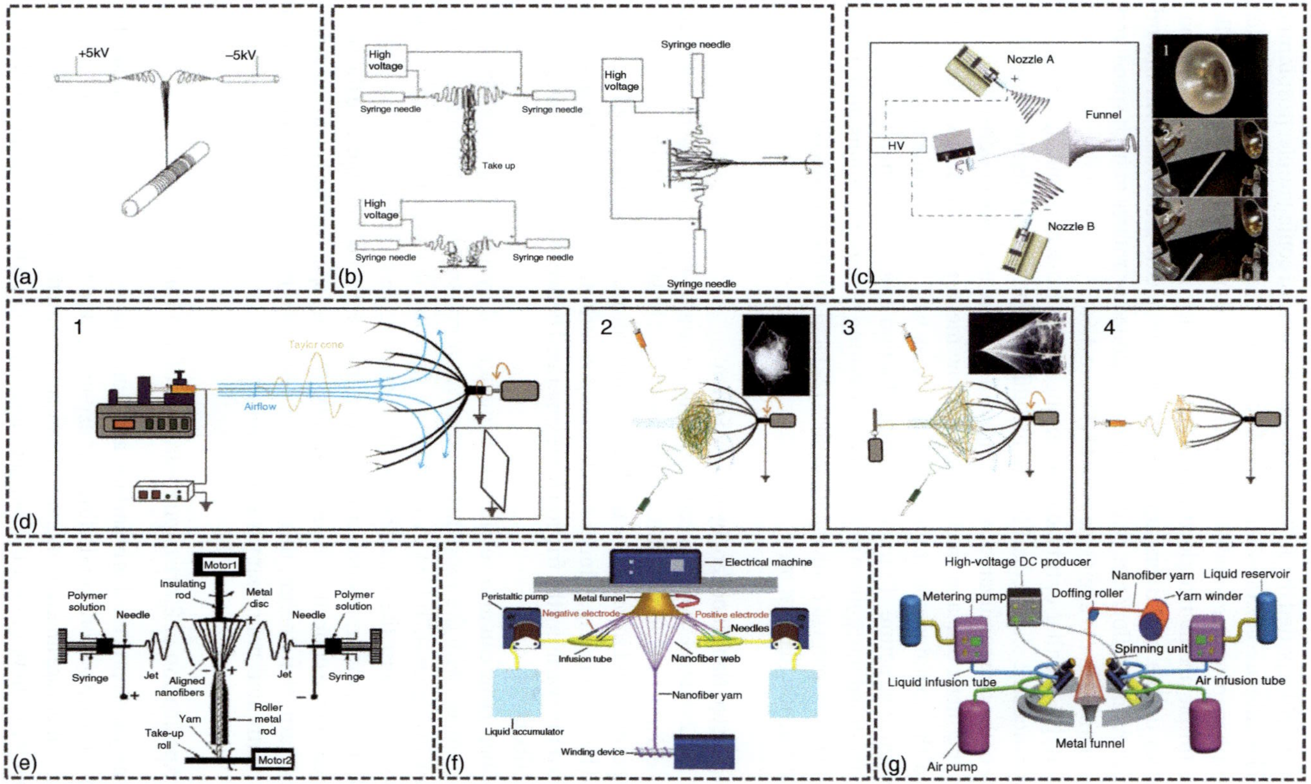

Figure 4.6 (a) Conjugate bundling device. Source: Pan et al. [17]. Reproduced with permission of Elsevier. (b) Conjugate bundling device. Source: Dabirian et al. [18]. (c) Conjugate electrospinning apparatus with a funnel collector. Source: Ali et al. [19]. Reproduced with permission of Wiley. (d) Airflow-assisted tip-induced conjugate electrospinning system. 1. schematic of the electrospinning device. 2–4. schematic for developing 3D fluffy mass, 1D yarn and 2D mats by the airflow-assisted tip-induced electrospinning system. Source: Joseph et al. [20]. Reproduced with permission of ACS Publications. (e) Conjugate electrospinning apparatus modified by a hollow metal rod and a metal disc. Source: Wu et al. [21]. Reproduced with permission of Elsevier. (f) Multiple conjugate electrospinning apparatus. Source: He et al. [22]. Reproduced with permission of John Wiley & Sons. (g) Multi-nozzle bubble electrospinning system. Source: He et al. [23]. Reproduced with permission of John Wiley & Sons.

generate a hollow nanofiber cone on the funnel collector, as shown in Figure 4.6c. Twisted nanofiber yarns can directly withdraw from the cone apex, with the rotation of the funnel collector. The twist of the nano yarn can reach 7400 rpm. Likewise, Joseph et al. [20] reported the airflow-assisted tip-induced conjugate electrospinning system. The unique collector was achieved by an array of equidistant point electrodes at the periphery of an open-ended frame made of flexible metallic rods. Each metallic rod has multiple needle-like electrodes at its head, which leads to the residual opposite charges on the nanofibers attracted toward the needle electrodes. An inward negative pressure airflow can be created as the collector rotates at high speed and drives the nanofibers to aggregate inward (Figure 4.6d).

Wu et al. [21] further optimized the conjugate electrostatic spinning system by adding a concentric sharp-edged hollow metallic rod and a metal disc in the vertical direction of the two conjugate nozzles (Figure 4.6e). The nanofibers were positioned between the metal disc and the tip of the hollow metal rod. The device effectively improved the spinning continuity and stability as well as the uniformity of yarn morphological structure. To increase the production efficiency of nanofiber yarns, He et al. modified the method of mechanical twisting of multiple conjugate bundles consisting of two pairs of conjugated needles [22] and bubble nozzles [23] as shown in Figure 4.6f and Figure 4.6g.

4.2.1.6 Twisted Nanofiber Yarn by Airflow System

Zhou et al. [24] developed an electrospinning apparatus by combining the stepped airflow field and traditional spinning (Figure 4.7a) to break the limitation of electrospinning output and improve the quality of nanofiber yarn. First, the nanofibers are collected by two dust cages under negative air pressure, and the arranged nanofibers are twisted into yarns by the friction force generated between the friction rollers and the nanofibers.

Nanofiber yarns can also be directly manufactured by a needleless melt electrospinning process with the assistance of negative air pressure and rotating collection, as shown in Figure 4.7b [25]. A negative air pressure device was used to bundle nanofibers and twist the nanofibers with a rotating disk. The needle-free nozzle greatly improves the production efficiency of nanofiber yarns, and adjusting the ratio of the speed of the collecting rollers to the disc rotating speed is the key to the twisting of the yarn. Nonetheless, this technology requires strict airflow stability control to keep continuous nanofiber yarn fabrication.

4.2.2 Application of Electrospun Pure Nanofiber Yarns

Nanofiber yarns process high surface area, and more functional groups can be exposed, providing more possibilities for expanding the functionalization of fibers. Also, nanofibers are an excellent matrix for loading functional particles. Meanwhile, nanofiber yarns have more interfiber gaps and channels, which can greatly enhance the wicking effect of the yarn. Unique characteristics of nanofiber yarn display irreplaceable advantages in many application fields.

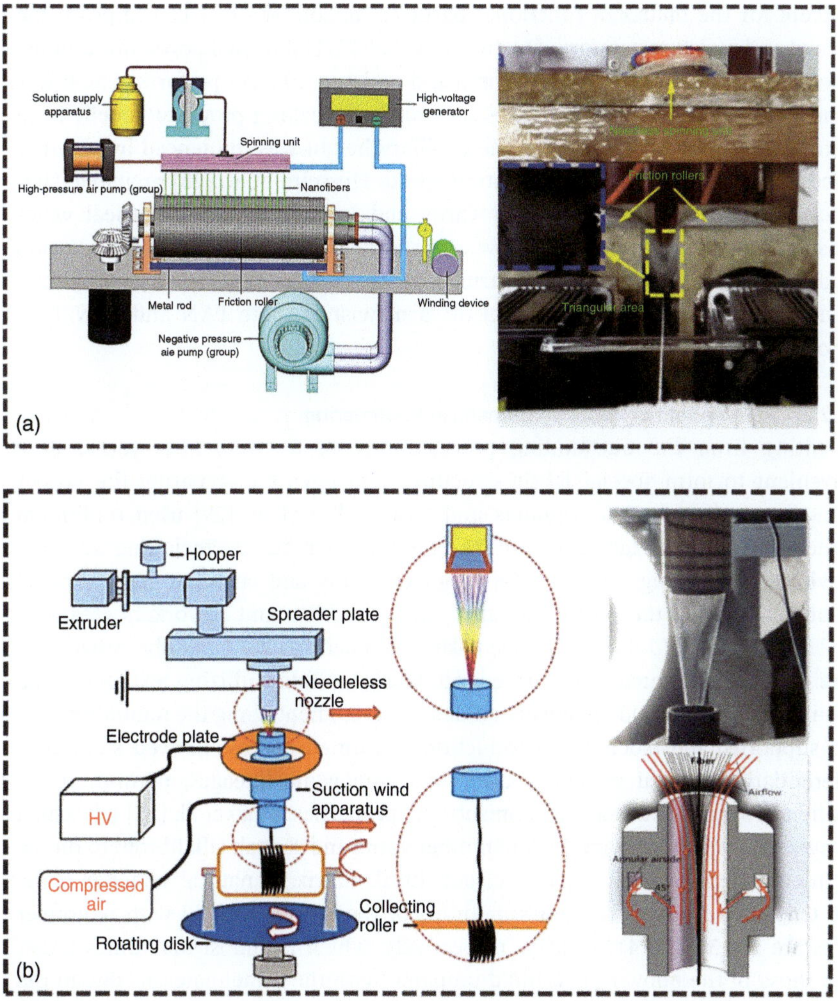

Figure 4.7 (a) Stepped airflow electrospinning and friction twisting-based electrospinning setup. Source: Zhou et al. [24]/Springer Nature. (b) Needleless melt electrospinning nanofiber yarn manufacturing device. Source: Ma et al. [25]. Reproduced with permission of John Wiley & Sons.

4.2.2.1 Pure Nanofiber Yarns in Functional Textiles

For antibacterial textiles, Hossein Barani [26] created antibacterial nanofibrous hybrid yarns comprising PVA and PLLA with antibacterial silver nanoparticles. The findings showed that the silver nitrate, which increases the conductivity of the polymer solution, has an impact on the surface fiber shape and fiber diameter. The results of the mechanical testing demonstrate that the hybrid yarn exhibits less stress and strain at the break when loaded than when unloaded. In addition, the AgNps-loaded nanofiber yarns present good bactericidal and bacteriostatic efficiency.

Except for the matrix of functional particles, nanofiber yarns can improve the thermal and moisture comfort of textiles. Jin et al. [27] fabricated polysulfone amide nanofiber yarns and studied the thermal and moisture transfer performance. It was found that nanofiber yarn possesses an excellent wicking property benefit from its unique nanofibrous structure. The capillary height was influenced by polymer concentration and the collector rotating speed. Hosseini et al. [28] prepared TiO_2 nanoparticle-loaded PAN nanofiber yarns and investigated their vertical water transport behavior. The capillary flow was observed by the rise of colored solution soaked up by the PAN nanofiber yarns. The result presented the lower capillary height was caused by a higher level of yarn twist for both PAN and PAN/TiO_2 nanofiber yarns.

4.2.2.2 Pure Nanofiber Yarns in Biomedical Engineering

Benefiting from the outstanding processability of the nanofiber yarns, it is convenient to form special fabric structures and sizes for preparing the proper biomedical materials, like scaffolds and tissues. Wu et al. [29] used traditional textile techniques to manufacture PAN nanofiber yarn fabrics, including weaving, weaving, and knitting, showing the re-processability and potential application of nanofiber yarns in the textile industry. After that, Wu and coworkers created a woven fabric scaffold for heart valve tissue engineering [30]. Nanofiber yarns that were uniaxially oriented were formed into fabric scaffolds utilizing textile-forming techniques. The scaffold effectively guided the alignment along the nanofiber yarn and supported the biochemical induction of human adipose-derived stem cells differentiation toward smooth muscle and osteogenic lineages, permitting the creation of synthetic tissues with anisotropic properties. Shao et al. [31] fabricated a woven scaffold by weaving PLA nanofiber yarns and Tussah silk fibroin to mimic the hierarchical architecture of collagen fibril matrix in natural lamellar bone. The tensile strength and Young's modulus of the woven scaffold with nanofiber yarns are 180.36 and 417.65 MPa, respectively, which is almost four times larger than those of the nonwoven scaffold prepared with the same materials due to the orientational arrangement of nanofibers. The woven scaffolds also promote the proliferation of mouse mesenchymal stem cells and biomineralization. Concurrently, it shows excellent mechanical properties, biocompatibility, osteoinductivity, and regenerative activity.

Joseph et al. [32] developed a woven fabric for artificial vascular by meticulously interweaving nanofiber yarns in the longitudinal and transverse directions. This nanofiber yarn woven textile shows remarkable enhancement of both physical and biological performance benefits from the multitude of hierarchically arranged nanofibers. The oriented and ordered arrangement of nanofibers enhances protein adhesion and promotes cell adhesion and proliferation. The in vivo testing result proved that the nanofiber woven conduit was robust, suturable, and thrombosis-free. Li et al. [33] developed a highly stretchable helical scaffold that can seed cells. The electrospun helical tissue presents a toughness of 57 GJ m^{-3} and stretchability of 15 times its length. The helical architecture of the electrospun tissue protects the seeding living cell by avoiding fiber deformation during the stretching process.

Abhari et al. [34] used an industrial braiding machine to produce multifilament electrospun yarns with complex yarn structures for medical sutures. The production speed is five meters of stretched filament per hour and with a linear density of nine tex. The electrospinning braided yarns possess a higher porosity and a lower tensile strength compared with commercial sutures.

4.2.2.3 Pure Nanofiber Yarns in Other Fields

Heterogeneous nanofiber yarn can achieve two or more functions independent of each other by conjugate electrospinning. Fan et al. [35] successfully prepared bifunctional heterogeneous nanofiber yarns by conjugating electrospinning for the first time. Two kinds of spinning solutions were loaded in the oppositely charged spinning injectors and electrospun to fabricate nanofiber yarns. The morphology, magnetism, and fluorescent performance were investigated. The separating strategy avoids the strong absorption of fluorescent by dark-colored Fe_3O_4. The intensities of fluorescent excitation of heterogeneous nanofiber yarns are about 25 times stronger than those of homogeneous nanofiber yarns with the same composition and concentration. Later, a tri-functional nanofiber yarn of superparamagnetism, electronic conductivity, and fluorescence was prepared by Fan and coworkers [36] in a similar method.

4.3 Electrospun Core-spun Yarns

4.3.1 Processing of Electrospun Core-spun Yarns

4.3.1.1 Single-needle Electrospun Core-spun Yarn-producing System

Compared with pure nanofiber yarns, electrospun core-spun yarns refers to those with a core of fiber bundles, filaments, or yarns covered by electrospun nanofibers on the surface. In 2009, Zhou et al. [37] reported a method for nanofiber coating on the filaments by electrospinning (Figure 4.8a). They collected the nanofibers by placing monofilaments or compound filaments horizontally in the nanofiber-deposited zone. The filament was placed perpendicular to the spinneret between the electrospinning nozzle and the collecting plate. When the filament is withdrawn from the bobbin and is placed directly toward the electrospinning zone, the nanofibers coat the filament surface. Zhou et al. [38] changed the collecting plate to a rotating round disc (Figure 4.8b). Su et al. [40] used metal cylindrical discs as nanofiber collectors to deposit nanofibers. However, the arrangement of nanofibers on the surface of the cored yarn was disordered. Yang et al. [41–43] used a similar method to deposit nanofibers on the core yarn, but the nanofibers prepared by this method were disorderly arranged, and the uniformity of the cladding structure was difficult to control. Liu [39] et al. added a set of grounded thin electrodes between the disc-shaped twisting devices, inducing the nanofibers to orient themselves vertically to the direction of movement of the core yarn, and twist under the action of disc rotation (Figure 4.8d). This improvement allowed the nanofibers to be uniformly wrapped around the surface of the core yarn.

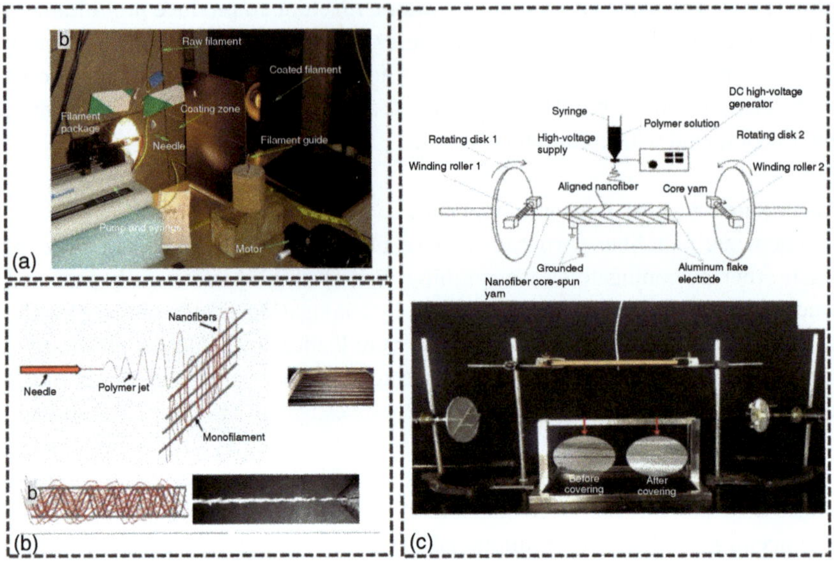

Figure 4.8 (a) Single-polymer jet coating nanofibers on the filament. Source: Zhou et al. [37]. Reproduced with permission of Elsevier. (b) Single-polymer jet coating nanofibers on multifilaments. Source: Zhou et al. [38]. Reproduced with permission of Elsevier. (c) Nanofibers deposited on a filament rotated by two rotating discs. Source: Liu et al. [39]. Reproduced with permission of Elsevier.

4.3.1.2 Conjugate Electrospun Core-spun Nanofiber Yarn-producing System

Conjugate electrospinning is the most common method to prepare the core-spun nanofiber yarn. The advantage of conjugate electrospinning is the uniformity of nanofiber wrapping and the controllable twist of the coated nanofibers. The difference between the various conjugate spinning devices lies mainly in the shape of the receiver. Dabirian et al. [44] reported a conjugate electrospinning system to produce core-spun nanofiber yarns (Figure 4.9a). A copper wire was introduced into the electrospinning zone, and electrospun nanofibers contact with the wire surface and form a spinning cone. The core-spun nanofiber yarn was twisted by the rotating twister. Mo et al. [45] prepared a kind of core-spun yarn by polylactic acid-glycolic acid microfibers covered with polylactic acid-glycolic acid nanofibers using conjugate electrospinning equipment with two oppositely charged nozzles (Figure 4.9b). The core-spun yarn shows great mechanical properties and strong capillary action. Padmakumar et al. [46] prepared a nanofiber core-spun yarn with poly-L-lactic acid micron fiber as the core, and poly-lactic-co-glycolic acid nanofibers as the sheath, using a tip-arc electrode array receiver device (Figure 4.9c). Mao et al. [47, 49, 50] prepared structurally stable nanofiber core-spun yarns by tip-induced conjugate electrospinning method. The core-spun yarn has a controllable twist and adjustive diameter, Figure 4.9d. However, low preparation efficiency is still a difficulty in the development of nanofiber core-spun yarn preparation technology.

He and his colleagues [48] created the multi-nozzle air jet electrospinning equipment to create the polyacrylonitrile/viscose nanofiber core-spun yarn, shown

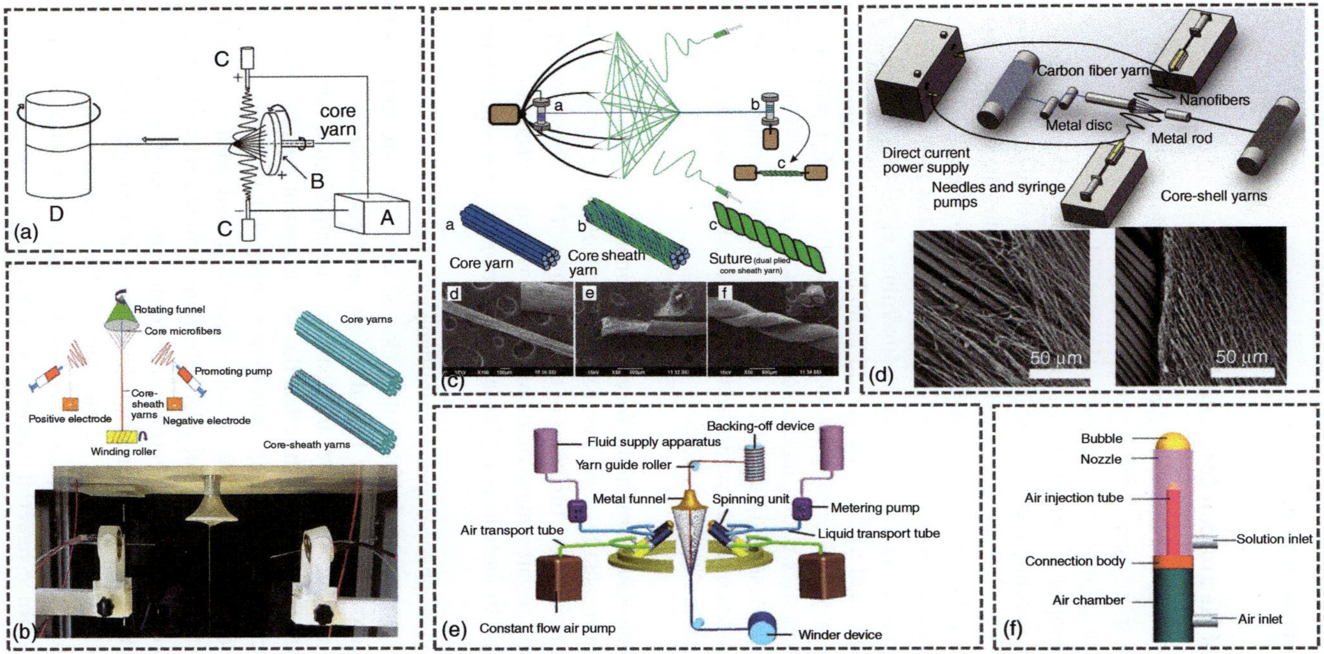

Figure 4.9 (a) Conjugate electrospinning collected by a disc. A. high voltage DC, B. twister plate, C. syringe needle, D. take up unit. Source: Dabirian et al. [44]. Reproduced with permission of John Wiley & Sons. (b) Conjugate electrospinning collected by a funnel. Source: Gu et al. [45]. Reproduced with permission of Elsevier. (c) Conjugate electrospinning collected by a tip-arc array. a. core yarn, b. core-sheath yarn, c. suture. d–f. SEM images of the core yarn, the core-sheath yarn and the suture. Source: Padmakumar et al. [46]. Reproduced with permission of ACS Publications. (d) Conjugate electrospinning collected by a metal disc and a metal rod. Source: Mao et al. [47]. Reproduced with permission of Elsevier. (e) Multi-nozzle air jet electrospinning apparatus. Source: He et al. [48]. Reproduced with permission of Springer Nature. (f) Structural illumination of the air jet electrospinning nozzle. Source: He et al. [48]. Reproduced with permission of Springer Nature.

in Figure 4.9e,f. By altering the electrospinning settings and the rotating speed of the funnel collector, they were able to manage the coating rate, winding speed, and twist level of the core-spun yarn. This technique results in orientated and tightly twisted electrospun core-spun yarns.

4.3.2 Application of Electrospun Core-spun Yarns

4.3.2.1 Electrospun Core-spun Yarns in the Biomedical Engineering Field

The electrospun core-spun yarns have higher breaking strength offered by the inner core yarn, which promotes its application in surgical sutures and muscles. Wang et al. [51] developed a core–shell scaffold that employed electrospinning nanofiber yarn as the core, and a photocurable hydrogel as the shell for simulating native skeletal muscle tissue. The core–shell scaffold was fabricated by dry–wet electrospinning and the hydrogel shell was adhered by photo-cross-linking. The core–shell scaffold demonstrated excellent biocompatibility and performance in inducing cellular alignment and elongation. Padmakumar et al. [46] prepared a central poly-L-polylactic acid and polylactic acid nanofiber core-spun yarn using a tip electrode array receiver for surgical sutures. This drug-loaded polymeric core-spun yarn demonstrates strong mechanical properties and long-time active drug release. The drug-release surgical suture enhanced cellular migration in the wound-healing assay. The core-spun yarn is an ideal template for loading drugs and controlling drug release as a surgical suture.

4.3.2.2 Electrospun Core-spun Yarns in the Wearable Electronics

1D wearable electronics usually need the assembling of different materials in multi-layers for reaching different functions in the electronics. The core-spun yarns with core-sheath structures are proper candidates for 1D wearable electronics and have already been used in electric skin [52], energy storage [47, 53], and energy harvesters [54, 55].

You et al. [52] coated graphene-doped polyurethane nanofibers on the surface of Ni-coated conductive cotton yarn for fabricating electric skin. The core-spun yarn was helically wound around an elastic thread. As-prepared composite yarns are woven into a fabric to sense external force by measuring the capacitance variations of the fabric electrode. The electric fabric has the advantages of a wide sensing range, high sensitivity, low detection limit, and short response time.

Mao et al. [47] reported a novel type of supercapacitor based on a core-spun polyaniline nanowire array growing on aligned carbon nanofibers/carbon fiber yarn electrodes. The core-spun yarn electrode, assembled into solid-state supercapacitors, shows a high specific capacitance of 234 mF/cm^2 at a current density of 0.1 mA/cm^2 in EMIMBF$_4$/PVDF/DMF gel electrolyte. The adequate pores in the core-spun yarn effectively improve the specific surface area of the yarn and the accessibility of the electrode material, showing a high energy density and cycle life. By wrapping PAN nanofibers around the core Ni-cotton yarn, Sun et al. [53] presented a core-spun yarn-based supercapacitor. The yarn/nanofiber/PEDOT:PSS nanoparticle hierarchical structured electrode material enhanced specific area and pseudocapacitance. The core-spun exhibits a high volumetric capacitance of 26.88 F cm^{-3} at 0.08 A cm^{-3} as well as high energy and power densities of 9.56 mWh and 830 mW cm^{-3}.

A wearable piezoelectric energy harvester based on core-spun yarns was initially electrospun by Ji et al [54]. The electrospun piezoelectric shell is coated on the inner conductive thread. Subsequently, the core-spun yarn electrode was braided with two strands of conductive thread to improve the deformation according to various body movements during the energy-harvesting process. The energy could be directly harvested by wearing garments like gloves and guards made from core-spun yarns. Ma et al. [55] coated the conductive silver wire with polyvinylidene fluoride/polyacrylonitrile core-spun nanofiber yarns and studied their triboelectric power generation performance. Textile-based triboelectric nanogenerators, which consist of this kind of electrospinning core-spun yarns, show a high energy-harvesting performance of 40.8 V, 0.705 μm cm^{-2}, and 9.513 nC cm^{-2} by applying 5 V, with 2.5 Hz mechanical drive.

4.3.2.3 Electrospun Core-spun Yarns in the Functional Textiles

The core-spun yarn also shows great potential in fabric moisture transfer because of enhanced water transport through the core and the sheath layer, led by the diameter difference between the inner and outer fibers. To examine the contact force between the nanofiber layer and microfiber layer of the core-spun yarn, Mao et al. [56] devised a novel experiment. The results display a favorable association between the contact force and the polarity and hairiness of the core yarns. Additionally, the wetting performance of these fabrics made from core-spun yarn shows improved moisture management and moisture-wicking capabilities. Afterward, Mao et al. [50] prepared a series of core-spun yarn with polymer nanofibers as the sheath and cotton yarn as the core and studied its wetting properties. Finite element simulation was carried out to simulate the velocity and pressure fields of the core-spun yarns. The simulation results show that the initial transport of water velocity at the outer boundary is greater than that in the conventional yarn structure. The water transport is accelerated in the core-spun nanofiber yarns by the push–pull effect and the differential capillary effect. As a result, the core-spun yarn fabric shows a high one-way transport index of 846%. Later, Mao et al. [49] studied the relationship between the surface wettability of nanofibers and the wetting properties of the core-spun yarn fabric. DTAB was added to the polymer solution to adjust the contact angle of the electrospun nanofibers. The result shows that the capillary pressure of nanofibers in the core-spun yarn can be changed by adjusting the surface wettability. Water transport between the sheath and the core layer is driven by the capillary pressure. The wettability of the fabrics fabricated by these core-spun yarns displays an ultrahigh one-way transport index of 1034.5% with an overall moisture management capability of 0.88.

4.3.2.4 Electrospun Core-spun Yarns in Gas Sensors

Liu et al. [57] prepared polypyrrole/polyacrylonitrile nanofiber core yarns and polyaniline/polyacrylonitrile nanofiber yarns for ammonia sensing by the conjugated beam method. On the one hand, the high specific surface area of the nanofibers exposes more gas contact sites; on the other hand, the orientation structure of the yarn and the effectiveness of carrier transport are improved, so that the nanofiber yarn shows good responsiveness and recovery. Similarly, Wu et al. [58] reported an ammonia sensor based on electrospinning core-spun polyaniline (PANI) polyacrylonitrile (PAN) nanofibers via combining in situ solution polymerization and conjugate electrospinning method. This ammonia sensor shows excellent sensitivity and fast response/recovery upon exposure to NH_3 of 10–2000 ppm at room temperature.

4.4 Micro-/nanofiber Composite Yarns

Functional textiles like antibacterial, moisture management, and one-way moisture transport are desirable for the needs of health and comfort of clothing. The major issue for electrospun nanofiber functionalized textiles is incorporating functional nanofibers into the textile substrates with strong adhesion. In this regard, many works have been done to embed nanofibers in the cotton web during the cotton yarn spinning process directly for enhancing the adhesion of functional nanofibers and yarn substrate. A new type of micro-/nanofiber composite yarn has been developed and the corresponding methods and devices are designed.

4.4.1 Processing of Micro-/Nanofiber Composite Yarns

By electrospinning nanofibers onto the carded cotton web and then spinning the cotton as usual, Jiang et al. [59] created antibacterial nanofiber/cotton hybrid yarns (Figure 4.10a). The results present that although the nanofiber/cotton hybrid yarns have high antibacterial activity against both Gram-positive and Gram-negative bacteria, they only contain a small amount of triclosan. Yang et al. [60, 62] investigated the processability, structure, and properties of micro-/nanofiber composite yarns fabricated by the trans-scale spinning method (Figure 4.10b). The fluorescence tracing method was used to show the distribution of the electrospinning nanofibers (Figure 4.10c). The results demonstrate that the electrospun nanofibers are uniformly distributed along the axial of yarns in all kinds of micro-nanofiber composite yarns. Besides, the nanofibers show shorter and more uniform segments distributed in the chemical fibers than cotton fibers. The addition of nanofibers provides excellent physical performance such as tensile properties, yarn hairiness, and yarn unevenness. Afterward, Chen et al. [61] combined traditional ring spinning and blend electrospinning to fabricate composite core-spun yarn of submicron-cotton/polyester (Figure 4.10d). The submicron-cotton fibers are made using a carding machine equipped with a free-surface electrospinning device, and the submicron-cotton/polyester composite core-spun yarn is made using a rolling frame with a modified bell mouth. The intermixing and entwining of the submicron-cotton fibers improve the mechanical characteristics of submicron-cotton/polyester core-spun yarn.

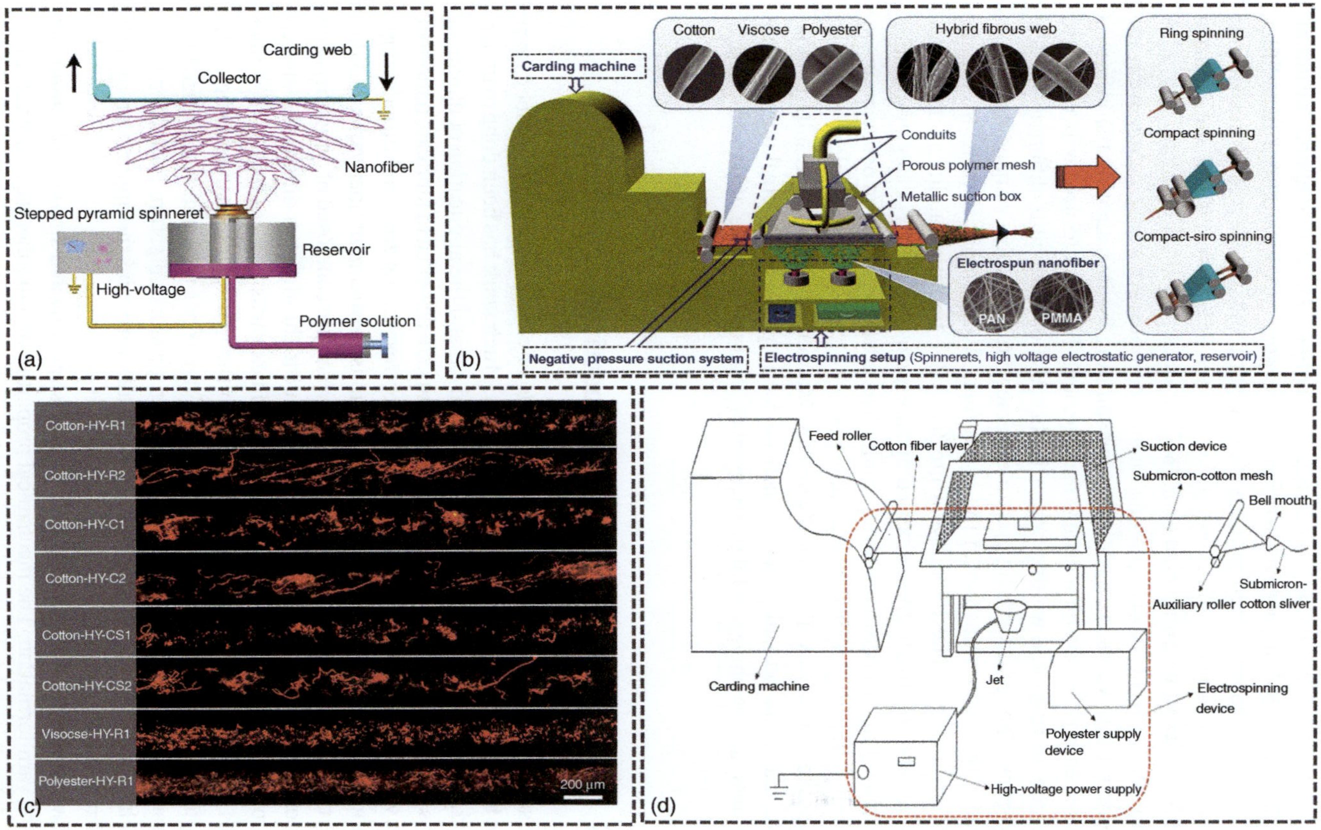

Figure 4.10 (a) Cotton/nanofiber composite yarn spinning system. Source: Jiang et al. [59]. Reproduced with permission of Springer Nature. (b) Trans-scale composite yarn spinning system. (c) Confocal microscopy images of micro/nanofiber composite yarns. Source: Yang et al. [60]/from SAGE Publications. (d) Trans-scale composite core-spun yarn spinning system. Source: Chen et al. [61]. Reproduced with permission of Taylor & Francis.

4.4.2 Application of Micro-/Nanofiber Composite Yarns

Qiu et al. [63] developed an antibacterial textile using the electrospinning blended yarn composed of antibacterial electrospun nanofibers and cotton fibers. The inhibition rate was 99.99% for both *Escherichia coli and Staphylococcus aureus*. In addition, this textile based on electrospinning blended yarn had a low cytotoxic response to the seeded cells in vitro and shows still more than 95% against *E. coli* and *S. aureus* after 35 cycles of washing. Mao et al. [64] prepared two kinds of micro/nano-fiber composite yarn by electrospinning nanofiber onto a carding cotton web and spinning it into yarns via traditional ring spinning for investigating the influence of the sheath polymers on moisture management. PAN and PS are the two selected polymers to study the wetting properties because of their quite different physical properties. The results exhibit that an extremely small number of nanofibers could enhance the water transport in the micro-/nanofiber composite yarns from finite element analyses. The PS shows a greater water transport ability in the fabric woven with the micro-/nanofiber composite yarns proved by water evaporation rate and capillary pressure analysis results. The effect of migration behavior during the roller drafting of electrospun nanofibers on the quality of micro/nanofiber composite yarns was also investigated by Mao et al. [65]. The results demonstrate that PAN nanofibers are beneficial for improving the breaking strength, hairiness, and evenness, while PS nanofibers had unfavorable effects on physical properties.

4.5 Conclusions and Future Perspectives

The structure design and production methods of electrospinning nanofiber yarns have been developing rapidly over the decades. Much work has been done to understand the nanofiber formation mechanism and expand the application of nanofiber yarns. Several kinds of electrospinning apparatus are designed to fabricate nanofiber yarn via controlling the electric field in the electrospinning zone and introducing a rotational collecting device to prepare continuous twisted nanofiber yarns. In addition, the study of composite nanofiber yarns, which are composed of nanofibers and microfibers, has drawn widespread attention due to the diameter difference between microfibers and nanofibers. Both pure nanofiber yarns and composite nanofiber yarns display broad application prospects with their structural characteristics. This chapter overviewed the developing history of nanofiber yarn and summarized the characteristics and shortcomings of each yarn production setup. Nanofiber yarns are classified by the type of fibers that constitute yarn, and the application of each kind of nanofiber-based yarn is also involved.

However, few studies have investigated the stretching and stressing of nanofibers during yarn formation. The mechanical properties of nanofiber yarns are still weak and difficult to withstand damage in the common processing methods of textile yarns, which limits the further application of nanofiber yarns in the textile industry. Moreover, most of the nanofiber yarn formation apparatuses are equipped with needle-spinnerets, which exhibit a severely slow production scale of nanofiber yarn. Needleless spinnerets should be considered when designing higher-throughput nanofiber yarn formation systems.

References

1 Dalton, P.D., Klee, D., and Möller, M. (2005). Electrospinning with dual collection rings. *Polymer* 46 (3): 611–614.

2 Yan, H., Liu, L., and Zhang, Z. (2011). Continually fabricating staple yarns with aligned electrospun polyacrylonitrile nanofibers. *Materials Letters* 65: 2419–2421.

3 Teo, W.E. and Ramakrishna, S. (2005). Electrospun fibre bundle made of aligned nanofibres over two fixed points. *Nanotechnology* 16 (9): 1878–1884.

4 Shuakat, M.N. and Lin, T. (2016). Direct electrospinning of nanofibre yarns using a rotating ring collector. *Journal of the Textile Institute* 107 (6): 238–239.

5 Bazbouz, M.B. and Stylios, G.K. (2008). Novel mechanism for spinning continuous twisted composite nanofiber yarns. *European Polymer Journal* 44: 1–12.

6 Ko, F., Gogotsi, Y., Ali, A. et al. (2003). Electrospinning of continuous carbon nanotube-filled nanofiber yarns. *Advanced Materials* 15 (14): 1161–1165.

7 Fennessey, S.F. and Farris, R.J. (2004). Fabrication of aligned and molecularly oriented electrospun polyacrylonitrile nanofibers and the mechanical behavior of their twisted yarns. *Polymer* 45 (12): 4217–4225.

8 Khil, M.S., Bhattarai, S.R., Kim, H.Y. et al. (2005). Novel fabricated matrix via electrospinning for tissue engineering. *Journal of Biomedical Materials Research, Part B: Applied Biomaterials* 72B: 117–124.

9 Smit, E., Bűttner, U., and Sanderson, R.D. (2005). Continuous yarns from electrospun fibers. *Polymer* 46 (8): 2419–2423.

10 Teo, W.E., Gopal, R., Ramaseshan, R. et al. (2007). A dynamic liquid support system for continuous electrospun yarn fabrication. *Polymer* 48 (12): 3400–3405.

11 Tian, L., Yan, T., and Pan, Z. (2015). Fabrication of continuous electrospun nanofiber yarns with direct 3D processability by plying and twisting. *Journal of Materials Science* 50 (21): 7137–7148.

12 Yousefzadeh, M., Latifi, M., Teo, W.E. et al. (2011). Producing continuous twisted yarn from well-aligned nanofibers by water vortex. *Polymer Engineering and Science* 51 (2): 323–329.

13 Liu, C.K., Sun, R.J., Lai, K. et al. (2008). Preparation of short submicron-fiber yarn by an annular collector through electrospinning. *Materials Letters* 62 (29): 4467–4469.

14 Dabirian, F., Hosseini, Y., and Hosseini Ravandi, S.A. (2007). Manipulation of the electric field of electrospinning system to produce polyacrylonitrile nanofiber yarn. *Journal of the Textile Institute* 98 (3): 237–241.

15 Lotus, A.F., Bender, E.T., Evans, E.A. et al. (2008). Electrical, structural, and chemical properties of semiconducting metal oxide nanofiber yarns. *Journal of Applied Physics* 103 (2): 024910.

16 Gu, B.K., Shin, M.K., Sohn, K.W. et al. (2007). Direct fabrication of twisted nanofibers by electrospinning. *Applied Physics Letters* 90 (26): 263902.

17 Pan, H., Li, L., Hu, L., and Cui, X. (2006). Continuous aligned polymer fibers produced by a modified electrospinning method. *Polymer* 47 (14): 4901–4904.

18 Dabirian, F. and Hosseini, S.A. (2009). Novel method for nanofibre yarn production using two differently charged nozzles. *Fibres & Textiles in Eastern Europe* 17 (3): 45–47.

19 Ali, U., Zhou, Y., Wang, X., and Lin, T. (2012). Direct electrospinning of highly twisted, continuous nanofiber yarns. *Journal of the Textile Institute* 103 (80–88): 10.

20 Joseph, J., Nair, S.V., and Menon, D. (2015). Integrating substrateless electrospinning with textile technology for creating biodegradable three-dimensional structures. *Nano Letters* 15 (8): 5420–5426.

21 Wu, S.H. and Qin, X.H. (2013). Uniaxially aligned polyacrylonitrile nanofiber yarns prepared by a novel modified electrospinning method. *Materials Letters* 106: 204–207.

22 He, J., Qi, K., Zhou, Y., and Cui, S. (2014). Multiple conjugate electrospinning method for the preparation of continuous polyacrylonitrile nanofiber yarn. *Journal of Applied Polymer Science* (8): 131, 40137.

23 He, J.X., Qi, K., Zhou, Y.M., and Cui, S.Z. (2014). Fabrication of continuous nanofiber yarn using novel multi-nozzle bubble electrospinning. *Polymer International* 63 (7): 1288–1294.

24 Zhou, Y., Wang, H., He, J. et al. (2018). Novel method for preparation of continuously twisted nanofiber yarn based on a combination of stepped airflow electrospinning and friction twisting. *Journal of Materials Science* 53 (22): 15735–15745.

25 Ma, X., Zhang, L., Tan, J. et al. (2017). Continuous manufacturing of nanofiber yarn with the assistance of suction wind and rotating collection via needleless melt electrospinning. *Journal of Applied Polymer Science* 134 (20): 44820.

26 Barani, H. (2014). Antibacterial continuous nanofibrous hybrid yarn through in situ synthesis of silver nanoparticles: preparation and characterization. *Materials Science & Engineering, C: Materials for Biological Applications* 43: 50–57.

27 Jin, S., Xin, B., and Zheng, Y. (2019). Preparation and characterization of polysulfone amide nanoyarns by the dynamic rotating electrospinning method. *Textile Research Journal* 89 (1): 52–62.

28 Hosseini Ravandi, S.A., Mehrara, S., Sadrjahani, M., and Khodaparast Haghi, A. (2020). Tunable wicking behavior via titanium oxide embedded in polyacrylonitrile nanofiber strings of yarn. *Polymer Bulletin* 77 (1): 307–322.

29 Wu, S., Zhang, Y., Liu, P., and Qin, X. (2016). Polyacrylonitrile nanofiber yarns and fabrics produced using a novel electrospinning method combined with traditional textile techniques. *Textile Research Journal* 86 (16): 1716–1727.

30 Wu, S., Duan, B., Liu, P. et al. (2016). Fabrication of aligned nanofiber polymer yarn networks for anisotropic soft tissue scaffolds. *ACS Applied Materials & Interfaces* 8 (26): 16950–16960.

31 Shao, W., He, J., Han, Q. et al. (2016). A biomimetic multilayer nanofiber fabric fabricated by electrospinning and textile technology from polylactic acid and Tussah silk fibroin as a scaffold for bone tissue engineering. *Materials Science & Engineering, C: Materials for Biological Applications* 67 (1): 599–610.

32 Joseph, J., Krishnan, A.G., Cherian, A.M. et al. (2018). Transforming nanofibers into woven nanotextiles for vascular application. *ACS Applied Materials & Interfaces* 10 (23): 19449–19458.

33 Li, Y., Guo, F., Hao, Y. et al. (2019). Helical nanofiber yarn enabling highly stretchable engineered microtissue. *Proceedings of the National Academy of Sciences of the United States of America* 116 (19): 9245–9250.

34 Abhari, R.E., Mouthuy, P.A., Vernet, A. et al. (2018). Using an industrial braiding machine to upscale the production and modulate the design of electrospun medical yarns. *Polymer Testing* 69: 188–198.

35 Fan, L., Ma, Q., Tian, J. et al. (2018). Conjugate electrospinning-fabricated nanofiber yarns simultaneously endowed with bifunctionality of magnetism and enhanced fluorescence. *Journal of Materials Science* 53 (3): 2290–2302.

36 Fan, L., Ma, Q., Tian, J. et al. (2017). Novel nanofiber yarns synchronously endued with tri-functional performance of superparamagnetism, electrical conductivity and enhanced fluorescence prepared by conjugate electrospinning. *RSC Advances* 7 (77): 48702–48711.

37 Zhou, F.L., Gong, R.H., and Porat, I. (2009). Nanocoating on filaments by electrospinning. *Surface and Coating Technology* 204 (5): 621–628.

38 Zhou, F.L., Gong, R.H., and Porat, I. (2010). Nano-coated hybrid yarns using electrospinning. *Surface and Coating Technology* 204 (21–22): 3459–3463.

39 Liu, C.K., He, H.J., Sun, R.J. et al. (2016). Preparation of continuous nanofiber core-spun yarn by a novel covering method. *Materials and Design* 112: 456–461.

40 Su, C.I., Lai, T.C., Lu, C.H. et al. (2013). Yarn formation of nanofibers prepared using electrospinning. *Fibers and Polymers* 14 (4): 542–549.

41 Tebyetekerwa, M., Xu, Z., Li, W. et al. (2018). Surface self-assembly of functional electroactive nanofibers on textile yarns as a facile approach toward super flexible energy storage. *ACS Applied Energy Materials* 1 (2): 377–386.

42 Shao, W., Tebyetekerwa, M., Marriam, I. et al. (2018). Polyester@MXene nanofibers-based yarn electrodes. *Journal of Power Sources* 396: 683–690.

43 Marriam, I., Wang, X., Tebyetekerwa, M. et al. (2018). A bottom-up approach to design wearable and stretchable smart fibers with organic vapor sensing behaviors and energy storage properties. *Journal of Materials Chemistry A* 6 (28): 13633–13643.

44 Dabirian, F., Hosseini Ravandi, S.A., Hinestroza, J.P., and Abuzade, R.A. (2012). Conformal coating of yarns and wires with electrospun nanofibers. *Polymer Engineering and Science* 52 (8): 1724–1732.

45 Gu, Z., Yin, H., Wang, J. et al. (2018). Fabrication and characterization of tgf-β1-loaded electrospun poly (lactic-*co*-glycolic acid) core-sheath sutures. *Colloids and Surfaces, B: Biointerfaces* 161 (1): 331–338.

46 Padmakumar, S., Joseph, J., Neppalli, M.H. et al. (2016). Electrospun polymeric core-sheath yarns as drug eluting surgical sutures. *ACS Applied Materials & Interfaces* 8 (11): 6925–6934.

47 Mao, N., Chen, W., Meng, J. et al. (2018). Enhanced electrochemical properties of hierarchically sheath-core aligned carbon nanofibers coated carbon fiber yarn electrode-based supercapacitor via polyaniline nanowire array modification. *Journal of Power Sources* 399: 406–413.

48 He, J., Zhou, Y., Wang, L. et al. (2014). Fabrication of continuous nanofiber core-spun yarn by a novel electrospinning method. *Fibers and Polymers* 15 (10): 2061–2065.

49 Mao, N., Peng, H., Quan, Z. et al. (2019). Wettability control in tree structure-based 1D fiber assemblies for moisture wicking functionality. *ACS Applied Materials & Interfaces* 11 (47): 44682–44690.

50 Mao, N., Ye, J., Quan, Z. et al. (2020). Tree-like structure driven water transfer in 1d fiber assemblies for functional moisture-wicking fabrics. *Materials and Design* 186: 108305.

51 Wang, L., Wu, Y., Guo, B., and Ma, P.X. (2015). Nanofiber yarn/hydrogel core-shell scaffolds mimicking native skeletal muscle tissue for guiding 3D myoblast alignment, elongation, and differentiation. *ACS Nano* 9 (9): 9167–9179.

52 You, X., He, J., Nan, N. et al. (2018). Stretchable capacitive fabric electronic skin woven by electrospun nanofiber coated yarns for detecting tactile and multimodal mechanical stimuli. *Journal of Materials Chemistry C* 6 (47): 12981–12991.

53 Sun, X., He, J., Qiang, R. et al. (2019). Electrospun conductive nanofiber yarn for a wearable yarn supercapacitor with high volumetric energy density. *Materials* 12 (2): 273.

54 Ji, S.H., Cho, Y.S., and Yun, J.S. (2019). Wearable core-shell piezoelectric nanofiber yarns for body movement energy harvesting. *Nanomaterials* 9 (4): 555.

55 Ma, L., Zhou, M., Wu, R. et al. (2020). Continuous and scalable manufacture of hybridized nano-micro triboelectric yarns for energy harvesting and signal sensing. *ACS Nano* 14 (4): 4716–4726.

56 Mao, N., Peng, H., Qin, X., and Yu, J. (2021). Contact force within electrospun nanofiber core-spun yarns and moisture management ability of their fabrics. *Journal of the Textile Institute* 10: 1–13.

57 Liu, P., Wu, S., Zhang, Y. et al. (2016). A fast response ammonia sensor based on coaxial PPy-PAN nanofiber yarn. *Nanomaterials* 6 (7): 121.

58 Wu, S., Liu, P., Zhang, Y. et al. (2017). Flexible and conductive nanofiber-structured single yarn sensor for smart wearable devices. *Sensors and Actuators, B: Chemical* 252: 697–705.

59 Jiang, G., Zhang, J., Ji, D. et al. (2017). A novel approach for fabricating antibacterial nanofiber/cotton hybrid yarns. *Fibers and Polymers* 18 (5): 987–992.

60 Yang, Y., Quan, Z., Zhang, H. et al. (2022). Investigation on the processability, structure and properties of micro-/nano-fiber composite yarns produced by trans-scale spinning. *Journal of the Textile Institute* 51 (3 suppl): 5409S–5426S.

61 Chen, C., Cao, J., Qin, X. et al. (2022). A novel concept to produce submicron-cotton/polyester composite core-spun yarn via modified apparatus. *Journal of Natural Fibers* 19 (16): 13232–13242. https://doi.org/10.1080/15440478.2022.2087132.

62 Yang, Y., Zhao, Y., Quan, Z. et al. (2019). An efficient hybrid strategy for composite yarns of micro-/nano-fibers. *Materials and Design* 184: 108196.

63 Qiu, Q., Chen, S., Li, Y. et al. (2020). Functional nanofibers embedded into textiles for durable antibacterial properties. *Chemical Engineering Journal* 384: 123241.

64 Mao, N., Qin, X., Wang, L., and Yu, J. (2021). Electrospun nanofiber/cotton composite yarn with enhanced moisture management ability. *Textile Research Journal* 91 (13–14): 1467–1477.

65 Mao, N., Qin, X., Wang, L., and Yu, J. (2021). The migration behavior of electrospun nanofibers within cotton slivers in roller drafting and their effects on composite yarn quality. *Textile Research Journal* 91 (13–14): 1555–1564.

5

Application of Electrospinning in Air Filtration

Yunpu Liu and Xiaohong Qin

Donghua University, Key Laboratory of Textile Science & Technology, Ministry of Education, College of Textiles, 2999 Renmin North Road, Songjiang District, Shanghai 201620, China

5.1 Introduction

As global industrialization continues to move forward, air pollution is becoming a problem that plagues many countries. The high dependence on and demand for fossil fuels in industries such as steel, power generation, metallurgy, automobiles, textile printing, and dyeing have led to huge emissions of waste dust and smoke, which have caused serious damage to the atmospheric environment and exacerbated the air pollution situation. Breathing polluted air for a long time will lead to various diseases and even a shorter life [1]. Controlling air quality is a vital initiative for the sustainable development of all mankind. With the concept of environmental protection increasingly emphasized, many countries seek to protect air and control air pollution.

The concentration of particulate matter (PM), SO_2, NO_2, O_3, and other pollutants is an important basis for measuring good or bad air quality. Among them, PM is the main component of haze, which is an important air pollutant [2]. PMs include solid particles and liquid droplets, and the main sources include thermal power generation, automobile emissions, and road dust. PM can be classified into dustfall and drifting dust according to the particle size range of the particles. Particles with size greater than 10 microns usually settles due to gravity. The finer particles below 10 microns can be suspended in the air for a long time and can be inhaled by the human body. $PM_{2.5}$ (fine PM with an aerodynamic equivalent diameter of 2.5 μm or less) is a more minute component of PM_{10}, which is not easily intercepted and easily adsorbed to carry toxic substances, and the higher its concentration, the greater the hazard.

PM can harm industrial production. The presence of fine PM can make the precision of operation in the precision processing workshop not guaranteed, making electronic machinery and communication equipment prone to failure. Particulate dust filled in the air or attached to the lines and equipment can produce fire, electricity, and other fire hazards.

Electrospinning: Fundamentals, Methods, and Applications, First Edition.
Edited by Liming Wang and Xiaohong Qin.

The finer PMs, in the absence of protection, can be inhaled directly into the human body and enter the lungs, predisposing to a variety of respiratory diseases and allergic reactions [3, 4]. Moreover, PMs suspended in the air are potentially benign carriers of bacteria, viruses, and small pollutants. Influenza virus and new coronavirus can be spread through tiny solid particles or droplets in the air to achieve transmission from triggering group infectious diseases. It can be said that PM poses a great threat to human health.

Taking the necessary measures to deal with the PM problem has become a top priority in combating air pollution. The prevention and control of PMs are important tools to deal with the air pollution problem. Reducing particulate emissions at the source is the fundamental solution, but it cannot be achieved in a short period at present. Therefore, passive filtration is regarded as an important and effective way to stop the deterioration of air quality.

Inhalable PM is a significant public health hazard, and suitable air filtration materials can play a good role in protection and management. The use of air filtration materials is an important way to combat respirable PM. Air filtration materials are mainly divided into two categories, porous membrane filtration and fiber filtration materials [5]. Among these, fiber material is the mainstream material currently used in the field of air filtration. The fiber filtration material with air filtration performance is made by randomly arranging and stacking the fibers into a mesh structure with a certain thickness. By changing the chemical composition of fibers, stacking structure, and molding process, various fibrous membranes with different filtration properties and functions can be adjusted to meet different usage requirements. Therefore, the preparation of convenient, good air filtration performance, easy to modify the innovative fiber-based air filtration materials not only become the mainstream direction of scientific research, but also the focus of market sales direction.

At present, fiber-based air filtration materials are mainly divided into the following categories: nonwoven spunbond materials, melt-blown materials, glass fiber materials, and electrospinning nanofiber materials [6]. The following materials are available [7].

There are many types of fiber materials prepared by the nonwoven process. The filtration performance of spunbond prepared by the common nonwoven process is average and can only intercept large particles visible to the naked eye. Therefore, spunbond is usually used as a secondary substrate in composite filter materials for preliminary filtration and mechanical reinforcement. Nonwoven melt-blown fibrous membranes prepared by polymer extrusion are widely used filtration materials [8]. The melt-blown process produces fibers of small fineness, usually less than 10 microns in diameter, and most of the fibers have a diameter distribution of 1–4 microns. Therefore, fibrous membranes have small pore sizes and can provide a good filtration effect. Melt-blown materials are usually used in the field of personal protection with spunbond as a composite material because of the shortcomings in mechanical properties. However, melt-blown materials can only intercept micron-scale particles due to the limitation of fiber diameter, and cannot effectively intercept nanoscale particles. Improved electret melt-blown materials have been fabricated to obtain the ability to intercept nanoscale particles, but the high pressure drop and charge dissipation problems are still the shortcomings that limit their development [9].

Glass fiber is an inorganic filtration material with a very wide range of applications and is used in construction, transportation, sports, aerospace, and other fields. Continuous raw filaments or short-cut raw filaments are nondirectionally bonded together by chemical binders or mechanical action to make fiber mats that can be used for air filtration. Glass fiber has excellent physical and mechanical properties, chemical stability, corrosion resistance, and high-temperature resistance, which can be used on some special occasions. However, glass fibers are brittle, have poor abrasion resistance and softness, and are made of bulky products, which makes them gradually replaced by organic synthetic fibers in the field of air filtration [10–12].

The electrospun nanofibrous membrane is an emerging filtration material. The nanofibers made by electrospinning technology have a small fineness and a large specific surface area. The membranes formed by nanofiber stacking have excellent characteristics such as light mass, small pore size, high porosity, high specific surface area, and good internal connectivity. They can intercept submicron-scale particles and have excellent air filtration performance [13–15]. Similar to melt-blown materials, the mechanical properties of fibrous membranes prepared by electrospinning are generally poor, and they also need to be laminated with micron-scale nonwoven fiber layers used as mechanical supports [16]. With excellent filtration performance, electrostatically spun nanofibrous membranes are now a hot material for air filtration research and are expected to be prepared in large quantities for industrialization into the market in the future. To achieve better filtration performance and wider application in the field of filtration, researchers have made a lot of efforts in the preparation of high-efficiency and low-resistance membranes and functionalized membranes.

5.2 Characterization of Filtration Effect of Electrospun Nanofibrous Membranes

5.2.1 Filtration Efficiency

Filtration efficiency is an indicator used to evaluate the ability of a filter to intercept the passage of PM. The filtration efficiency is expressed as the ratio of the particle concentration before and after filtration to the particle concentration before filtration, and the unit is %. The filtration efficiency reflects the filter's ability to intercept PMs, and the larger the filtration efficiency, the stronger the filter's ability to intercept PMs. Therefore, improving filtration efficiency is one of the directions pursued by related researches. The filtration efficiency of the same fibrous membrane is different for different particles. The larger the particle size, the easier it is to intercept and the more efficient the membrane is for that particle. The smaller the particle size, the easier it is to pass through the filter, and the lower the efficiency of the membrane for that particle. The nature of the particles and the flow rate of the test air also affect the filtration efficiency value, so it is important to specify the particles used in the test and the test conditions to measure the filtration efficiency of the fibrous membrane. For electrostatic spun nanofibrous membranes, the longer the spinning time, the thicker the membrane, the smaller the pore size, and the higher the filtration efficiency, all other conditions being equal. The diameter of the fibers also significantly affects the filtration efficiency of the membrane after a certain duration of spinning. Membranes with finer diameter fibers generally have a higher filtration efficiency and can intercept fine particles more easily [16–18].

5.2.2 Pressure Drop

Pressure drop is an indicator to evaluate the ease of airflow through the filter. The pressure drop comes from the filter's obstructing effect on the normal passage of air. It is expressed by the air pressure difference between the upper and lower sides of the filter at the rated air flow rate, and the unit is Pa. The higher the pressure drop, the greater the resistance encountered by the airflow through the filter, and the more likely the filter will become clogged. Masks with high resistance can lead to poor breathing, and commercial air filters with high resistance can increase energy consumption [19]. Therefore, the filter material is expected to have a smaller pressure drop.

All other conditions being equal, the longer the spinning time, the thicker the membrane, the smaller the pore size, the greater the resistance to airflow through the membrane, and the greater the pressure drop across the membrane. The denser and tighter the fiber stacking, the higher the density of the stacking, and the higher the pressure drop of the membrane. In addition, the pressure drop results are different depending on the test conditions. It has been shown that there is a positive relationship between pressure drop and face velocity [20].

5.2.3 Quality Factor

Based on the two indicators of filtration efficiency and pressure drop, the quality factor (QF) was introduced to evaluate the filtration performance of the filter matrix comprehensively. The formula for calculating the quality factor is expressed as

$$\mathrm{QF} = -\frac{\ln(1-\eta)}{\Delta P} \tag{5.1}$$

where η represents the filtration efficiency and ΔP represents the pressure drop across the filter. The unit of QF is Pa^{-1}. The higher the filtration efficiency and the lower the pressure drop of the filter material, the higher the QF and the better the overall filtration performance of the filter. However, the general measures to increase the filtration efficiency will also increase the pressure drop, and the methods to reduce the pressure drop will also reduce the filtration efficiency. How to find a way to achieve a balance between the two and get a high QF value is the basic problem of air filtration material research.

5.2.4 Dust Holding Capacity

The dust holding capacity is a measure of the filter's ability to hold PM. When filtration continues for some time, the particles are partly intercepted on the filter surface and partly penetrated into the fibrous membrane structure. As filtration continues, a cake layer will be formed on the surface of the filter media, and the filter efficiency and pressure drop of the filter will increase due to the appearance of the cake layer. When the pressure drop is too high, the filtration performance of the filter is significantly affected and the filtration effect cannot meet the requirements.

The dust capacity of the filter is generally defined as the amount of dust that accumulates when the resistance reaches a specified value (generally two times the initial resistance) due to dust accumulation under the action of a certain air volume. The dust holding capacity determines to a certain extent the single service life of the filter. The construction of a special membrane structure allows particles to be gradually trapped inside the filter material, thus increasing the dust holding capacity of the filter.

5.3 Filtration Mechanism of Electrospun Nanofibrous Membranes

5.3.1 Single-Fiber Filtration Mechanism

The filtration process is theoretically divided into two stages [21]. The first is the stabilization phase. The inherent properties of the fibrous membrane, the nature of the PM, and the airflow conditions determine the filtration performance of the membrane. The filtration efficiency and pressure drop of the membrane remain constant during this stage and do not change significantly over time. The second stage is the instability stage. As filtration proceeds and PM is gradually deposited, the filter's performance in capturing PM changes over time. Most of the applied filtration materials have a negligible thickness of the filter cake layer resulting from the deposition of PM over a considerable period of time. Therefore, the process of fibrous membrane filtration can be considered as being within a stable phase.

The classical filtration theory is based on the analysis of single fibers. This theory reduces the fibrous membrane to a collection of fibers consisting of several unconnected cylinders, which allows the study of the airflow state around the fibers and the particle interception process. According to classical filtration theory, there are five mechanisms of particle trapping by fibers: inertial impaction, interception, Brownian motion, gravitational settling, and electrostatic deposition [22, 23].

Inertial impaction. The flow line of the air stream will be bent when passing through the fiber, and the particles approaching the fiber with the air stream will be captured by the fiber because the mass is much larger than the air molecules, so the significant inertia effect will make the particles advance along the original trajectory. The larger the mass of the PM inertia effect is more obvious, the more easily captured.

Interception. When the PM is close to the fiber, the PM will be mutually attracted to the fiber due to the van der Waals force effect, and when the radius of the PM is greater than or equal to the streamline to the fiber surface, it will be intercepted and captured by the fiber. The larger the PM, the greater the van der Waals force effect, and the more obvious the interception effect.

Brownian motion is the constant irregular motion of particles of small diameter suspended in the air. This property causes some particles to move out of the direction of the original airflow and to be captured by the fibers. This effect becomes more pronounced when the size of the particles is below 100 nm.

Gravitational settling is the effect of gravity on particles of larger masses, which are intercepted by the filter. In practice, high-efficiency filtration by nanofibrous membranes is often directed at small, light particles that are less affected by gravity, and the gravitational effect can be largely ignored.

Electrostatic deposition. PMs are attracted to fibers by electrostatic forces due to its charged nature.

At the microscopic scale, the filtration behavior of a single fiber can be described as a process in which PM is transported to the fiber surface with the airflow and is adhered to by van der Waals forces due to the abovementioned mechanisms. PM is trapped as a result of the combined action of several mechanisms. There have been several theoretical models to study the process of interception of PM by nanofibers, but the predictions of these models often differ significantly from the actual situation because of the specific nature of the airflow on the surface of nanofibers and the complexity of several mechanisms.

Based on the average free range of air molecules and the diameter of nanofibers, that is, the Knudsen number of fibers, the flow state of gases can be classified into four types: free molecular flow regime, transition flow regime, slip flow regime, and continuum flow regime [24]. The Knudsen number is calculated as

$$\mathrm{Kn} = \frac{\lambda}{d_\mathrm{f}} \tag{5.2}$$

where λ is the average free range of air molecules, which is about 65.3 nm, and d_f is the diameter of the nanofiber. When the fiber diameter is above the micron level, the airflow around the fiber is in a continuum flow regime, and the direct collision between air molecules and the fiber surface is the main reason for the filtration resistance. The behavior of the airflow around the fiber changes as the fiber diameter decreases to the nanometer scale. Nanofibers prepared by electrospinning reach the nanoscale in diameter, so the continuum flow assumption is no longer applicable to the flow field on the nanofiber surface. The airflow on the nanofiber surface is usually in the transition flow regime or slip flow regime. When the fiber diameter is close to the mean free range of the air molecules, a portion of the air molecules will slip flow on the fiber surface to avoid direct collision with the fiber surface, which will greatly reduce the filtration resistance of the air through the membrane [16, 25–27]. It is this effect that gives nanofibers the advantage of being significantly different from micron-scale materials suitable for filtration. The aerodynamic behavior of some air molecules bypassing the fibers allows for a reduction in the macroscopic pressure drop across the fibrous membrane.

5.3.2 Fibrous Membrane Filtration Mechanism

A fibrous membrane is a large collection of single fibers. PMs are captured by the fibers during filtration and deposited on the fiber surface, while the voids between the fibers provide a path for airflow. Fibrous membranes also rely on a combination of single-fiber filtration mechanisms to achieve filtration. Depending on the location of the captured PM, the membrane filtration mechanism can be divided into a surface interception and internal interception [21, 28].

Surface interception refers to the capture of PM by fibrous membranes through gravitational sedimentation, electrostatic action, inertial effects, and the deposition of PM on the outer surface of the membrane. In addition to the role of several single-fiber filtration mechanisms, filtration screening is also one of the ways to achieve surface interception. For filter screening, the pores produced between the fibers are large and small. PM with a larger diameter approaching the fibrous membrane with the airflow cannot pass through the pores with a smaller size than itself and will be intercepted on the surface of the fibrous membrane [29]. This is the most basic mechanism of filtration achieved by membranes and is usually applied to explain how large particle size PMs are intercepted.

Internal interception means that the fibrous membrane relies on various effects to retain PM within the pore structure of the fibrous membrane. The structure and adsorption properties of the fibrous membrane are important factors affecting internal interception. The more zigzagging the fibrous membrane thickness direction is stacked, the higher the porosity, the more three-dimensional (3D) the structure, and the stronger the electrostatic effect, the better the internal interception effect [30].

5.4 Electrospun Nanofibrous Membranes for Air Filtration

Fibrous membranes prepared by electrospinning are considered to be a better choice for use in filtration, compared to other filtration materials. Many of the earliest studies revolved around the relationship between spinning solution concentration, spinning solution advancement rate, spinning voltage, spinning time, receiving distance, environmental conditions, and other factors and filtration performance in order to investigate the optimal parameters for obtaining ideal electrospun nanofibrous membranes [31].

However, adjusting the abovementioned influencing factors often increases the pressure drop while improving the filtration efficiency of the fibrous membrane. The ideal filtration material should be characterized by high efficiency and low resistance, but it is impractical to try to obtain high QF by modifying only these parameters of the spinning process. Membranes prepared by conventional electrospinning techniques are no longer able to meet the higher demands for filter performance.

With increasing research on air filtration, more and more new types of fibrous membranes are being designed and prepared. By improving electrospinning methods, modulating spinning fluid components, adding nanoparticles, modulating fibrous membrane structure, electrostatic electret, and post-treatment (impregnation, coating, and lamination), researchers have been able to significantly change the nanofiber morphology, form new structured fibrous membranes, or change the fibrous membrane charge properties. These membranes with special properties have higher filtration efficiency, lower pressure drop, and better overall performance. Electrospun air-filtration nanofibrous membranes can be classified into three categories depending on the principle of achieving improved filtration performance: fibrous membranes based on fiber morphology, fibrous membranes based on membrane structure, and fibrous membranes based on electrostatic electret technology.

5.4.1 Fiber-Morphology-Based Membranes

5.4.1.1 Beaded Fibers

Typically, electrospinning with a suitable concentration of polymer solution produces nanofibers with smooth surfaces, approximately circular cross sections, and solid structures. If the spinning parameters are changed, using different compositions or concentrations of polymer solution for spinning, it is possible to obtain products with a different morphology from ordinary nanofibers: microsphere particles and beaded nanofibers [32].

Beading is the special structure of fibers made by electrospinning when the concentration of the spinning solution is low. The jet generated by the low-concentration solution is broken in the high-voltage electrostatic field, and the polymer chains slip between each other so that the string bead structure is created on the fiber. The beads are generally spherical or spindle-shaped and have a larger diameter than the diameter of the rope fibers. Beading is considered in some studies to be a defect in the fiber due to improper spinning concentration [33]. However, other studies found that the presence of beads optimized the stacking structure of the membrane and increased the permeability of the filter, and that fibrous membranes composed of beaded fibers had a lower pressure drop than ordinary fibrous membranes.

Yun prepared nanofiber (NF), beaded nanofiber (BF), and composite particle/nanofiber (CF) in three different morphologies using different concentrations of polyacrylonitrile (PAN) and polymethyl methacrylate (PMMA), respectively. NF and BF filter materials are made directly by electrospinning into a fibrous membrane structure in one step. CF filter materials consist of a composite structure of polymer particles and nanofibers. The polymer particles in it are made by electrospinning technique using a lower concentration of PMMA solution. Performance tests of the three different filters revealed that the diameters of the main fibers in all three materials were relatively close to each other, distributed around 400 nm, but the BF had a wider diameter distribution. The average particle size of the beads in the beaded fibers and the polymer particles in the CF were relatively close, but the former had a more dispersed particle size distribution and the latter had a more concentrated particle size distribution. The pressure drop of beaded nanofibers is significantly lower than the other two materials, and the quality factor is higher. Overall, the filtration performance of CF is slightly better than that of NF, and the difference between the two is not much, and both are far less than that of BF [32]. This study shows that the morphological characteristics of nanofibers have a great influence on their filtration performance. The special morphology of beaded fibers makes it easier for air to pass through the filter. Further, it was demonstrated that the beaded fibers produced by electrospinning are not a simple combination of microspheres and ordinary nanofibers.

The viscosity of the spinning solution is an important parameter that affects the fiber morphology. By adjusting the viscosity, fibers with some more interesting morphologies can be made based on ordinary beaded fibers. Kim et al. finely regulated the mass concentration of polyethylene oxide (PEO) solution and prepared the following three types of filter materials by electrospinning: nanofiber, ordinary beaded nanofiber, and branched nanofiber. The nanofiber (362 mPa s) was smooth and straight with an average fiber diameter of 172 nm. The average diameter of ordinary beaded nanofibers (109 mPa s) is 74 nm, and there are uniformly dispersed beads with a spindle shape and a thickness of about 500 nm. Branched nanofiber (45 mPa s) is similar to the beaded type. There are many micron-scale beads similar to the shape of zongzi, a traditional Chinese food. The average fiber diameter of the branched nanofiber is 45 nm. The filtration performance of the three fibrous membranes was measured and then compared. The filtration efficiencies of samples 1, 2, and 3 were 66.7%, 86.2% and 94.1%, and the initial pressure drops were78.96, 84.69 and 90.42 Pa, respectively, and the quality factors were 0.014, 0.023, and 0.031, respectively. It can be seen that although the pressure drops of the three materials showed an increasing trend, the increase was not significant, and the filtration efficiency did increase significantly and the overall quality factor achieved an increase. Beaded fibers (common beaded fibers and branching fibers), especially branching fibers, show obvious filtration advantages. Beaded fibers have lower air resistance due to finer fiber diameters, longer distances between fibers, and the bead-supported structure. Branching fibers, on the other hand, have a higher filtration efficiency due to the additional nanofibers spun out of each of their beads [34].

In another study, GAO et al. investigated the suitable process parameters for the preparation of polysilsesquioxane (PSQ)-immobilizing polylactic acid (PLA) (PLA-PSQ) as feedstock for beaded fibers using an electrospinning technique [35]. It was found that PSQ was immobilized and uniformly distributed in PLA, making the nanofibers more difficult to form yet easier to fabricate beaded nanofibers and providing denser and uniformly distributed beads. The formation mechanism of PLA-PSQ beaded fibers is proposed, and beads are formed for two reasons: the improvement of immobilized PSQ for the even distribution of beads and the improvement of the crystallization process due to PSQ immobilization in PLA. By comparing PLA-PSQ beaded fibers with PLA/PSQ (a dispersion system formed by adding PSQ nanoparticles into PLA spinning solution) beaded fibers, it was found that the beads in PLA/PSQ beaded fibers were not uniform in size, not uniformly distributed in linear fibers and less in number, which was explained as a result of PSQ agglomeration in some parts. In contrast, PLA-PSQ beaded fibers have a very concentrated bead size and a uniformly dispersed positional distribution because PSQ is uniformly immobilized in PLA. In addition, for semicrystalline polymers like PLA, the rate of fiber formation can be controlled by the crystallization process, which makes it possible to prepare beaded structures with larger polymer molecular weights and more entangled molecular chains. (Usually, concentrated solutions of polymers with higher molecular weights and more entangled molecular chains are not easily made into beaded fibers.) Immobilization of PSQ increases the rate of PLA crystallization, which accelerates the rate of formation of specially shaped fibers. Thus, PLA-PSQ has the property of being more prone to bead formation than PLA.

The filtration performance of three different fibrous membranes, PLA-PSQ, PLA/PSQ, and plain PLA was analyzed in comparison. The lowest pressure drop was observed for PLA-PSQ fibrous membranes, followed by PLA/PSQ fibrous membranes, and the highest pressure drop was observed for plain PLA fibrous membranes when 100% filtration conditions were reached. This indicates that the presence of beads enlarges the distance between the fibers, resulting in a lower pressure drop across the fibrous membrane. Also, the beads make the pores of the filter more uniformly distributed, which inhibits the accumulation of PM in the pores. More, PLA-PSQ fibrous membranes show more outstanding filtration performance because of having a larger number of appropriately sized and uniformly distributed beads.

By adding other components, beaded fibers can also be obtained. Graphene oxide (GO) is a material with easy preparation and excellent performance, which is widely used [36]. The GOPAN prepared by mixing GO in PAN also showed beaded morphology and exerted the effect of forming a more fluffy fiber film with reduced pressure drop [37].

The means of preparing beaded fibers to achieve improved overall filtration performance of fibrous membranes have been utilized by many scholars. However, the specific effects of bead size and distribution on filtration as well as the mechanism remain to be investigated in depth, which is an important direction for further research on bead-based fiber filtration materials.

5.4.1.2 Rough Surface Fibers

Specific surface area is important for fiber filtration. Reducing the diameter is a convenient way to increase the specific surface area [38], but on the one hand, it may lead to undesired higher filtration resistance, and on the other hand, electrospinning means cannot consistently reduce the fiber diameter to an arbitrary degree. In fact, making the fiber diameter less than 100 nm is not very easy to achieve. So researchers have opened another way to achieve an increase in the specific surface area of fibers, which is to rough up the fiber surface. Fibers with a smooth surface have a low coefficient of friction and are less likely to trap PM [39]. Rough surface fibers have a higher specific surface area than smooth fibers, and the change in surface morphology increases the friction on the fiber surface, making it easier to trap PMs. At the same time, the rough surface reduces the filter resistance by increasing the airflow around the fiber.

A common method for preparing rough surface fibers is to add nanoparticles to polymer solutions to make composite fibers. The distribution of nanoparticles in the solution system will change the conductive properties of the solution and give the fibers unique morphological characteristics. Inorganic oxides, metals and their compounds, carbon nanotubes, and semiconductor materials are often used as auxiliary materials to make nanofibrous membranes [37].

In a study, silica nanoparticle (SNP) was added to PAN to make a composite bimodal fibrous membrane [40]. The morphology of the nanofibers obtained by spinning changed significantly, and the surface changed from smooth to rough. The rough surface brought a larger specific surface area to the fibers. The addition of carbon nanotubes to the polymer solution can improve the electrical conductivity of the system. It was also demonstrated that carbon nanotubes can construct the rough structure of fiber surfaces and increase the electrical conductivity and specific surface area of fibers. Wrinkled Fibers.

Wrinkled morphology refers to the formation of many pleated rough structures on the fiber surface. This roughness is no longer a slight change in surface roughness, but a substantial change in fiber cross section and appearance that is significantly different. Similar to rough surfaces, wrinkled morphology achieves efficient PMs interception by increasing the contact area between fibers and PMs [41].

The template method, microemulsion method, and chemical etching method are all methods for forming wrinkles. The template method is described here. The template method uses a templating agent as an auxiliary material to spin the fiber and then removes the templating agent by using the difference in properties between the main material and the templating agent so that the residual part shows a wrinkled state. The wrinkled fiber prepared by the template method has a unique morphology with a wrinkled structure, and the specific surface area of the fiber is greatly increased. At the same time, the fineness of the fiber can be further reduced because the template agent component is removed, which will be more conducive to the filtration performance improvement.

Researchers have reported a polyimide (PI) fibrous membrane with excellent filtration performance and a wrinkled shape [42]. The appearance of the wrinkled morphology was achieved by the template method. PI is insoluble in organic solvents, and one-step preparation is difficult, so PI nanofibrous membrane was prepared by a two-step method. A certain proportion of PAN is added to polyamic acid (PAA) to make a well-mixed precursor spinning solution. Then, PAA/PAN fibrous membrane was made by electrospinning using a precursor spinning solution. In the normal preparation process, the PAA fibrous membrane is subjected to a gradient temperature treatment to imidate the PAA to produce PI, and the subtlety of this study using the template method lies in the direct use of the original process of temperature treatment to remove the PAN because PI is a high-temperature-resistant material, but PAN is not, and PAN will melt and decompose under high temperature and then disappear in the fibrous membrane. Therefore, after the PAA/PAN fibrous membrane is warmed up, it becomes a PI fibrous membrane. And the PI membrane exhibits very rough folds. This fold is not a slight surface roughness structure, but a huge change from inside to outside all over the fiber. The improved morphology directly gives PI a more powerful removal efficiency. Deng et al. prepared polyvinyl alcohol/sodium alginate/hydroxyapatite (T-PVA/SA/HAP) nanofibers and further heat-treated the fibers to obtain a rare wrinkled helical structure [43]. PVA is a biodegradable and environmentally friendly material, but its water-soluble physical properties can cause the fibrous membrane to collapse and become useless under high humidity conditions. HAP is added as an electret material to improve the PM capture ability. The T-PVA/SA/HAP fibers combine the advantages of fiber structure with the synergy of electrostatic electret (highlighted later in this chapter) technology to demonstrate highly satisfactory filtration performance, which is potential in the field of air filtration. In addition, some researchers have also made fibrous membranes with wrinkled structures by adding salt to alter certain properties of the polymer system that can affect the electrospinning process [44]. Attabi et al. added tetraethyl orthosilicate (TEOS) to PAN, and the electrostatically spun nanofibers developed a pleated structure. The pleated structure provided a nonslip surface for the nanofibers (twice the specific surface area of bare PAN fibers) that could attach more particles. The TEOS/PAN fibrous membranes made by this strategy possess higher filtration efficiency than bare PAN.

5.4.1.3 Porous Fibers

Porous fibers are fibers with tiny pores on their surface. A simple way to prepare porous fibers is to change the environmental parameters. When electrospinning is performed at low relative humidity, the solvent in the jet drawn by the electric field can evaporate at a faster rate, which leads to a porous structure in the final fiber [45]. Dense pores appear on the fiber surface, bringing higher porosity to the fiber, which results in higher filtration efficiency and lower pressure drop in the fiber-stacked membrane. However, the tiny pores all over the fiber surface may not last long because they are very easy to clogged and thus counter-productive. However, in any case, the preparation of porous fibers is an alternative way to improve the ability of fibers to capture PMs.

Han and his colleagues made polyetherimide (PEI) fibers using the aforementioned template method. PEI fibers are engineering plastics with good thermal stability [46]. A safe and nontoxic poly(3-hydroxybutyrate-*co*-3-hydroxyvalerate) (PHBV) was used as a templating agent, and the PEI/PHBV fibrous membrane was first spun. The significant difference in thermal decomposition temperature between PEI and PHBV (the difference is around 200 °C) was then used to remove PHBV under the action of high temperature. The relative ratio of PEI and PHBV and the effect of heat treatment temperature on fiber morphology were specifically investigated. It was found that when the heat treatment temperature was set to 210 °C. Both PEI/PHBV (75/25) and PEI/PHBV (50/50) fibers exhibited porous structures.

In addition, researchers also made poly(L-lactic acid) (PLLA) nanofibers with porous structures by post-processing [47]. The preparation of these porous fibers is based on the principle of non-solvent-induced phase separation, and the solubility of PLLA differs between dichloromethane (DCM) and *N*, *N*-dimethylformamide (DMF). The DMF component, which is not soluble in PLLA, is allowed to occupy a certain space inside the formed fibers by pre-configuration of the solution. Then, after the DMF evaporates, a certain amount of pore space is formed. At this time, the PLLA fiber is then treated by soaking in acetone. The fiber will be swollen, and when the acetone is removed, the fiber will have a complex pore structure.

5.4.1.4 Curly Fibers

In the electrospinning process, the nanofibers formed by jet deposition at the receiving end are mostly flat and straight, with a small number of bends. Although the improvement of the fiber surface morphology can serve to enhance the filtration efficiency, the pores formed by such straight fibers stacked into a membrane are difficult to improve further, thus keeping the resistance of the membrane to air high. Curly fibers are nanoscale polymers formed in a wool-like form. The convoluted structure allows the fibrous membrane to achieve further stacking, thus exhibiting higher porosity to reduce the pressure drop across the membrane.

During the electrospinning process, the motion of the jet shows a stable and then unstable process. In the stabilization process, the trajectory of the jet is close to a straight line. At the end of the motion, the jet undergoes an unstable spiral motion due to the reduction of electric field strength and environmental factors. In response to this motion characteristic of the jet, the parameters were adjusted to control the helical motion to make self-supporting 3D nanofibers with a curled structure [48]. The curled nanofibers formed a 3D structure with low stacking density and higher porosity, which led to a lower pressure drop in the fibrous membrane. Li et al. improved the spinning conditions and adjusted the exchange rate between water molecules and DMF solvent so that the trajectory of polyvinylidene difluoride (PVDF) jets varied as expected, resulting in a curly morphology of the fibers deposited on the receiver device [49]. "Non-solvent-induced phase separation" is the theoretical basis for this preparation. The preparers call these curly fibers nano wool. Indeed, the curled nanofibers not only have a morphology similar to that of wool, but they also exhibit fluffy properties. The nano wool stacked fibrous membrane has an ultra-high porosity of 98.7%, which directly reduces the filtration resistance by more than 30% compared to the equivalent non-curled structure. Yang et al. [50] reported a curly structured fiber made by the synergistic effect of electric field and airflow. The appearance of the curl pattern is related to the external airflow introduced during the jet flight. Based on the curl property of the fiber, the fibrous membrane showed good dust holding properties (maximum dust holding capacity of 168 $g\,m^{-2}$ for $PM_{2.5}$) while ensuring high efficiency and low resistance (99.9% $PM_{2.5}$ removal efficiency and 56 Pa pressure drop) characteristics.

However, the limitations of the preparation conditions have led to few reports and applications of crimped fibers in the field of air filtration. There are also some other areas of research using improved electrospinning processes [51–53], spinning solution components, and post-treatment [54]. There are also several other areas of research that have used improved electrospinning processes, improved spinning fluid components, and post-treatment to make micro-/nanofibers that are also curled or spiraled, which will not be discussed here.

There are also many researchers who have prepared fibers with morphological or structural differences from ordinary nanofibers in various ways and have used such changes in fiber morphology and structure to improve filtration performance; for example, shaped cross-section fibers. The cross section of fibers is associated with various physical properties of fibers, mechanical properties, color, and moisture absorption. In order to obtain fabrics with different properties, various chemical fibers with heterogeneous cross sections are manufactured to meet the diverse needs of people. But these fibers are quite coarse compared to the nanofibers made by electrospinning. The cross section of nanofibers formed by electrospinning was considered to be approximately circular for a long time, and few studies have been conducted for noncircular cross sections (mainly because it is difficult to make nanofibers made by the electrospinning process with as diverse cross sections as industrially produced chemical fibers). However, studies have shown that noncircular cross sections will change the specific surface area of the fibers and some of their properties against airflow, and this change may help the fiber aggregates to achieve better mechanical filtration properties.

Since there are few examples of truly anisotropic cross sections being produced in practice, only a superficial illustration is given here by means of a simulation study carried out by Hosseini and Tafreshi [25]. Several different cross sections of single fibers in the no-slip and slip flow regimes were simulated for filtration. The cross section of nanofibers with a significant slip effect has a greater influence on filtration than fibers with diameters in the micron range. The ability of a single fiber to capture PM is more related to the specific surface area of the fiber when only the shape of the fiber is considered, ignoring the influence of factors such as electrostatic forces. The difference in specific surface area between several geometries is not significant, so the effect of the cross section on the ability of fibers to capture particles is not significant. However, the shape of the cross section has an important influence on the ease of airflow through the fiber, which is commonly known as the filtration resistance of the fiber. The flowline models of several different cross-sectional fibers can be found that the more the fiber has a cross section adapted to the streamlined shape, the more it can exert the slip effect and the easier the airflow can pass through. The slip effect is less pronounced for fibers with a distinct large stagnation region in the front.

5.4.2 Structure-Based Membranes

5.4.2.1 Bimodal Structure

By bimodal diameter nanofibrous membranes, we mean nanofibrous membranes with two distinctly different peak ranges of fiber diameter distribution. In general, the distribution of fiber diameters in fibrous membranes prepared using electrospinning is consistent with a normal distribution with a single peak. It is only the range of the diameter distribution that is inconsistent. According to the general pattern between diameter and filtration performance of single-peaked fibrous membranes, coarse diameter fibers lead to low pressure drop with low filtration efficiency, and fine diameter fibers lead to high filtration efficiency with high pressure drop. Single-peak fibrous membranes with a narrow diameter distribution are difficult to address in a uniform manner to improve filtration resistance and reduce filtration resistance. It has been found that the diameter distribution can affect the nature of airflow through the membrane, which has an impact on the performance of the filter. There are two ranges of fibers in bimodal membranes, coarse and fine fibers. This coarse and fine are generally defined subjectively based on the pattern of diameter distribution. This type of fibrous membrane has an improved stack structure due to the introduction of coarse fibers into the fine fibers. Fine fibers assume the role of maintaining high filtration efficiency, while coarse fibers perform the function of reducing the pressure drop, which improves the filtration performance of the bimodal membrane [55].

Wang et al. reported an efficient and low-resistance double-peaked diameter nanofibrous membrane. Electrospinning was performed using two combined single-needle nozzles. Nanoscale silica solid particles were added to the polyacrylonitrile solution, and by setting different ratios of the polyacrylonitrile spinning solution combination, and then spinning layer by layer, two different diameters of fibers were stacked on the receiver device to form a nanofibrous membrane with a bimodal size distribution. The filtration efficiency of this bimodal diameter fibrous membrane can reach 99.989% and its filtration resistance is only 117 Pa [40]. Zhang et al. designed and fabricated a highly efficient and low-resistance bimodal nano-fibrous membrane following these two criteria: first, there should be fine fibers with small enough diameter in the fibrous membrane to create small pores; second, the fibrous membrane should have high porosity and low filling density [56]. The bimodal fibrous membranes were prepared from PAN solution and polysulfone (PSU) solution by a multi-jet co-blending spinning process. The effects of different PAN/PSU jet ratios on the performance of bimodal fibrous membranes were investigated. In this study, the PAN fibers prepared by electrospinning had a nanoscale diameter with an average diameter of 270 nm, while the PSU fibers had a micron-scale diameter with an average diameter of 1.3 μm. This more pronounced difference in fiber diameter (almost five times the average diameter difference) resulted in a randomly arranged 3D structure of the fibrous membranes. This structure creates a number of zigzagging channels inside the membrane to achieve the goal of blocking PM while facilitating the passage of air. The results of the filtration performance tests show that the PAN/PSU composite membrane with a 3/1 PAN/PSU jet ratio (81.211% filtration efficiency, 4.8 Pa pressure drop, and QF value of 0.348) has a slightly reduced filtration efficiency compared to the pure PAN nanofibrous membrane (82.414% filtration efficiency, 7 Pa pressure drop, and 0.248 QF), while the pressure drop decreased significantly and the quality factor increased significantly. It can be seen that the addition of PSU micron fibers resulted in improved filtration performance of the fibrous membranes.

In addition, Fu et al. reported a protein-based bimodal structured fibrous membrane. Gelatin fibers and zein fibers were prepared by electrospinning with two single-needle nozzles. The protein-blend membrane with a bimodal structure exhibited a filtration efficiency of 99.67% for $PM_{2.5}$ with a pressure drop of only 38 Pa and maintained a high filtration efficiency of over 96% after filtering 1000 l of polluted air for a certain period of time. The high filtration performance of 96% of separation efficiency after a certain period of operation [57].

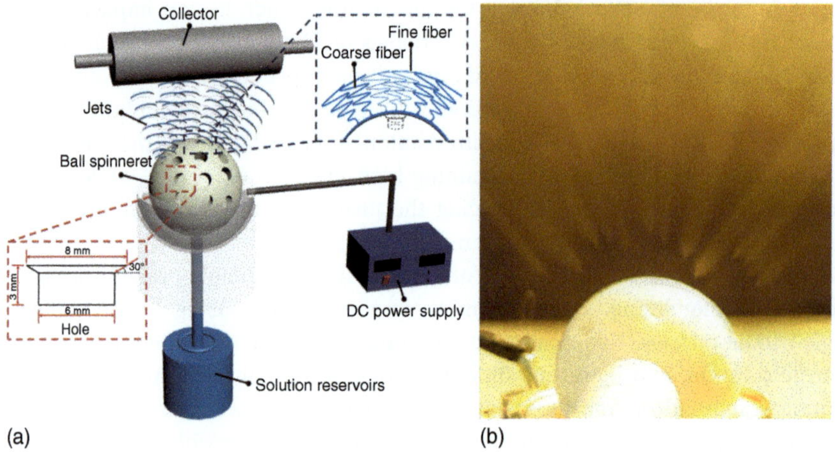

Figure 5.1 (a) Schematic diagram of needleless electrospinning setup using a multi-hole curved surface rotary spinneret. (b) An image of multiple jets from a digital camera during needleless electrospinning. Source: Quan et al. [58]. Reproduced with permission of Elsevier.

The successful functioning of the bimodal structure depends on the ability to make two different diameters of fibers. The general method of preparing bimodal fiber films is to use multiple single-needle spinnerets loaded with different solutions in a certain spatial configuration or time interval for spinning so that multiple jets generated by different spinning solutions are stretched and deposited in an electric field to produce a single-fiber film. Unlike this, Quan et al. developed a multi-hole curved surface rotary spinneret, as shown in Figure 5.1. Using this electrospinning device, fibers with two different diameters of coarse and fine fibers can be made in one step. In this study, finite element analysis of this spinneret was carried out to explain the principles of coarse and fine jet generation from the viewpoint of the electric field. The quadratic general rotary unitized design (QGRUD) was also performed to investigate the effect of different factors on the performance of the bimodal fibrous membrane. The spinning experiments verified that this multi-hole curved surface rotary spinneret is able to produce bimodal fibrous membranes in a simple and efficient way by virtue of its sophisticated design [58].

5.4.2.2 Bonding Structure

In the conventional electrospinning process, the jet is stretched by an electric field, the solvent evaporates, and the fibers are deposited to form a nanofiber film. The fibers are in contact with each other but do not form a fixed connection. By controlling the spinning fluid components, adjusting the spinning process parameters, or post-treating the fiber film, adhesions can be created between the fibers. The bonding points allow the contacting fibers to be immobilized, resulting in a more stable and durable structure and stronger mechanical properties. The presence of the adhesive structure does not directly contribute to the filtration performance of the membrane but rather ensures that filtration is sustained. In order to achieve high efficiency and low resistance, fibrous membranes are designed with various fluffy structures. However, the mechanical properties of nanofibrous membranes made by common processes are not good and can be easily damaged by external forces. In the case of increased airflow and deposition of large amounts of particles, the special structure of the membrane is easily weakened or even lost due to the mechanical defects of the membrane being squeezed and compacted.

Zhang et al. prepared fibrous membranes with adhesive structure in order to make the filtration material with better resistance to deformation to maintain the filtration advantages given by the fluffy porous structure of the membrane [56]. Based on the prepared PAN/PSU bimodal fibrous membranes, PEO@PAN/PSU fibrous membranes (PAN/PSU jet 3:1) were prepared by multi-jet co-blending spinning with PAN/PEO spinning solution and PSU spinning solution as raw materials. It was found that PEO@PAN/PSU fiber films form the bonding structures after heating and annealing due to the incorporation of PEO, a low melting point polymer (melting point of 60–70 °C). This does not damage the porous surface morphology of PSU. This bonding structure makes the 3D stacking structure of the bimodal fibers more stable, which enhances the mechanical properties of the fibrous membrane and ensures that effective filtration can be sustained.

5.4.2.3 Nano-Spider Web Structure

Physical screening is an important form of particle interception by fibrous membranes. Efficient physical screening is dependent on the pore size of the fibrous membrane. Reducing the diameter of the fibers is the key to reducing the pore size. However, it is difficult to control the diameter of nanofibers below 50 nm. Prof. Bin D's group has prepared a series of spider web-like nano-spider webs based on electrospraying–netting technique. The diameter of the fibers in the nano-spider webs can be controlled to less than 20 nm, which is able to perform very high filtration efficiency. Unlike the electrospinning process, the electrostatic spraying process generates both jets and droplets from the solution by the electric field force. The jets are stretched and cured to form nanofibers, while the charged droplets expand to form a film and then undergo phase separation to form nano-spider webs. The ionization ability of salt is an important factor in the formation of spider webs, and the addition of salt to the polymer provides a greater charge density to the charged droplets, which will facilitate the formation of nano-spider webs [59, 60].

Two-dimensional (2D) spider-web-structured fibrous membranes are produced simultaneously with ordinary nanofibers and are therefore often combined with nanofibers to make composite fibrous membranes for air filtration. For example, it is combined with beaded fibers to form a nano-spider web membrane with a cavity structure [61, 62].

5.4.2.4 Gradient Structure

Similar to the bimodal structure, the gradient structure is also used to build an unusual membrane structure by combining fibers of different diameters to balance separation efficiency and pressure drop. However, coarse and fine-diameter fibers are evenly distributed in the bimodal structure fibrous membrane. The arrangement of fiber diameters does not form a pattern in the direction of airflow passage.

Gradient structured fibrous membranes show clear gradient distribution of fiber diameters in the direction of airflow passage.

In the real environment, the particles that filters have to deal with are not concentrated in a single diameter range but have different sizes. However, filters with different fibers or membrane structures as described above do not filter PMs differently, so the filter is prone to clogging due to the deposition of different sizes of PMs on the surface. Completely similar pore sizes and coupling structures do not make better use of the 3D structure of the fibrous membranes to perform the filtration function. The use of electrostatically spun fibers of different diameters to build a layered structure provides the fibrous membrane with the opportunity to intercept PMs of different sizes. This multilayer membrane composition filtration material with different fiber diameters, stacking densities, and pore sizes are called gradient structured fibrous membrane. This gradient structure makes the fibrous membrane more 3D and also reduces filtration resistance to a certain extent. At the same time, the thickness is increased, and the special structure of the membrane increases the dust holding capacity of the membrane, each fiber layer of the membrane can effectively intercept PMs of different sizes, and the dust holding capacity and service life of the membrane are increased.

A three-layer (coarse [C], medium [M], and fine [F] layers) structure containing fibers of different diameters was efficiently prepared by electrospinning using a self-developed free-liquid surface spinning device (Figure 5.2) [63]. The arrangement of the fiber layers was designed to investigate the effect of the arrangement of

Figure 5.2 Schematic of the phase compensation free surface electrospinning setup with mushroom-head-shaped spinnerets and frame diagrams of the fibrous membranes with FCM, CMF, and MFC structures. Source: Zhou et al. [63]. Reproduced with permission of Royal Society of Chemistry.

the fiber layers on the filtration performance of the composite membrane. The PAN gradient membranes were fabricated using pure PAN at controlled concentrations (10.5%, 13%, and 16%) with fiber diameters of 419 ± 24, 214 ± 23, and 115 ± 19 nm for the F, M, and C layers, respectively.

The mean flow pore (MFP) sizes of FMC fibrous membrane (1.533 μm) were larger than those of a single F layer (1.398 μm) but smaller than those of M layer, C layer, and all bilayer combinations. The pore size range of the FMC fibrous membrane was also the smallest. The filtration performance tests showed that the FMC membrane had better filtration performance than the other combinations, with 99.28% filtration efficiency, 55.86 Pa pressure drop, and 0.088 QF. The effect of different arrangements on the filtration performance of the membranes was further investigated, and the quality factors of the FMC and CMF membranes were found to be significantly better than those of other combinations (MCF, MFC, CFM, FCM). When filtration occurs from the C layer through the M layer to the F layer, the pressure drop of the fibrous membrane is significantly lower than the other combinations. This is related to the arrangement of the different size pores formed by the fiber layers. It is easier for the airflow to pass through the large pore size and exit from the small pore size than to pass through the fibrous membrane with a small pore size distributed on the surface first.

Further, in order to enhance the performance of the filter, methods of reducing the fiber diameter and improving the roughness of the fiber surface were combined with the construction of the gradient structure of the fibrous membrane, respectively. More work has been carried out on the basis of PAN gradient fibrous membranes. These works specifically include the modification of the finest layer fibers by doped titanium dioxide nanoparticles, the modification of the finest layer fibers by doped single-walled carbon nanotubes, and the improvement of the gradient structure by adding PMMA to improve the composition of the spinning solution to modulate the fiber fineness. It was found that all these behaviors could improve the filtration performance of the gradient fibrous membranes to different degrees and better play the role of high efficiency and low resistance. The filtration efficiency of the gradient fibrous membrane (FMC structure) with PAN/titanium dioxide (PAN/TiO_2) fibers containing TiO_2 nanoparticles reached 99.74%, which is an improvement over pure PAN, but the pressure drop also increased to 147 Pa. The doping of single-walled carbon nanotubes (SWCNTs) into PAN reduced the average diameter of F-layer fibers to 78 nm and gave a rougher surface structure. The PAN/SWCNTs fibrous membranes with FMC gradient based on this modification achieved 99.99% filtration efficiency with a pressure drop of only 92 Pa. While the micro-gradient PAN/PMMA fibrous membranes with improved polymer composition by adding PMMA had an average diameter of 200, 110, and 70 nm, respectively. Compared with the micro-gradient structure of pure PAN, the fiber fineness of each layer in the PAN/PMMA gradient fibrous membrane (FMC structure) was further reduced, and the overall filtration effect was significantly improved, with a filtration efficiency of 99.97% and a pressure drop of 106.8 Pa.

Based on the above studies, the gradient structure filtration mechanism is described. The small fiber diameter of the fine fiber layer of the gradient fibrous membrane is close to the average free range of air molecules and has a significant slip effect, which gives the fibrous membrane a low resistance. The relationship between pore structure and air resistance is generally expressed by the equation of the equilibrium factor τ

$$\tau = \frac{d_f}{d^2} \tag{5.3}$$

where d_f is the fiber diameter and d is the distance between fibers.

It can be seen that the smaller the fiber diameter and the larger the distance between the fibers, the more the pore structure can have lower airflow resistance. For gradient fibrous membranes with FMC triple-fiber layers, the 3D structure is constructed to help build the pore structure and thus reduce air resistance. In fact, research results have shown that the gradient structure, constructed by decreasing the fiber diameter from coarse to fine or increasing the diameter from fine to coarse, can provide a much better filtration effect than the normal single-diameter fibrous membranes.

5.4.2.5 Multilayer Composite Structure

As air filtration has been more intensively studied, a wide variety of atypical morphological fibers and special membrane structures have been developed. However, a single application of a certain morphological fiber or a certain membrane structure is not the best choice for the preparation of air filtration materials. Combining various approaches to enhance filtration performance to make multilayer composite structures of nanofibrous membranes is a better solution to meet the demand for high-efficiency and low-resistance performance.

Quan et al. designed a sandwich structure based on a bimodal structure [26]. The optimal single nanofiber diameter with the lowest pressure drop under the influence of the slip effect was first determined by simulation and experiment. The optimal single nanofiber diameter with the lowest pressure drop under the influence of the slip effect is 70 nm, and then the optimal bimodal structure with fiber diameters of 70 and 200 nm is determined by further simulations and experiments.), thus making a composite fibrous membrane with a three-layer structure. The bimodal fiber layer provides a 3D support structure, which helps to increase the distance between the fibers and reduce the pressure drop. The lamination of three layers of fibers with different scale diameters creates gradient channels in the membrane, which helps to meet precise filtration and reduce air resistance. The composite sandwich membrane achieves a filtration efficiency of 99.984% for $PM_{0.26}$ with a pressure drop of 85.02 Pa and a QF up to 0.1028.

Yang et al. reported a PA-6/PAN/PA-6 (PA-6, polyamide-6) fibrous membrane consisting of PAN bead-on-string fibers and a 3D nano-spider web composite structure [61]. The bead-on-string fibers were prepared by precise control of spinning voltage and PAN mass concentration, and the effect of beads on filtration was systematically investigated. In addition, sandwich composite structures with different mass ratios of beaded fiber layers and nano-spider web layers were designed and prepared, and the pore structure and filtration performance of the composite fibrous

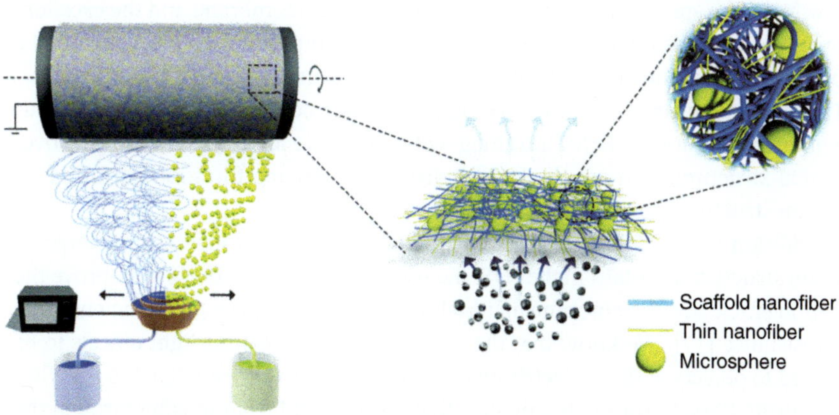

Figure 5.3 Schematic illustrating the fabrication of low filtration resistance PAN 3D composite membrane composed of ternary structures via free surface electrospinning. Source: Gao et al. [64]. Reproduced with permission of Royal Society of Chemistry.

membranes were analyzed in depth. As a result, PA-6/PAN/PA-6 fibrous membranes with a mass 1:8:1 showed excellent filtration performance for 300–500 nm airborne particle (99.9998% filtration efficiency, 117.5 Pa filtration resistance) with improved mechanical properties. This is due to the synergistic effect of the nano-spider web structure and the beaded structure. Ultrathin nanowires form a nano-spider web structure on both sides of the composite membrane to provide a strong guarantee of high filtration efficiency, while the beaded fibers wrapped in the middle layer of the composite membrane promote the formation of cavities inside the fibrous membrane.

As shown in Figure 5.3, based on the research of our group on free surface spinning, an embedded microsphere nanofibrous membrane with a 3D structure was prepared in one step by an electrospinning device with a pyramid-shaped copper spinneret as the core component, using different concentrations of PAN solution [64]. The microspheres were embedded in a network formed by interspersing scaffold nanofibers with fine nanofibers. The cavities appeared on fibrous membrane were due to the presence of microspheres, and the distance between fibers was further due to the presence of microspheres. The 3D structure gives the fiber membrane easy to allow airflow through the property, which allows the membrane to maintain a high filtration efficiency of 99.99% with a pressure drop of only 126.7 Pa. More meaningfully, the membrane is easy to prepare and is produced in one step without repeated stacking and assembly. The production efficiency is 120 times higher than that of a normal single-needle membrane.

Similarly, a composite fiber combining beaded fibers with a gradient structure was prepared and characterized for performance [65]. The PA-6 ultrafine nanofiber layer with an average fiber diameter of about 60 nm was placed on the upper side to deal with PM first, and the polystyrene (PS) fiber layer with a beaded structure was placed in the middle to form the fluffy cavity structure of the fibrous membrane, and the thermoplastic polyurethanes (TPU) submicron fiber layer was placed below. With the special structural design of the fibrous membrane and the excellent charge-bearing performance of the PS fibers, the composite membrane exhibits a superior removal efficiency of 99.99% and a low pressure drop of 57 Pa.

The abovementioned studies show that multilayer composite structured fibrous membranes have unparalleled advantages of high efficiency and low resistance over common structures and are the best choice for optimizing filtration performance based on structure.Electrostatic-Effect-Based Membranes.

In addition to the aforementioned modulation of fiber and membrane morphology and structure to obtain higher QF, the use of electrostatic effects to improve the filtration efficiency of membrane materials is a means to achieve high efficiency and low resistance [17]. We know that the electrostatic effect causes light objects to be attracted to objects with an electric charge. This effect is very beneficial to the filter in capturing PMs. Filtration is a process that prevents PM from passing through the filter material. When a fibrous membrane exhibits a certain charged nature, particles with opposite charges are attracted by Coulomb forces as they pass through the membrane with the air stream. For most particles, a fibrous membrane with electrostatic force can perform better. This electrostatic force is very beneficial to help the membrane achieve higher efficiency filtration [66].

The electrostatic effect does not dominate the filtration mechanism in classical filtration theory. This is because the charge generated by nanofibers due to friction and other factors cannot exist stably with the filtration medium for a long time and will soon decay to such a low level that it can hardly affect the process of PM separation by fibrous membranes. For example, ordinary fibrous membranes that have just been removed from the spinning device will carry a certain charge due to the spinning process and will exhibit a strong ability to adsorb PMs at this time. However, the amount of charge carried by the fibers decays with time and environmental changes, so the electrostatic effect cannot be sustained by simply charging the fibrous membrane through the spinning process. However, by artificially and proactively placing an electrical charge inside the fiber and trying to keep it in a stable state, the fibrous membrane will not only perform mechanical filtration through the original filtration mechanism but will also have higher filtration efficiency due to the electrostatic force generated. Satisfactorily, this approach hardly increases filtration resistance. Therefore, nanofibrous membranes based on electrostatic effect are considered a very promising filtration material. The research related to electrospinning air filters is also gradually focusing on the combination of structure optimization and electrostatic electret methods.

A typical approach to enhance the electrostatic effect of filter materials is electret technology. An electret is a dielectric that maintains its charged properties for a long time. In the presence of a strong external electric field, the electret can be polarized and remain polarized for a long time. Filters made by inserting and storing charges in a fibrous membrane by electret technology are called electret filters. Typical methods to prepare electret materials are corona charging, triboelectric charging, and electrospinning [67, 68].

Corona charging is a common method used to manufacture melt-blown electret materials. The main mechanism is that a high voltage is applied to the needle, causing the surrounding air to be ionized to produce positive and negative ions. The ions generated by ionization are deposited onto the filter media by electric field forces, which charge the surface of the material. While this method allows the filter media to have a high surface potential, the charge is usually difficult to distribute evenly and tends to decay [69].

Traction charging is a method that uses the friction of two substances with different electrical properties to generate an electric charge. This electret technique is usually not applied to nanofibrous membranes because the friction process is difficult to achieve for nanofibrous membranes prepared by electrospinning, which do not have outstanding mechanical properties.

Electrospinning is an important method for manufacturing electret filtration materials. Common polymers made by the electrospinning process are capable of some degree of polarization. The polarization of polymer dielectrics is achieved by adjusting the alignment of the electric dipoles inherent in the dielectric material. The degree of alignment of the electric dipoles depends on the chemical structure of the dielectric, so the electrostatic adsorption properties of fiber filters made by high-pressure electrospinning vary from polymer to polymer. Among several common polymers, PAN exhibits better filtration efficiency due to its high dipole moment and the high surface potential of the fibers made by electrospinning [70]. However, the electrostatic effect of polymer-oriented dipole electret filters formed simply by the spinning process is limited, and the charge given to the fibers by the electrospinning process is quickly dissipated without action, especially in high-humidity environments. The addition of nanoparticles and/or the use of multiple polymer composites to modulate the components of the spinning solution followed by electrospinning prepares electret nanofibrous membranes with the advantages of high charge storage capacity and long-lasting charge maintenance time. Therefore, some inorganic nanoparticles such as silicon dioxide (SiO_2), titanium dioxide (TiO_2), and barium titanate ($BaTiO_3$) are often added as charge enhancers in studies to obtain higher surface potential of the fibrous membranes. The use of polar polymers in combination with nonpolar polymeric materials also contributes to the enhancement and stabilization of the electrostatic effect of fibrous membranes.

Improvements to electrostatic electret nanofibrous membranes are based on both enhancing the surface potential and slowing down the degradation of electrical properties. The addition of charge enhancers to assist the polymer to store more charge is a common way to obtain higher surface potential. Titanium dioxide is often used in the electrospinning process to prepare filter fibrous membranes, not only because of the change in fiber morphology due to the presence of titanium dioxide but also because titanium dioxide allows the fiber to be injected with more charge during the electrospinning process, thereby increasing the surface potential.

The molecular space configuration of $BaTiO_3$ determines its high dipole moment and high polarizability. Kilic et al. reported a study, in which $BaTiO_3$ particles with high dielectric constants were added to PP fibers, giving electret fibers a denser surface charge with little effect on fiber morphology [71].

Previous studies have tended to focus on improving the ability to create change by trying different electret particles as charge enhancers but neglected the interaction between the electret particles and the fiber material.

According to a study, PVDF electret nanofibrous membranes were fabricated by adding HAP nanoparticles on top of the prepared convoluted structured fibers [49]. The effect of HAP is dependent on the hydrogen bonding interaction between HAP and PVDF to promote the regular arrangement of electric dipoles, which leads to the polarization of PVDF. In addition, the Maxwell–Wagner effect was introduced for the first time to explain the electret mechanism in the preparation of electret filter materials by electrospinning. The difference in electron transport capacity between the two materials allows charge accumulation at the interface resulting in space charge formation rather than transfer.

To cope with the high relative humidity of the filtration environment and the high number of oily components, Pan Jiang et al. reported an efficient low-resistance electret fibrous membrane made of polyvinyl butyral (PVB) combined with hydro-oleophobic fluorinated polyurethane (FPU) added with Si_3N_4 NPs (silicon nitride nanoparticles) (the fibrous membrane has a high filtration efficiency of 99.950% and a pressure drop of 55 Pa), and a dissipation mechanism for the charge injected into the fibers was proposed [72]. The optimal ratio of Si_3N_4 NPs and FPU incorporation was determined. The Si_3N_4 NPs composition effectively increased the dielectric constant of the material, and the use of FPU gave the fibrous membrane hydrophobic and oleophobic properties. The final PVB/Si_3N_4-FPU membranes were prepared with multiple functional advantages.

In addition, the difference in properties between polymers can also be used to create electret composite fibrous membranes with abundant electrostatic charges and solid electrostatic adsorption properties [73]. Combining micro-PS fibers with nano-PAN fibers to form a sandwich structure, the filter with membrane structure and electrostatic effect showed excellent filtration performance (99.96% filtration efficiency and 54 Pa pressure drop). The PS fibers placed at the top and bottom of the fibrous membrane are highly resistive nonpolar polymeric materials, and the PAN fibers in the middle layer are strongly polarized. This spatial configuration allows the electrostatic charge to be stably stored within the fibrous membrane and not easily dissipated. The effect of ambient relative humidity on the amount of charge held by the spun-prepared PAN material was also investigated. When the humidity was low, the charge was less likely to be accumulated within the fibrous membrane on the receiver device. As the humidity increased, the amount of charge accumulation also increased until a maximum value. After that, the humidity is increased again and the charge stored in PAN decreases. This is due to the fact that many of the charges stored in PAN due to high-voltage static places will be very easily dissipated in high-humidity environments.

Liu also prepared PVDF electret fibrous membranes using in situ electret electrospinning/netting technology [74]. The high-voltage electrospinning process promoted the neat alignment of the electric dipoles of PVDF molecules. Thanks to the chemical structure of PVDF and the 2D nano-spider web structure, the fibrous membrane has a surface potential of up to 6.8 KV and exhibits excellent filtration performance (99.998% filtration efficiency and 93 Pa pressure drop for $PM_{0.3}$).

Changing the composition of the spinning solution and preparing fibrous membranes by spinning a mixed solution containing multiple polymers can also exhibit good electrostatic effects. A PAA/PAN composite membrane was prepared from PAA and PAN. The addition of PAA gives the composite membrane material a negative charge characteristic, which enables the separation of PM from the airflow through electrostatic interaction with the PM [75].

All of the abovementioned studies have shown that the electret technique enabled the electrospun nanofibrous membranes to obtain improved air filtration performance, with little change in the filtration resistance of the material. However, the fibrous membranes prepared by the electret technique also have disadvantages. In high-temperature and high-relative humidity environments, it is difficult for the fibrous membrane to maintain a stable charged state, and the filtration efficiency will be greatly reduced. For this reason, a lot of work has been done to ensure that the charge in the electret membrane filter can be stored in the fiber in a stable manner and that the electrostatic force can exert a high adsorption effect on the particles. Unfortunately, the theory of electret application to electrospun nanofibrous membrane filters is complex and many problems are still not well explained and solved. The mechanism of charge dissipation is also not particularly clear. Further research is needed to minimize the charge dissipation of electret filters and to improve their ability to withstand high-humidity and high-temperature environments.

5.5 Functional Nanofibrous Membranes for Filtration

Nanofibrous membranes prepared by electrospinning have been regarded as the best material for achieving effective air filtration. As laboratory research on air filtration continues to intensify, people are no longer limited to the pursuit of high efficiency and low resistance of this excellent filtration material alone and are beginning to try more research in other directions.

Nanofibers are expected to have more practical functions, such as high-temperature resistance, antibacterial properties, biocompatibility, harmful gas adsorption, hydrophobicity, and oleophobic properties, for more special applications. Moreover, we do not expect these functions to be completely independent of each other, but to achieve many functions in one fibrous membrane at the same time is the goal we are pursuing.

5.5.1 Heat-Resisting Membranes

We know that to deal with PMs, we can use air filters to achieve human and indoor protection. The filtration that plays a protective role is passive filtration, which can only prevent PM but not eliminate it, and is also called "treating the symptoms but not the root cause." An important source of PMs is factories. The PM problem can be greatly alleviated if factories can use filters to intercept the PM generated and make the airflow meet the requirements. The PM generated by industrial production is often accompanied by the release of high-temperature gases, such as thermal power generation, which emits gases at 260–300 °C. Fibrous membrane materials must withstand the challenge of high temperatures in order to perform their filtration role. Although ordinary polymer filtration materials have excellent filtration performance, they will undergo significant mechanical property degradation and thermal decomposition under high-temperature conditions for a long time, and cannot be directly applied to high-temperature scenarios. High-temperature-resistant air filtration materials are an important direction to explore for the treatment of PM.

High-temperature-resistant materials are defined as materials that maintain their original properties for a long time under high temperature (above 200 °C) conditions. The achievement of thermal stability does not depend on the spinning method or the structure of the fibrous membrane but is related to the chemical composition of the fibrous membrane. Suitable high-temperature-resistant polymers are the key to preparing high-temperature-resistant nanofibrous membranes. Thermal stability is an important reference for polymer selection. Materials such as PI, polysulfone amide (PSA), polytetrafluoroethylene (PTFE), PMIA (meso-aramid), polyphenylene sulfide (PPS), and polybenzimidazole (PBI) have been proven to have good thermal stability in many studies. However, polymers such as PTFE, PPS, and PI are difficult to be directly applied to electrospinning because of their lack of good solubility, which greatly limits the preparation of high-temperature-resistant nanofibrous membranes. Currently, polymer blending is a simple modification method that combines the advantages of two materials. This method specifically operates by blending a polymer into a polymer with good spinnability to make a blending solution then used for electrospinning.

Specifically, many high-temperature-resistant air filtration materials have been prepared. PI exhibits excellent chemical stability, high strength, radiation resistance, and high-temperature resistance because of the action of the imide bond and benzene ring in the molecular structure. PI, a high-temperature-resistant polymer, is not easily soluble in solvents, so it is difficult to make nanofibrous membranes directly. Researchers have solved this problem by making precursor PAA nanofiber film first and then heat-treating it to generate PI. There are also studies on the chemical modification of PI to make soluble PI.

PMIA with excellent mechanical properties and PSA with stronger thermal stability properties were blended for electrospinning to produce composite membranes that combined the advantages of each other and compensated for their weaknesses [76]. In a study conducted by Tian et al, PMIA/PSA blended fibrous membranes with excellent overall performance were prepared. The PSA component enhanced the thermal stability of the composite membrane, and the PMIA/PSA (3/7) nanofibrous membrane maintained a separation efficiency of more than 99% for $PM_{2.5}$ at 200 °C for 120 hours. The spinnability of the polymers is also particularly important, which will directly determine whether they can be electrostatically spun into nanofibrous membranes. Some studies [77] showed that as PSA is not easy to electrospinning characteristics, using PSA/PAN mixed solution electrospinning to prepare composite fibrous membrane, the composite fibrous membrane also showed good thermal stability. After being calcined at 400 °C, the PSA/PAN fibrous membrane still maintained a high filtration efficiency of 99.95% for $PM_{2.5}$. In addition, the blending of PMIA and PAN can also produce high-temperature-resistant composite fibrous membranes, and the mechanical properties and filtration properties of the fibrous membranes remain stable after high-temperature treatment [78].

Polymer poly(arylene sulfide sulfone) (PASS), based on a modified PPS, is a thermally stable and soluble polymer. Based on a special three-layer structure stack, an M-PASS fibrous membrane has both good mechanical and filtration properties (99.97% filtration efficiency, 44.3 Pa pressure drop) [79]. Even at 200 °C, the filter maintains good PM removal capacity.

Using POSS-NH_2 (octa(amino-propylsilsesquioxane)) as the linkage structure, NH_2-ZIF-8 (amino-functionalized zeolitic imidazolate framework-8) is immobilized on the surface of PI fibers to make PI-POSS@ZIF fibrous membrane with special surface properties [80]. This treatment enables and significantly improves the filtration performance of the membrane. Filtration efficiency of 99.95% at 25 °C and 99.28% at 280 °C is achieved. And there is no significant loss of mechanical properties after the fibrous membrane is subjected to high-temperature treatment at 300 °C. In particular, the PI-POSS@ZIF fibrous membrane significantly improved the adsorption capacity of harmful gases compared with the ordinary PI fibrous membrane as well as the PI@ZIF fibrous membrane without the POSS-NH_2 connection.

5.5.2 Harmful Gas Adsorbing Membranes

The quality of air is not only related to PM concentration but also depends on the concentration of various harmful gases. Formaldehyde, nitrogen dioxide, carbon monoxide, and other harmful gases are all serious threats to human health. However, the traditional fibrous membrane can only filter PMs and cannot stop the passage of harmful gases. Air purification devices usually set an additional layer of activated carbon to remove harmful gases such as formaldehyde in addition to the fiber layer that filters PMs. However, this method leads to a bloated filter size, a dramatic increase in air resistance, and increased energy consumption [81]. Therefore, trying to make the fibrous membrane not only intercept PMs but also adsorb harmful gases is also an interesting direction to prepare functional filters. The fiber material itself usually does not have the property of adsorbing gases, and additional functional materials need to be applied to separate these harmful gases. However, the pore space and high specific surface area resulting from the porous characteristics are conducive to better adsorption of hazardous gases by fibrous membranes.

Metal–organic frameworks (MOFs), a kind of hybrid material that can be effectively used for gas separation, have been developing rapidly in recent years and have become an important research direction in the field of chemistry. MOFs combined with electrospinning technology to prepare nanofiber composites that can separate harmful gases and remove PM is a new research trend.

A ZIF-67@PAN nanofibrous membrane preparation process and morphological characteristics. Firstly, $Co(AC)_2$/PAN composite membrane was made by electrospinning, and then the fibrous membrane was immersed in a 2-methylimidazole-ethanol solution to generate uniformly distributed ZIF-67 nanocrystals on the fiber surface by chemical reaction. This semi-insertion structure makes the MOFs and the fibers closely connected and not easy to fall off, and the ZIF-67@PAN fibrous membrane not only improves the filtration efficiency due to the change of surface structure but more importantly, the fibrous membrane has 84% formaldehyde removal efficiency because of the presence of MOFs, which will be extremely beneficial to indoor air purification [82].

5.5.3 Antimicrobial Membranes

PMs in the air are good carriers of bacteria and viruses, and the spread of many diseases is directly related to the inhalation of PMs that carry these microorganisms. Although nanofibrous membranes can separate PMs from the air, PMs that are intercepted on the outside and inside of the fibrous membranes accumulate for a long time, which can easily spread bacteria. Therefore, nanofibrous membranes with antibacterial function have gained the favor of researchers.

The realization of the antimicrobial properties of fibrous membranes depends on the addition of functional nanoparticles and salts. Metal nanoparticles such as silver and copper were often used in conjunction with electrospinning to give antimicrobial properties to fibrous membranes. Silver nanoparticles (AgNPs) were produced by the liquid phase reduction method [83]. AgNPs were added to PVDF solutions in a certain ratio as raw material for electrospinning, and the AgNPs/PVDF nanofibrous membrane with a special dendritic morphology was fabricated. It was found that the addition of AgNPs components not only made the AgNPs/PVDF nanofibrous membrane have good and long-lasting antibacterial properties, but the dendritic structure of the fiber also improved the filtration performance of the material.

Titanium dioxide has antimicrobial properties not only in crystalline form but also in amorphous form. Amorphous titanium oxide (mTiO_2) was incorporated into PVP to make PVP-mTiO_2 nanofibrous membranes. This membrane with a composite of organic and inorganic components exhibited strong and long-lasting filtration performance (99% filtration efficiency was maintained after 160 hours of filtration) and showed significant antibacterial properties against *Escherichia coli* and *Staphylococcus aureus* [84].

N-halamine compounds are nonpolluting and nontoxic to humans. Chitosan (CS) is a natural biodegradable polysaccharide. Both materials exhibit strong antimicrobial activity and are suitable to be introduced as additives to exert antimicrobial efficacy. In a study, P (ADMH-NVF) biopolymer containing *N*-halamine and CS were combined with PVA to make PVA/CS and PVA/P (ADMH-NVF) films, respectively. The PVA/P(ADMH-NVF) was used as an intermediate layer and the PVA/CS was sequentially assembled on both sides as a mechanically reinforced layer to form a composite structure. This ecofriendly PVA composite filter based on *N*-halamine compounds and CS also exhibits excellent antimicrobial function [85].

5.5.4 High Humidity and Greasy Smoke Environment-Resistant Membranes

The environment in which the filter is placed determines the corresponding features that must be present at the same time to achieve filtration. For example, in high humidity conditions, the fibrous membrane may need to be hydrophobic in nature. In a highly oily smoke environment, the fibrous membrane may need to have a good oil absorption effect. In order to avoid morphological or even physical changes in the membrane due to water vapor or oily particles in the environment that may lead to filtration failure, the membrane needs to have some additional properties in addition to its excellent filtration performance. The selection of suitable polymer materials, the addition of special nanoparticles, and the post-treatment of fibrous membranes have all been used as ways to provide filters with a more desirable affinity for water or oil.

Polar polymers, which are considered to be favorable for PMs adsorption, also have hydrophilic properties, which cause the fibrous membrane to be damaged by the absorption of water molecules under high humidity conditions [84]. This leads to the destruction of the fibrous membrane by absorption of water molecules under high humidity conditions. Fibrous membranes are expected to possess hydrophobic properties in order to adapt to the actual application environment. Therefore, the preparation of hydrophobic polar polymeric filtration materials is the key to avoiding the destruction of fibrous membranes with high filtration performance by water molecules, and PMMA exhibits hydrophobicity due to the ester group at the end of the molecule. Fibrous membranes made with PMMA exhibit good hydrophobic characteristics and are not easily covered by water films. Large-size water droplets do not adhere to the hydrophobic fibrous membrane, so they can quickly slip off by gravity. However, making the filter material hydrophobic does not directly mean that the problem of filtration under high humidity is solved. In high-humidity environments, liquid water molecules may also clog the pores of hydrophobic membranes and make filtration unsustainable.[5] . Therefore, choosing the right material and fiber structure to meet the high humidity challenge is also an important concern. Hydrophobic materials can be combined with hydrophilic materials to conduct moisture and prevent clogging of the membrane. Superhydrophilic PAN/SiO_2 fibers combined with hydrophobic PVDF fibers make a composite membrane that can filter properly under high humidity conditions. The PAN/SiO_2 fibers perform the function of transporting water vapor, while PVDF prevents capillary water formation. This composite membrane achieves a high moisture–vapor transmission rate due to the difference in the affinity of the two materials for water [86]. The hydrophobic function is achieved by the inherent properties of the fibers, which allows for a longer-lasting effect than simply coating or impregnating the membrane.

In addition to the more common high-humidity environment, the fume environment is also an important challenge for filtration. Food processing, such as cooking, is often accompanied by large amounts of fumes, which can have a negative impact on the health of those working indoors. The use of air filters is extremely important in places like kitchens. Unlike water, which can be easily removed, oily particles can quickly clog filters after being trapped, so trapping and breaking down oily substances is a function that filters need to have. A study has prepared a filter to cope with oily fumes using TiO_2 with photocatalytic degradability and good lipophilic CNTs based on Ni foams. The oily substances will be adsorbed on the filter media and then degraded [87]. This may be an approach that can be learned from fibrous membrane type filtration materials. The oleophilic fibrous membrane combined with the component of decomposing oily PM can be a solution to cope with high oily smoke environment filtration.

5.5.5 Biodegradable Membranes

Today, the concept of environmental protection has been deeply rooted in people's minds, and the preparation of green and environmentally friendly air filtration materials has gradually become the target of attention in air filtration research.

Many researchers are obsessed with improving the filtration performance of nanofibrous membranes and use a lot of chemicals in the spinning process of fibrous membrane preparation. Most of these chemicals are polluting the environment, and some are even toxic and harmful to humans. For example, many organic solvents, which are not good for the environment, are used in large quantities in the spinning process. At the same time, the solvent components remaining in the fibrous membrane pose a threat to human health. In addition, nanofibrous membranes for air filtration are faced with the problem of how to dispose of them after they have finished their function. Like many plastic products for apparel and industrial use, fibrous membranes are composed of polymers that are difficult to degrade naturally. Due to the contamination of the filtration process, the complex composition of the membranes is difficult to recycle into other products and can only be incinerated or disposed of in landfills. If discarded indiscriminately, these chemical materials will remain in the soil in their original state for a long time. Incineration of the polymer will again create air pollution problems (which the nanofibrous membranes for air filtration are designed to solve). If these nanofibrous membranes are indeed used in large quantities as air filtration materials in the future, they will cause a lot of harm to the environment. Therefore, green biocompatible nanofibrous membranes are becoming more acceptable as environmentally friendly filtration materials and are expected to replace conventional filtration materials in the future.

Polyvinyl pyrrolidone (PVP), which exhibits good degradability, has a high dipole moment and is one of the raw material choices for the preparation of biodegradable filters. Mikheev et al. reported a fibrous membrane made of PVP for the collection of aerogel particles. The proper configuration of the solvent allowed us to obtain finer diameters of nanofibers, which led to improved filtration performance of PVP fibrous membranes. Unusually, PVP fibrous membranes with good filtration can be easily soluble in water, which is more convenient to handle than polymers such as PAN [88]. Of course, water solubility is somewhat detrimental to membrane filtration. This is because moisture in the humid air can dissolve the membrane and no longer block the passage of PM.

PVA is a good environmentally friendly material, but water solubility is the main factor that restricts the application of PVA in air filtration. In order to avoid damage to the fibrous membrane caused by moist environments or direct water contact, the addition of cross-linking agents to modify PVA is a common means of research. Zhang et al. prepared biodegradable green air-filtration membranes using PVA and cellulose nanocrystals (CNCs) without the use of cross-linking agents. The crystallinity of PVA was improved by adding CNCs and heating the membrane to obtain resistance to moisture and humidity. The filters can be washed several times to achieve reuse [89].

PLA is a bio-based polymer with nontoxic properties and biodegradability, which is made from starch extracted from renewable plant resources by treatment. Due to its green properties, PLA is gradually becoming the choice of filtration substrate, and Wang et al. made a composite fibrous membrane of nanoscale PLA (PLA-N) interwoven with porous PLA (PLA-P) from PLA. Porous PLA has a micron-scale diameter and the surface is covered with densely packed small pores, which serve as scaffold fibers in the composite membrane. The average diameter of nano-PLA fibers is less than 150 nm, which can form a network of fibers with good removal efficiency. The combination of two PLA fibers with different morphological characteristics can be decomposed in the natural state while ensuring good filtration performance, which makes it an attractive and environmentally friendly green filter [90].

Natural polymer materials, such as proteins, cellulose, starch, and lignin are good green materials that have received more attention in recent years and are being used more frequently in electrospinning. Natural proteins, which have better moisture resistance compared to general bio-based polymer materials, and the functional groups possessed by protein molecules show a stronger ability to remove PMs, may become an ideal substitute for synthetic polymers in the future. Liu et al. reported a Zein-based protein fibrous membrane. Aqueous ethanol solution was used as a solvent and acted as a denaturing agent for the protein [91]. By improving the spinning method, Zein spinning solutions with different concentrations were prepared as bimodal fibrous membranes. The protein-based fibrous membranes exhibited excellent and stable filtration performance in both dry and humid environments. Although natural polymers are good bio-based degradable materials, they generally have the disadvantages of insufficient solubility and poor mechanical properties [92]. Researchers usually take the approach of blending natural materials with other polymers to improve the weaknesses of natural polymer materials.

The abovementioned studies provide promising candidates for the subsequent development of air filtration materials. In general, a considerable number of green filter materials have been prepared and studied, and nonpollution to the environment and degradability are gradually becoming important considerations for the preparation of air filters by electrospinning.

5.6 Summary and Prospect

Nanofibrous membranes prepared by electrospinning have greater advantages than conventional air filtration materials. The nanoscale diameter of fibers not only enables the membrane to have high filtration efficiency due to the huge specific surface area and pore structure but also, the nanofibers enable a significant air slip effect, which allows the nanofibrous membrane to have lower pressure drop. These advantages make electrospinning-based nanofibrous membranes more competitive than micron-scale conventional air filtration materials and will gradually make them mainstream filtration materials.

Achieving the unity of high efficiency and low resistance remains an important goal for nanofibrous membranes. The combination of fiber morphology changes, construction of atypical membrane structures, and the use of electrostatic electret techniques will be the key to research to achieve better filtration performance of nanofibrous membranes. More importantly, the continuous discovery of nanofibrous membranes with special functionalities such as high-temperature resistance, antibacterial properties, and biodegradability for air filtration requirements in different environmental conditions plays an important role in expanding the application of air filtration nanofibrous membrane research results. In addition, the electrospinning process determines the low yield of nanofibrous membranes, which is currently the key difficulty preventing the large-scale application of nanofibrous membranes in commercial high-efficiency airers. We have reason to believe that once the difficulties are overcome, nanofibrous membranes will take over the market at a rapid pace and truly become the obvious choice in air filtration.

References

1 Sharma, A.K., Baliyan, P., and Kumar, P. (2018). Air pollution and public health: the challenges for Delhi, India. *Reviews on Environmental Health* 33 (1): 77–86.

2 Yue, H., He, C., Huang, Q. et al. (2020). Stronger policy required to substantially reduce deaths from PM2.5 pollution in China. *Nature Communications* 11 (1): 1462.

3 Zhang, Q., Jiang, X., Tong, D. et al. (2017). Transboundary health impacts of transported global air pollution and international trade. *Nature* 543 (7647): 705–709.

4 Doiron, D., de Hoogh, K., Probst-Hensch, N. et al. (2019). Air pollution, lung function and COPD: results from the population-based UK Biobank study. *The European Respiratory Journal* 54 (1): 1802140.

5 Liu, H., Huang, J., Mao, J. et al. (2019). Transparent antibacterial nanofiber air filters with highly efficient moisture resistance for sustainable particulate matter capture. *iScience* 19: 214–223.

6 Hung, C.-H. and Leung, W.W.-F. (2011). Filtration of nano-aerosol using nanofiber filter under low Peclet number and transitional flow regime. *Separation and Purification Technology* 79 (1): 34–42.

7 Joubert, A., Laborde, J.C., Bouilloux, L. et al. (2011). Modelling the pressure drop across HEPA filters during cake filtration in the presence of humidity. *Chemical Engineering Journal* 166 (2): 616–623.

8 Patel, R.M., Martin, J., Claasen, G., and Allgeuer, T. (2017). Advances in polyolefin-based spunbond and binder fibres, Chapter 6. In: *Polyolefin Fibres*, (2)e (ed. S.C.O. Ugbolue), 157–185. Woodhead Publishing.

9 Wehmann, M. and McCulloch, W.J.G. (1999). Melt blowing technology. In: *Polypropylene: An A-Z reference* (ed. J. Karger-Kocsis), 415–420. Dordrecht: Springer Netherlands.

10 Chen, Z., Wei, B., Mo, X. et al. (2009). Mechanical properties of electrospun collagen–chitosan complex single fibers and membrane. *Materials Science and Engineering: C* 29 (8): 2428–2435.

11 Modarresifar, F., Bingham, P.A., and Jubb, G.A. (2016). Thermal conductivity of refractory glass fibres. *Journal of Thermal Analysis and Calorimetry* 125 (1): 35–44.

12 Gupta, P.K. (2002). Strength of glass fibers. In: *Fiber Fracture* (ed. M. Elices and J. Llorca), 127–153. Oxford: Elsevier Science Ltd.

13 Jiang, Q., Jiang, Z., Ma, K. et al. (2014). Development of cytocompatible antibacterial electro-spun nanofibrous composites. *Journal of Materials Science* 49 (19): 6734–6741.

14 Xiang, H., Tan, S., Yu, X. et al. (2011). Sound absorption behavior of electrospun polyacrylonitrile nanofibrous membranes. *Chinese Journal of Polymer Science* 29 (6): 650–657.

15 Mirjalili, M. and Zohoori, S. (2016). Review for application of electrospinning and electrospun nanofibers technology in textile industry. *Journal of Nanostructure in Chemistry* 6 (3): 207–213.

16 Leung, W.W.-F., Hung, C.-H., and Yuen, P.-T. (2010). Effect of face velocity, nanofiber packing density and thickness on filtration performance of filters with nanofibers coated on a substrate. *Separation and Purification Technology* 71 (1): 30–37.

17 Tang, M., Chen, S.-C., Chang, D.-Q. et al. (2018). Filtration efficiency and loading characteristics of PM2.5 through composite filter media consisting of commercial HVAC electret media and nanofiber layer. *Separation and Purification Technology* 198: 137–145.

18 Ardkapan, S.R., Johnson, M.S., Yazdi, S. et al. (2014). Filtration efficiency of an electrostatic fibrous filter: studying filtration dependency on ultrafine particle exposure and composition. *Journal of Aerosol Science* 72: 14–20.

19 Lee, H.-J. and Choi, W.S. (2020). 2D and 3D bulk materials for environmental remediation: air filtration and oil/water separation. *Materials* 13 (24): 5714.

20 Xia, T., Bian, Y., Zhang, L., and Chen, C. (2018). Relationship between pressure drop and face velocity for electrospun nanofiber filters. *Energy and Buildings* 158: 987–999.

21 Qin, X.-H. and Wang, S.-Y. (2006). Filtration properties of electrospinning nanofibers. *Journal of Applied Polymer Science* 102 (2): 1285–1290.

22 Li, P., Wang, C., Zhang, Y., and Wei, F. (2014). Air filtration in the free molecular flow regime: a review of high-efficiency particulate air filters based on carbon nanotubes. *Small* 10 (22): 4543–4561.

23 Kim, M.-W., An, S., Seok, H. et al. (2019). Electrostatic transparent air filter membranes composed of metallized microfibers for particulate removal. *ACS Applied Materials & Interfaces* 11 (29): 26323–26332.

24 Gad-el-Hak, M. (1999). The fluid mechanics of microdevices - the freeman scholar lecture. *Journal of Fluids Engineering* 121 (1): 5–33.

25 Hosseini, S.A. and Tafreshi, H.V. (2011). On the importance of fibers' cross-sectional shape for air filters operating in the slip flow regime. *Powder Technology* 212 (3): 425–431.

26 Quan, Z., Zu, Y., Wang, Y. et al. (2021). Slip effect based bimodal nanofibrous membrane for high-efficiency and low-resistance air purification. *Separation and Purification Technology* 275: 119258.

27 Choi, H.-J., Kumita, M., Seto, T. et al. (2017). Effect of slip flow on pressure drop of nanofiber filters. *Journal of Aerosol Science* 114: 244–249.

28 Zhang, S., Rind, N.A., Tang, N. et al. (2019). Electrospun nanofibers for air filtration, Chapter 12. In: *Electrospinning: Nanofabrication and Applications* (ed. B. Ding, X. Wang, and J. Yu), 365–389. William Andrew Publishing.

29 Winans, J.D., Smith, K.J.P., Gaborski, T.R. et al. (2016). Membrane capacity and fouling mechanisms for ultrathin nanomembranes in dead-end filtration. *Journal of Membrane Science* 499: 282–289.

30 Liu, G., Xiao, M., Zhang, X. et al. (2017). A review of air filtration technologies for sustainable and healthy building ventilation. *Sustainable Cities and Society* 32: 375–396.

31 Huang, Z.-M., Zhang, Y.-Z., Kotaki, M., and Ramakrishna, S. (2003). A review on polymer nanofibers by electrospinning and their applications in nanocomposites. *Composites Science and Technology* 63 (15): 2223–2253.

32 Yun, K.M., Suryamas, A.B., Iskandar, F. et al. (2010). Morphology optimization of polymer nanofiber for applications in aerosol particle filtration. *Separation and Purification Technology* 75 (3): 340–345.

33 Kong, L. and Ziegler, G.R. (2014). Rheological aspects in fabricating pullulan fibers by electro-wet-spinning. *Food Hydrocolloids* 38: 220–226.

34 Kim, S.C., Kang, S., Lee, H. et al. (2020). Nanofiber filter performance improvement: nanofiber layer uniformity and branched nanofiber. *Aerosol and Air Quality Research* 20 (1): 80–88.

35 Cao, Q., Meng, X., Tan, S. et al. (2019). Electrospun bead-in-string fibrous membrane prepared from polysilsesquioxane-immobilising poly(lactic acid) with low filtration resistance for air filtration. *Journal of Polymer Research* 27 (1): 5.

36 Park, J.H., Yoon, K.Y., Na, H. et al. (2011). Fabrication of a multi-walled carbon nanotube-deposited glass fiber air filter for the enhancement of nano and submicron aerosol particle filtration and additional antibacterial efficacy. *Science of the Total Environment* 409 (19): 4132–4138.

37 Li, J., Zhang, D., Yang, T. et al. (2018). Nanofibrous membrane of graphene oxide-in-polyacrylonitrile composite with low filtration resistance for the effective capture of PM2.5. *Journal of Membrane Science* 551: 85–92.

38 Podgórski, A., Bałazy, A., and Gradoń, L. (2006). Application of nanofibers to improve the filtration efficiency of the most penetrating aerosol particles in fibrous filters. *Chemical Engineering Science* 61 (20): 6804–6815.

39 Cui, J., Wang, Y., Lu, T. et al. (2021). High performance, environmentally friendly and sustainable nanofiber membrane filter for removal of particulate matter 1.0. *Journal of Colloid and Interface Science* 597: 48–55.

40 Wang, N., Si, Y., Wang, N. et al. (2014). Multilevel structured polyacrylonitrile/silica nanofibrous membranes for high-performance air filtration. *Separation and Purification Technology* 126: 44–51.

41 Lu, T., Cui, J., Qu, Q. et al. (2021). Multistructured electrospun nanofibers for air filtration: a review. *ACS Applied Materials & Interfaces* 13 (20): 23293–23313.

42 Xie, F., Wang, Y., Zhuo, L. et al. (2020). Electrospun wrinkled porous polyimide nanofiber-based filter via thermally induced phase separation for efficient high-temperature PMs capture. *ACS Applied Materials & Interfaces* 12 (50): 56499–56508.

43 Deng, Y., Lu, T., Cui, J. et al. (2022). Morphology engineering processed nanofibrous membranes with secondary structure for high-performance air filtration. *Separation and Purification Technology* 294: 121093.

44 Al-Attabi, R., Morsi, Y., Kujawski, W. et al. (2019). Wrinkled silica doped electrospun nano-fiber membranes with engineered roughness for advanced aerosol air filtration. *Separation and Purification Technology* 215: 500–507.

45 Xing, Q., Dong, X., Li, R. et al. (2013). Morphology and performance control of PLLA-based porous membranes by phase separation. *Polymer* 54 (21): 5965–5973.

46 Han, S.O., Son, W.K., Cho, D. et al. (2004). Preparation of porous ultra-fine fibres via selective thermal degradation of electrospun polyetherimide/poly(3-hydroxybutyrate-*co*-3-hydroxyvalerate) fibres. *Polymer Degradation and Stability* 86 (2): 257–262.

47 Song, J., Zhang, B., Lu, Z. et al. (2019). Hierarchical porous poly(l-lactic acid) nanofibrous membrane for ultrafine particulate aerosol filtration. *ACS Applied Materials & Interfaces* 11 (49): 46261–46268.

48 Zheng, G., Shao, Z., Chen, J. et al. (2021). Self-Supporting three-dimensional electrospun nanofibrous membrane for highly efficient air filtration. *Nanomaterials* 11 (10): 2567.

49 Li, Y., Cao, L., Yin, X. et al. (2020). Ultrafine, self-crimp, and electret nano-wool for low-resistance and high-efficiency protective filter media against PM0.3. *Journal of Colloid and Interface Science* 578: 565–573.

50 Yang, Z., Zhang, X., Qin, Z. et al. (2022). Airflow synergistic needleless electrospinning of instant noodle-like curly nanofibrous membranes for high-efficiency air filtration. *Small* 18 (14): 2107250.

51 Zhao, Y., Miao, X., Lin, J. et al. (2017). Coiled plant tendril bioinspired fabrication of helical porous microfibers for crude oil cleanup. *Global Challenges* 1 (3): 1600021.

52 Xiong, C., Wang, L., Yu, J. et al. (2022). Continuous self-crimped micro-nanofiber via one-step electrospinning. *Composites Part A: Applied Science and Manufacturing* 158: 106959.

53 Li, C., Wang, J., and Zhang, B. (2012). Direct formation of "artificial wool" nanofiber via two-spinneret electrospinning. *Journal of Applied Polymer Science* 123 (5): 2992–2995.

54 Liu, W., Lipner, J., Moran, C.H. et al. (2015). Generation of electrospun nanofibers with controllable degrees of crimping through a simple. *Plasticizer-Based Treatment. Advanced Materials* 27 (16): 2583–2588.

55 Xiong, R., Hua, D., Van Hoeck, J. et al. (2021). Photothermal nanofibres enable safe engineering of therapeutic cells. *Nature Nanotechnology* 16 (11): 1281–1291.

56 Zhang, S., Liu, H., Yin, X. et al. (2016). Anti-deformed polyacrylonitrile/polysulfone composite membrane with binary structures for effective air filtration. *ACS Applied Materials & Interfaces* 8 (12): 8086–8095.

57 Fu, X., Liu, J., Ding, C. et al. (2021). Building bimodal structures by a wettability difference-driven strategy for high-performance protein air-filters. *Journal of Hazardous Materials* 415: 125742.

58 Quan, Z., Wang, Y., Zu, Y. et al. (2021). A rotary spinneret for high output of electrospun fibers with bimodal distribution. *European Polymer Journal* 159: 110707.

59 Wang, N., Wang, X., Ding, B. et al. (2012). Tunable fabrication of three-dimensional polyamide-66 nano-fiber/nets for high efficiency fine particulate filtration. *Journal of Materials Chemistry* 22 (4): 1445–1452.

60 Barakat, N.A.M., Kanjwal, M.A., Sheikh, F.A., and Kim, H.Y. (2009). Spider-net within the N6, PVA and PU electrospun nanofiber mats using salt addition: novel strategy in the electrospinning process. *Polymer* 50 (18): 4389–4396.

61 Yang, Y., Zhang, S., Zhao, X. et al. (2015). Sandwich structured polyamide-6/polyacrylonitrile nanonets/bead-on-string composite membrane for effective air filtration. *Separation and Purification Technology* 152: 14–22.

62 Zuo, F., Zhang, S., Liu, H. et al. (2017). Free-standing polyurethane nanofiber/nets air filters for effective PM capture. *Small* 13 (46): 1702139.

63 Zhou, M., Fang, M., Quan, Z. et al. (2019). Large-scale preparation of micro-gradient structured sub-micro fibrous membranes with narrow diameter distributions for high-efficiency air purification. *Environmental Science: Nano* 6 (12): 3560–3578.

64 Gao, H., Yang, Y., Akampumuza, O. et al. (2017). A low filtration resistance three-dimensional composite membrane fabricated via free surface electrospinning for effective PM2.5 capture. *Environmental Science: Nano* 4 (4): 864–875.

65 Chen, J.-P., Chen, S.-C., Wu, X.-Q. et al. (2021). Multilevel structured TPU/PS/PA-6 composite membrane for high-efficiency airborne particles capture: preparation, performance evaluation and mechanism insights. *Journal of Membrane Science* 633: 119392.

66 Thakur, R., Das, D., and Das, A. (2013). Electret air filters. *Separation and Purification Reviews* 42 (2): 87–129.

67 Goel, M. (2002). An integrated view of applications of electret technology in energy and health sectors. *Proceedings. 11th International Symposium on Electrets* 169–172.

68 Nifuku, M., Zhou, Y., Kisiel, A. et al. (2001). Charging characteristics for electret filter materials. *Journal of Electrostatics* 51–52: 200–205.

69 Giacometti, J.A. and Oliveira, O.N. (1992). Corona charging of polymers. *IEEE Transactions on Electrical Insulation* 27 (5): 924–943.

70 Liu, C., Hsu, P.-C., Lee, H.-W. et al. (2015). Transparent air filter for high-efficiency PM 2.5 capture. *Nature Communications* 6: 6205.

71 Kilic, A., Shim, E., and Pourdeyhimi, B. (2015). Electrostatic capture efficiency enhancement of polypropylene electret filters with barium titanate. *Aerosol Science and Technology* 49 (8): 666–673.

72 Jiang, P., Zhao, X., Li, Y. et al. (2017). Moisture and oily molecules stable nanofibrous electret membranes for effectively capturing PM2.5. *Composites Communications* 6: 34–40.

73 Cai, R.-R., Li, S.-Z., Zhang, L.-Z., and Lei, Y. (2020). Fabrication and performance of a stable micro/nano composite electret filter for effective PM2.5 capture. *Sci. Total Environ.* 725: 138297.

74 Liu, H., Zhang, S., Liu, L. et al. (2020). High-performance PM0.3 air filters using self-polarized electret nanofiber/nets. *Advanced Functional Materials* 30 (13): 1909554.

75 Liu, Y., Park, M., Ding, B. et al. (2015). Facile electrospun polyacrylonitrile/poly(acrylic acid) nanofibrous membranes for high efficiency particulate air filtration. *Fibers and Polymers* 16 (3): 629–633.

76 Tian, X., Zhang, F., Xin, B. et al. (2019). Electrospun meta-aramid/polysulfone-amide nanocomposite membranes for the filtration of industrial PM2.5 particles. *Nanotechnology* 31 (5): 055702.

77 Zhang, A., Li, H., Zhang, A. et al. (2021). High-temperature bearable polysulfonamide/polyacrylonitrile composite nanofibers for high-efficiency PM2.5 filtration. *Composites Communications* 23: 100582.

78 Zhang, H., Xie, Y., Song, Y., and Qin, X. (2021). Preparation of high-temperature resistant poly(m-phenylene isophthalamide)/polyacrylonitrile composite nanofibers membrane for air filtration. *Colloids and Surfaces A: Physicochemical and Engineering Aspects* 624: 126831.

79 Su, Q., Wei, Z., Zhu, C. et al. (2022). Multilevel structured PASS nanofiber filter with outstanding thermal stability and excellent mechanical property for high-efficiency particulate matter removal. *Journal of Hazardous Materials* 431: 128514.

80 Xie, F., Wang, Y., Zhuo, L. et al. (2021). Multiple hydrogen bonding self-assembly tailored electrospun polyimide hybrid filter for efficient air pollution control. *Journal of Hazardous Materials* 412: 125260.

81 Zaatari, M., Novoselac, A., and Siegel, J. (2014). The relationship between filter pressure drop, indoor air quality, and energy consumption in rooftop HVAC units. *Building and Environment* 73: 151–161.

82 Bian, Y., Wang, R., Wang, S. et al. (2018). Metal-organic framework-based nanofiber filters for effective indoor air quality control. *Journal of Materials Chemistry A* 6 (32): 15807–15814.

83 Xiao, Y., Wang, Y., Zhu, W. et al. (2021). Development of tree-like nanofibrous air filter with durable antibacterial property. *Separation and Purification Technology* 259: 118135.

84 Lakshmanan, A., Sarngan, P.P., and Sarkar, D. (2022). Inorganic-organic nanofiber networks with antibacteria properties for enhanced particulate filtration: the critical role of amorphous titania. *Chemosphere* 286: 131671.

85 Zhang, L., Li, L., Wang, L. et al. (2020). Multilayer electrospun nanofibrous membranes with antibacterial property for air filtration. *Applied Surface Science* 515: 145962.

86 Zhao, X., Li, Y., Hua, T. et al. (2017). Cleanable air filter transferring moisture and effectively capturing PM2.5. *Small* 13 (11): 1603306.

87 Xu, C., Xie, W., Yu, Y. et al. (2019). Photocatalytic and filtration performance study of TiO_2/CNTs-Filter for oil particle. *Process Safety and Environment Protection* 123: 72–78.

88 Mikheev, A.Y., Kanev, I.L., Morozova, T.Y., and Morozov, V.N. (2013). Water-soluble filters from ultra-thin polyvinylpirrolidone nanofibers. *Journal of Membrane Science* 448: 151–159.

89 Zhang, Q., Li, Q., Zhang, L. et al. (2020). Preparation of electrospun nanofibrous poly(vinyl alcohol)/cellulose nanocrystals air filter for efficient particulate matter removal with repetitive usage capability via facile heat treatment. *Chemical Engineering Journal* 399: 125768.

90 Wang, Z. and Pan, Z. (2015). Preparation of hierarchical structured nano-sized/porous poly(lactic acid) composite fibrous membranes for air filtration. *Applied Surface Science* 356: 1168–1179.

91 Liu, J., Ding, C., Dunne, F.O. et al. (2020). A bimodal protein fabric enabled via in situ diffusion for high-performance air filtration. *Environmental Science & Technology* 54 (19): 12042–12050.

92 Adeli, H., Khorasani, M.T., and Parvazinia, M. (2019). Wound dressing based on electrospun PVA/chitosan/starch nanofibrous mats: fabrication, antibacterial and cytocompatibility evaluation and in vitro healing assay. *International Journal of Biological Macromolecules* 122: 238–254.

6

Electrospun Nanofibers for Separation Applications in Oil–Water Systems

Chengdong Xiong and Rongwu Wang

Donghua University, Key Laboratory of Textile Science & Technology, Ministry of Education, College of Textiles, 2999 Renmin North Road, Songjiang District, Shanghai 201620, China

6.1 Introduction

The rapid development of industrial modernization and the dramatic improvement in the quality of human life have been accompanied by the discharge of a large number of production pollutants into the water and the serious pollution of water bodies. Water quality deterioration is a common problem worldwide, and the water pollution will become a long-term bottleneck limiting the sustainable development of human economy and society. Although oily water is less carcinogenic compared to pollutants such as heavy metals and textile dyes, the total amount of oil-containing wastewater is huge. For the year 2021, one large spill (>700 tonnes) and five medium spills (7–700 tonnes) were recorded. The large spill occurred in Asia and the medium spills occurred in Africa, Asia, and North America. This is a small increase on 2020, when four spills ≥7 tonnes were recorded, but on a par with the annual average for the 2010s [1]. As the spilled oil also contains toxic and flammable chemicals including polycyclic aromatic hydrocarbons, toluene, and benzene [2], it can cause irreversible damage to marine ecology. At the same time, the casual discharge of daily oil also makes the inland offshore water surface oil pollution increasingly serious, bringing great harm to the human living environment [3–5].

In the face of different types of oil pollution in different areas, it is necessary to consider the viscosity of oil, the amount of oil, and the combination form of oil–water, so as to take different methods and materials to remove oil. Generally, the methods of removal of oil pollution mainly include adsorption and separation, the former dealing with separating state of oil–water and the latter dealing with the combining state of oil–water.

6.2 Current Situation of Oily Wastewater

Oil and water are mixed together in a certain state to form a liquid system called oil–water mixture system [6]. Oil pollution in water bodies (oily wastewater) arises

Electrospinning: Fundamentals, Methods, and Applications, First Edition.
Edited by Liming Wang and Xiaohong Qin.

in various industrial fields such as petrochemical industry, metal smelting industry, pharmaceutical engineering, food processing and production, and other industrial fields. According to the different particle sizes of oil droplets, the oil phase in the polluted water body is mainly divided into the following states: floating oil, dispersed oil, emulsified oil, and dissolved oil [7]. Floating oil, whose particle size is greater than 100 μm, is stratified with the water phase and is a stable and continuous phase that floats on the water in the form of an oil layer. Dispersed oil, whose particle size is 10–100 μm, is evenly dispersed in water and is thermally and mechanically unstable state that has the tendency to agglomerate to form floating oil under the action of external force. The particle size of emulsified oil is between 0.1 and 10 μm, and the oil–water mixture system with water content greater than 30% can usually form a stable "oil-in-water" emulsion under the condition of containing surface active substances or mechanical action. The oil beads are not easy to agglomerate with each other and can exist stably. As a result, such emulsions are often difficult to separate by traditional gravity sedimentation methods, while the use of centrifugation and other methods have high energy consumption, special application conditions, and other restrictions. Therefore, the separation of emulsified oil from water is the focus of current research in oily wastewater [8]. The particle size of dissolved oil is extremely small, between a few nanometers and 0.1 microns, and the oil substance is usually present in the water body in the form of molecules, making it difficult to remove under normal circumstances.

6.2.1 Source of Oily Wastewater and Its Hazards

In the previous section, the different combined forms of oil and water in oily water were introduced, so the specific sources of these polluted oil droplets include the following three points:

(1) A large amount of wastewater is generated in several aspects of oilfield development, such as produced water, reinjection water, and drilling water, among which, produced water is the main pollutant in oil development, accounting for 98% [9].
(2) The problem of uneven reserves and consumption of oil around the world inevitably makes it necessary to achieve through transportation. Oil transportation methods include pipeline, railroad, and maritime transport, and oil spills occur in marine transportation frequently [10].
(3) Oily wastewater discharge from petroleum and its products in the process of utilization. Steel rolling process in iron and steel enterprises produces a large amount of oily wastewater. Additionally, emulsified oil pollution caused by the use of lubricating oil and emulsified cutting fluid in mechanical processing, as well as in the chemical industry, textile, paper, food, catering, leather, and other industries also produce the oily wastewater [11].

The substances in these oily wastewaters are normally hard to degrade and are inherently very toxic, which can be teratogenic or carcinogenic to aquatic organisms, thus causing damage to the ecosystem. During the Deepwater Horizon

Figure 6.1 (a) Deep-sea oil platform explosion of GoM oil spill in 2010. Source: US Coast Guard/Wikimedia/Public domain. (b) Oil-tanker spills in GoM of 1990. Source: NOAA/Wikimedia Commons/Public domain. (c) The South Korean Army helped to clean up beaches slicked with oil from the Hebei Spirit in 2007. Source: Wan et al. [10] Reproduced with permission of Springer. (d) Oil-spill-threaten seabirds. Source: Louisiana GOHSEP/Wikimedia Commons/Public domain/CC BY-SA 2.0.

(DWH) blowout in 2010, which is the largest oil spill ever recorded in US waters (Figure 6.1a), approximately 168 million gallons of oil were leaked in the Gulf of Mexico (GoM) [12]. At some point, 88 522 square miles of GoM was covered by oil slicks and vast areas of the Gulf were closed for fishing. Although satellite detection found no more oil in large area of GoM, the invisible and toxic oil made the spill much worse than some satellite images have showed. Berenshtein et al. found that the residual oil pollution in fish near the oil spill area exceeded the standard, especially the concentration of oil in yellow-rimmed grouper increased by more than 800% [13]. In 2013, the "Qingdao pipeline explosion" released about 2000 tons of oil into the environment, and Gao et al. studied oil composition, bacterial diversity, and biotoxicity and found that three years after the event, it was still affecting the environment and influencing bacterial communities in the sediment [14]. Additionally, human errors are behind most oil-tanker spills (Figure 6.1b) [10, 15].

On the one hand, the large amount of oily wastewater causes damage to the structure of the soil covered by it due to its high viscosity [16]. As the viscosity increases, the air permeability of the soil becomes poor, which affects the respiration and nutrient transmission of plants. On the other hand, the highly viscous oily wastewater will not only stick to fish eggs and shrimp, but will also cause the feathers of marine creatures to be covered with oil, making them unable to maintain their body temperature and freezing to death (Figure 6.1d) [17]. These oily wastewaters lead to the reduction of marine products or a large number of pollutants [18] that even enter the human body through the food chain, causing human tissue lesions. The direct human impact of these oil contaminations is the direct economic loss of fishermen, as well as the arduous treatment process and the high economic cost (Figure 6.1c) [10, 19]. Therefore, the removal of these oily wastewaters and the reduction or even elimination of their toxicity are crucial for ecological and human health development.

6.2.2 Treatment Methods for Oily Wastewater

Throughout history, scientists and engineers have worked to improve methods of dealing with oily wastewater. Some of the most widely used methods are fence oil absorption [20], controlled combustion [21], chemical dispersion [22], biological oxidation [23], and flotation method [24]. However, no single method is successful in removing oily wastewater effectively, even integrated systems are less effective and efficient [5, 25]. For oil spills on the water, physical sorbent, chemical dispersants, in situ burning and bioremediation are the conventional and commercially used oil spill remediation methods. Each oil spill response method has its own advantages and limitations and is often used in combination with mechanical, biological, and chemical methods to reduce costs and increase efficiency [16]. For oil spills under the water, the underwater oil spills need to be treated systematically in terms of location, method, and efficiency. The mass balance of oil spills marine for oil snow (MOS) settlement was ~41% evaporated, ~15% ashore and in nearshore sediments (areas <10 m deep), ~3% burned or mechanically removed, ~38.4% in the water column (partially degraded; 29% shallower and 9.4% deeper than 40 m), and 2.6% sedimented by MOS and oil-mineral aggregates (OMA) deeper than 10 m [26]. Faced with the complexity of underwater oil spills, we need to consider the oil composition and bulk characteristics, release depth, environmental conditions, and droplet size distribution to design the cleanup strategy in segments.

This chapter introduces the state of the oil–water structure presenting the emulsification state, the dispersion state and its stability is the underlying logic of the study of oily wastewater. Among them, the behavior of the particles in oil–water depends mainly on their functional characteristics as well as on their size and porosity, in the form of emulsification and adsorption. According to the wettability, shape, and size of droplet particles, different separation principles are applied to achieve the decontamination of oily wastewater, which mainly includes physical separation, chemical separation, combined physical/chemical treatment, and biological treatment methods [27]. The mechanisms of these methods are roughly as follows:

(1) The first one is to use the difference in density, electrical conductivity, viscosity, and other physical properties of the two phases of oil–water, which mainly

includes gravity separation [28], centrifugal separation [29], electromagnetic separation [11], and membrane separation methods [30]. Different from others, membrane separation can be selected according to the particle size of oil–water emulsion, offering a wide size range of separation from molecular to coarse scale. Additionally, the separation process is a physical process without phase change. There is no secondary pollution because of no added chemical agent. Furthermore, the products can be separated on both sides and the liquid can be collected easily. Therefore, it is expected to realize the wide application of membrane separation technology in the field of oily wastewater treatment due to the advantages of energy saving and environmental protection.

(2) The second one is the addition of chemical agents to the oil–water mixture to achieve the purification of oily wastewater based on the principle of chemical reaction. For example, the flocculation method utilizes the flocculant to form flocs by the interaction force between the positive charge formed in the water and the negative charge carried by the oil, and then the flocs carrying the oil float to the surface or settle to the bottom [31]. These chemical methods have excellent separation effect, fast dispersion, and are easy to use. However, there are many types of chemical agents that are highly targeted and narrowly applicable, which are inevitably introduced during use and can be difficult to remove. Thus, this method is suitable for oil droplets with larger particle size, but it is more challenging to treat emulsified oil in water.

(3) The third one is to use a combination of the physical and chemical methods, which can be classified into air flotation [32], adsorption method [33], and coalescence methods [34]. Among them, adsorption method has been proven to be an efficient and facile solution for severe oily wastewaters [35]. New types of materials, preparation methods, structural configurations, and wettability modifications are being investigated to enhance their pore volume and hydrophilic properties in order to break the limits of their development.

(4) The last one, biological treatment, is based on the decomposition of the dissolved state or colloidal state of oil in water into carbon dioxide, water, and other nonpolluting substances by microorganisms through metabolism [36]. Common biological treatment methods include activated sludge, biofiltration, biofilm, and contact oxidation [37, 38]. However, due to the many types of components and complex states of the oil in water, it is necessary to develop various strains. The special cultivation of various bioengineering bacteria is not only expensive but also has biological safety problems, requiring cautious handing during the utilization process.

Currently, membrane separation and adsorption methods are relatively straightforward, efficient, green, and economical to deal with the purification of different oil–water combinations. So this chapter will focus on these two methods to introduce their oil removal mechanism, material selection, and oil removal performance for providing a reference for the study of oily wastewater purification. Fiber-based materials will be selected for the material substructure because their one-dimensional mechanism can flexibly build multidimensional structural materials that can adjust their pore size and pore distribution and facilitate post-treatment

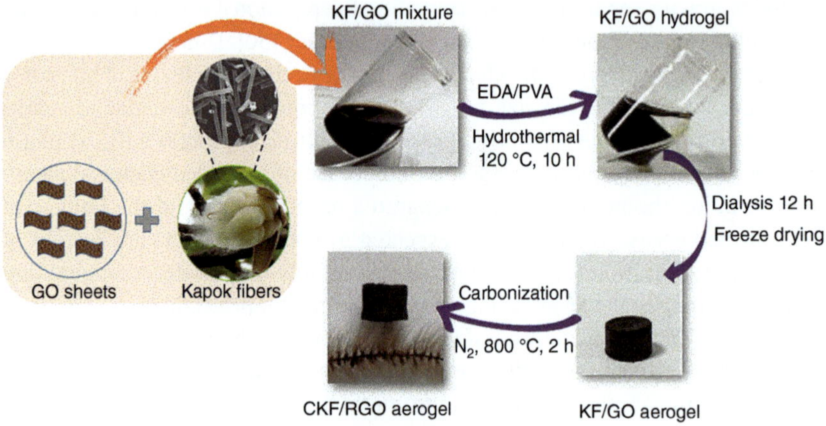

Figure 6.2 Scheme of preparation process of a CKF/RGO aerogel. Source: Zhou et al. [39] Reproduced with permission of Wiley.

modification to adapt to various types of oily wastewaters. As shown in Figure 6.2, a carbonized composite aerogel (CKF/RGO) fabricated based on kapok fibers (KFs) and graphene oxide (GO) through hydrothermal and carbonizing reactions was able to adsorb oil liquids up to 110 times of its own weight, and was still higher than 90% of the initial sorption capacity after eleven sorption-combustion cycles of n-hexane solvent. Generally, the natural fiber-based absorbents have excellent adsorption capacity [39]. Compared with synthetic absorbents, natural absorbents have excellent adsorption capacity, biodegradability, and nontoxicity. However, their low hydrophobicity increases the water adsorption and reduces the oil absorption capacity. Therefore, the design of a suitable synthetic fiber structure and the construction of desirable fiber wettability are the focus of research on oily wastewater treatment.

6.3 Electrospun Nanofibrous Membranes for Oil–Water Separation

Based on the advantages of large specific surface area, small pore size, high porosity, good pore connectivity, simple preparation process, low cost and high process controllability, electrospun nanofibrous membranes have been widely used in oil–water separation research [40–42]. The nanofibrous membrane materials produced by organic material blending, inorganic doping blending, or membrane post-treatment (grafting, polymerization, cross-linking, plasma etching, surface physicochemical deposition) can have strong hydrophobicity, which is very beneficial for the application in oil–water emulsion separation [43, 44]. But the resulted fibrous membranes need to have selective application of oil–water separation and adsorption because the separation is suitable for low-viscosity oil–water mixture and adsorption is better for high-viscosity oil removal [45].

6.3.1 Preparation Technology of Electrospun Nanofibrous Membrane

Electrospinning technology is a unique type of spinning technology with simple operation and special manufacturing process, which can effectively prepare diameter in the nano–micrometer range. It is considered to be one of the most important academic and technological activity in the field of materials science and technology in the last decade [46, 47]. In Figure 6.3, electrospinning device is simple, consisting of only three parts: high-voltage DC power supply, solution supply device, and collecting device [48a]. Different from the traditional spinning method that relies on mechanical forces such as traction and stretching, electrospinning utilizes the electrostatic repulsion generated by the surface charge under high pressure to stretch and control the diameter of the fiber. Therefore, the prepared fibrous membrane has a variety of unique and excellent properties with thinner fiber diameter and higher specific surface area.

In the preparation process of nanofibrous membranes by electrospinning process, intrinsic polymer properties, the operating parameters, and the surrounding conditions can be used to adjust the nanofibrous membranes with different compositions and structures. The regulation of intrinsic polymer properties includes polymer material selection as well as the regulation of relative molecular weight, solution concentration, viscosity, surface tension, electrical conductivity, and density, where the relative molecular weight and viscosity of the polymer must reach a certain value so that the molecular chains are entangled to avoid electrostatic atomization. And changing the surface tension of the polymer solution can achieve the regulation of the fiber diameter. The control of the operating parameters includes voltage, feed speed, collecting distance, nozzle size and shape, and collecting substrate. The surrounding conditions are mainly influenced by the ambient temperature and humidity. The ambient temperature and humidity can influence the solvent volatilization during fiber stretching to obtain fibers with different diameters and morphologies. In summary, there are many factors affecting electrospun nanofibers, and how to regulate suitable parameters is one of the keys to prepare nanofibrous membranes with different application properties.

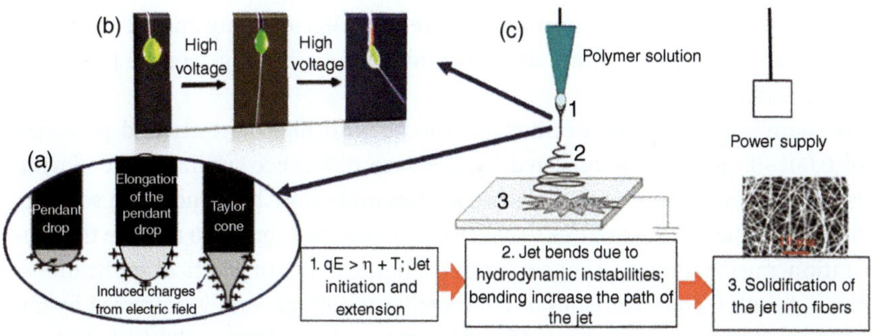

Figure 6.3 Schematic diagram of nanofiber formation during electrospinning. Source: Zhu et al. [48a] Reproduced with permission of Wiley.

The common electrospinning devices are roughly divided into spinneret modification and collector modification to obtain nanofibrous membranes with different morphological structures. These devices are summarized and discussed as follows: (a) a multi-spinneret electrospinning system can be applied to prepare nanofibrous membranes with designed thicknesses [49]; (b) electrospinning system with a coaxial spinneret can generate novel nanofibers with core–sheath and hollow structures, exhibiting gradient functionality [50]; (c) an optical chopper motor is used to achieve one-dimensional nanofibers that can assemble into extremely porous nanofibrous membranes [51]; (d) a cylindrical collector with appropriate rotating speed is used to result in membrane with aligned nanofibers [52]; (e) a rotating and traversing mandrel-type collector can be used to fabricate tubular-shaped nanofibrous membrane [53]; and (f) the compositions of polymeric dopes are optimized to generate spider-web-like nanonets with ultrafine fiber diameters less than 20 nm [54]. In summary, a variety of nanofibrous membrane products have been developed for oily wastewater treatment with their unique advantages by modifying electrospinning components.

6.3.2 Design Mechanism of Nanofibrous Membrane for Oil–Water Separation

6.3.2.1 Oil–Water Separation Membranes Based on Different Pore Sizes

Based on the difference in membrane pore size, commercially available membranes are divided into four main types: microfiltration (MF), ultrafiltration (UF), nanofiltration (NF), and reverse osmosis (RO) (Figure 6.4a) [48b].

(1) The pore size of the MF is generally between 0.01 and 10 μm. Based on the sieve effect, the particles larger than the pore size of the membrane are trapped on one side of the membrane, while the substances smaller than the pore size penetrate into the other side of the membrane under the action of external force. The pore size distribution is highly uniform and the porosity is high, so the MF membrane has high permeability and can be applied to the situation of large processing capacity [55]. Compared with the other three membranes, MF membranes prepared by electrospinning are suitable for treating oily wastewater containing micro-size particles. They exhibt higher separation capability and oil resistance at the same driving pressure, demonstrating great advantages in the treatment of oil–water emulsions.

(2) UF membrane is a kind of filter membrane with high porosity and pore size of 0.001–0.1 μm. UF is powered by pressure difference, utilizing the sieving mechanism to retain large-molecule solutes and allow high-molecule solvents or small-molecule substances to pass through the membrane to achieve the separation, purification, and concentration of liquids [56]. In practical applications, the separation characteristics of UF membranes are generally expressed by the relative molecular mass retained, which ranges from 1000 to 300 000, and the treatment effect is directly related to the size of the membrane pore size, which can retain proteins, bacteria, colloids, and other macromolecules. In oil–water

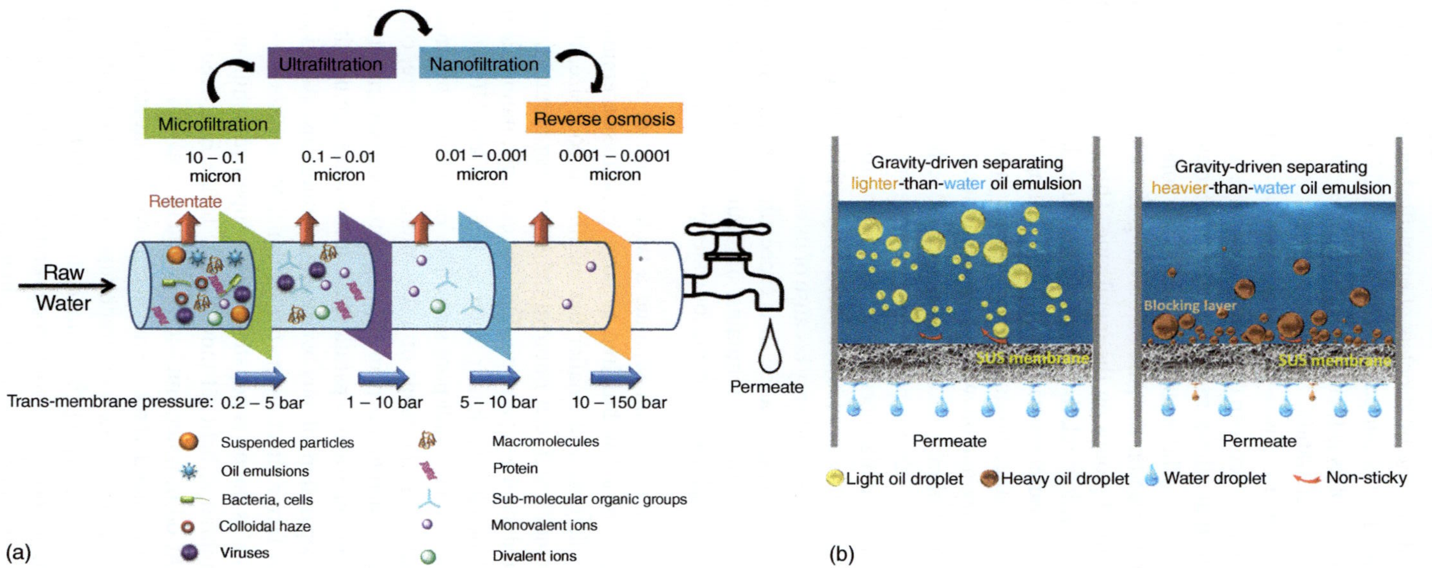

Figure 6.4 (a) Schematics of membrane water treatment system. Source: Selatile et al. [48b] Reproduced with permission of The Royal Society of Chemistry. (b) Schematic illustration of separation behavior of different oil–water mixtures with lighter-than-water oil and heavier-than-water oil. Source: Baig et al. [43] Reproduced with permission of Wiley.

separation, UF membrane prepared via electrospinning is generally used to separate emulsified oil. The oil droplets of emulsified oil can produce a good emulsion-breaking effect when passing through the UF membrane, and then make the small oil droplets gather and become larger to form dispersed oil or floating oil to achieve separation.

(3) NF membrane is a pressure-driven membrane material with a pore size between RO and UF membranes, between 0.01 and 0.001 μm. It is a functional semipermeable membrane that allows the permeation of certain solvent molecules, small-molecular-weight solutes, and low-valued salt ions [57]. The separation principle is mainly based on the pore size sieving effect and the charge effect formed by the charged groups on the membrane. This separation technology has the advantages of low relative operating pressure, large flux, diversified functions, strong pressure resistance, and anti-fouling. It is widely used in seawater desalination, juice production in the food industry, filtration in the pharmaceutical industry, and oil–water deep separation.
(4) The pore size of RO membrane is 0.001–0.0001 μm. RO technology uses chemical potential difference and osmotic pressure to separate substances with molecular weight less than 500 at high operating pressure, which can effectively remove organic compounds from water [58]. In recent years, this technology is widely used in the water treatment industry because of its easy operation and high efficiency in separation and purification. It is especially prevalent in applications such as the household drinking fountains, ultrapure water instruments, and other industries with high requirements for water purity.

Membrane separation method can separate a wide range of small molecules with molecular weights of a few hundred to large particle-size emulsions with particle sizes of ten microns. The smaller the membrane pore size, the smaller the separated compound, which can improve the separation accuracy, but at the same time will also reduce the separation flux. For example, nanoscale pore-size membranes prepared by electrospinning such as RO, NF, and UF often require external pressure due to their small pore sizes. The particle size of emulsions in oily wastewater is usually in the range of a few microns. Therefore, considering the combined effect of membrane pore size on emulsion separation flux and efficiency, gravity-driven oil–water separation MF membranes are usually selected to achieve efficient separation of oily wastewater emulsions. Additionally, oil density is another key factor affecting the conventional MF membrane separation process. In Figure 6.4b, as the oil is not sticky to the superhydrophilic-underwater superoleophobic membranes, light oil moves easily to the bulk feed during the separation of the light oil-in-water emulsions and generates less problem to the membranes. On the other hand, the accumulated heavy oil may start moving with the water permeation apart from the membrane superhydrophilicity and reduce the rejection rate of the membranes [43]. In summary, it is notable that the design of MF membranes with high performance should focus on the following factors: excellent antifouling properties, exceptional wettability, high permeate flux, and controlled pore size.

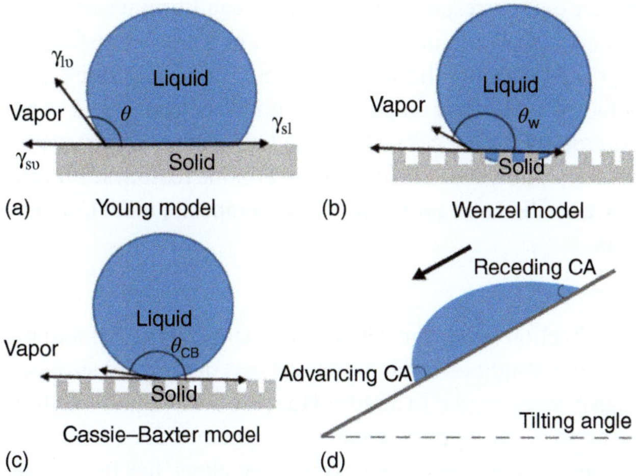

Figure 6.5 Wetting states of droplets on different solid surfaces in air and their CA models: (a) Young's, (b) Wenzel's, (c) Cassie–Baxter's, (d) Schematics of advancing CA, and droplet moving along an inclined surface with a tilting angle. Source: (a–c) Wang et al. [63] Reproduced with permission of MDPI; (d) Beedasy and Smith [62] Reproduced with permission of MDPI.

6.3.2.2 Oil–Water Separation Membranes Based on Different Wettability

Wettability is a macroscopic process in which a liquid replaces another fluid from the solid surface. In nature, the hydrophobic lotus leaf surface and water strider walking freely on water are the most common superwetting phenomena [59, 60]. The surfaces of these plants and animals in contact with water all contain a waxy substance with low surface energy or micro-nano-protruding structures. Wettability is an important physical and chemical property of solid materials, and there are two main factors affecting the wettability of solid surfaces: surface free energy and microscopic geometry [61–63].

Meanwhile, we can use three classical models to interpret the mechanism of surface wettability (Figure 6.5a–c): Young equation [64], Wenzel equation [65], and Cassie–Baxter equation [66]. Young equation is used to explain the mechanism of contact angle (CA, θ) and wettability of superhydrophobic surfaces:

$$\cos\theta = \frac{\gamma_{SV} - \gamma_{SL}}{\gamma_{LV}} \tag{6.1}$$

where γ_{SV}, γ_{SL}, and γ_{LV} are the surface tensions at the solid–vapor, solid–liquid, and liquid–vapor interfaces, respectively. Young equation is mainly used to explain the mechanism of wettability of smooth planes. The degree of CA is mainly determined by four factors: the intrinsic property of the substance, the roughness of the surface, the adsorption of the substance, and the inhomogeneity of the interface. For rough surfaces, Young equation cannot well explain the mechanism of superhydrophobicity of the substrate surface. Therefore, scientists have developed the Wenzel equation by modifying it on the basis of the Young equation:

$$\cos\theta_W = r\cos\theta \tag{6.2}$$

where θ_W is the apparent CA that corresponds to the stable equilibrium state; r is the roughness factor, defined as the ratio of the actual area of the solid surface to the apparent surface area ($r = 1$ for a smooth surface and $r > 1$ for a rough one); and θ is the Young's CA. Obviously, it is based on the assumption that the liquid penetrates into the roughness grooves. In the case where the liquid does not penetrate into the grooves, the Wenzel equation does not apply [67]. For this reason, scientists Cassie and Baxter et al. have studied and proposed a new composite model, called the Cassie–Baxter equation:

$$\cos\theta_{CB} = r_f f_{SL} \cos\theta + f_{SL} - 1 \tag{6.3}$$

where f_{SL} is the solid–liquid fraction under the contact area. Unlike the total roughness factor r, r_f is defined as the roughness ratio of the wet part of the solid surface, and r_f is always smaller than r. It shows that in addition to r, the solid–liquid fraction (solid–gas fraction) is also an influencing parameter of the apparent CA.

The difference between the three wetting states mentioned above lies in the CA hysteresis phenomenon, which reflects the heterogeny of the surface. CA hysteresis phenomena are common on rough and chemically nonhomogeneous surfaces. CA hysteresis, representing the difference between advancing and receding CA (Figure 6.5d), is completely different for Wenzel and Casey–Baxter wetting states [68]. Generally, droplets on the same surface may exhibit the effects of both models, that is, they may be in the Wenzel state and may also show the Cassie–Baxter state, but the CA of the former is smaller than that of the latter [69]. For the Cassie–Baxter state, it is sub-stable, so if it is disturbed by external conditions, the gas-phase structure between the solid–liquid phase may be disrupted to a Wenzel state, and the apparent CA of the droplet will be reduced. The wetting mechanism of superwetting surfaces is a combination of the Wenzel and the Cassie–Baxter model, which allows solid surfaces to exhibit two different states of CA simultaneously.

6.3.3 Oil–Water Separation Modes

Based on these basic wetting properties, Figure 6.6 illustrates superhydrophobic, superhydrophilic, superoleophobic, superoleophilic, and other types of surface functions that can be obtained by any combination of the two [70, 71]. When superhydrophobic/superoleophobic or superhydrophilic/superoleophobic are simultaneously available, separation of oil–water can be accomplished with premium efficiency. In addition to surfaces with these static wetting properties, in some cases the surface chemistry or geometry of a rough surface can be dynamically tuned. In such cases, it is possible to obtain a smart surface whose wettability can be reversibly switched between superhydrophobic and superhydrophilic or superoleophobic and superoleophilic. As mentioned earlier, the wettability of a solid is determined by its surface free energy and surface geometry. Therefore, dynamically changing one of these two factors can be used to regulate the surface wettability [72, 73].

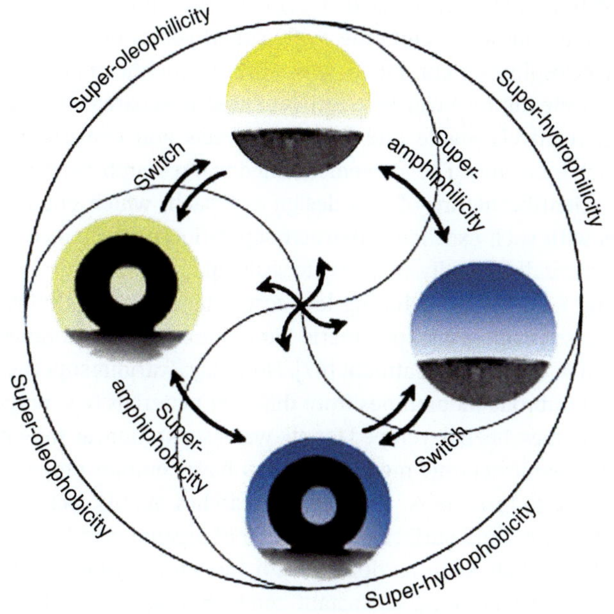

Figure 6.6 Illustration of the relationship between the four basic superwetting properties and further special surface superwetting functions obtained by combining two basic properties. Source: Feng and Jiang [70] Reproduced with permission of Wiley.

6.3.3.1 Hydrophobic and Oleophilic Membranes

Hydrophobic membrane materials are generally oleophilic, and oil droplets can quickly infiltrate and spread on the membrane surface to form an oil layer during the separation process, while the water phase is repelled from the upper surface of the membrane, thus effectively removing oil from water. Hydrophobic oil–water separation membrane materials are characterized by high surface roughness and low surface energy. So the main methods to obtain superhydrophobic properties on the membrane surface are: first, doping or coating the high roughness surface with substances of low surface energy, and second, constructing micro–nano rough structures on the low surface energy surface via various process methods [45]. Deng et al. fabricated ecofriendly composite nanofibers composed of poly(L-lactide) (PLLA), poly(D-lactide) (PDLA), and maghemite nanoparticles γ-Fe_2O_3 nanoparticles. Simultaneously, they achieved a nanoscale porous structure and stereocomplex crystallites in fibers by regulating the processing parameters and introducing additional annealing treatment [41]. The electrospun stereocomplex polylactide (PLA) porous nanofibrous membrane had good oil–water separation ability with 6824.4 Lm^{-2} h^{-1} of gravity-driven oil flux and good stability during the cycling measurements. Ma et al. successfully prepared a new flexible and self-supporting Fe^{3+}-phytic acid (PA)/octadecyltrimethoxysilane (OTMS)/polyimide (PI) nanofibrous membrane with superhydrophobicity and superoleophilicity by electrospinning combined with surface modification technology [74]. The prepared membranes can be effectively used to separate various oil–water mixtures with relatively stable and high flux levels and consistently higher than 99% separation efficiency. The abovementioned two research examples fully demonstrated the scientific nature of this design approach, which enables nanofibrous membranes with such excellent oil–water separation performance to be cost-effective, environmentally friendly, and easy to scale up.

Owing to the significant advantages including low cost, outstanding flexibility, good processability, and ease of operation, polymeric nanofibrous MF membranes have been extensively used in oily water treatment [67]. Nowadays, various superhydrophobic electrospun nanofibrous membranes from different materials (e.g. polymers, ceramics, and carbon) have been developed for oil–water separation, as shown in Table 6.1 [90]. Since inorganic ceramic membrane often have complex preparation processes and are prone to brittleness, the actual researches mainly focus on building surface rough structures and surface modification of organic membranes. This is achieved through methods such as doping with inorganic hydrophobic/hydrophilic materials, chemical grafting, surfactant modification, plasma technology, high-energy irradiation grafting, and others. These methods aim to impart superwetting to polymeric nanofibrous membranes for efficient oil-water separation.

6.3.3.2 Hydrophilic and Oleophobic Membranes

Hydrophilic and oleophobic membranes have oleophobic properties in air or water. When oily wastewater passes through the membrane surface, water can quickly permeate downward through the membrane. The oil phase is intercepted on one side of the membrane to achieve the effect of oil-water separation. These membranes can mostly avoid contamination by oil and can be reused many times with high separation effect, antipollution, and long service life. Hydrophilic-oleophobic membranes can be divided into organic and inorganic ones. Inorganic ones require calcination for preparation. The process is relatively complex, the repeatability of membrane preparation is poor, and it is difficult to prepare membranes with small pore sizes. In addition, it is costly to prepare. Therefore, its use is limited to a certain extent. Comparatively, the preparation of organic polymeric ones is simpler and more cost-effective. However, it is usually difficult for ordinary polymer membranes to reach the superhydrophilic and superoleophobic states, thus a lot of researches have focused on hydrophilic modification of polymeric membrane, which employs both chemical and physical modification methods. Chemical modification is to enhance the hydrophilicity of polymer membrane surface through chemical action, such as creating hydrophilic groups (carboxyl, hydroxyl, amide, etc.) on the membrane surface by chemical grafting, and hydrolysis. In addition, the design of surface roughness structure can also meet the requirements of enhancing the hydrophilicity and oleophobicity of organic membrane surface. Common methods to build roughness include

Table 6.1 Summary of hydrophobic-oleophilic electrospun nanofibrous membranes for oil–water separation.

Polymer	Additive	WCA/OCA (°)	Separation efficiency (%)	Filtration flux (l (m^{2}h)$^{-1}$)	References
PVDF	—	>150/0	>99.9	12994 (SFE) 2994 (SSE)	[75]
	SiO_2 NPs	150/0	97	6900 (mixture)	[76]
	CNC	144/0	97	5842 (SSE)	[77]
	P(MMA-*r*-FDMA)	140/0	—	6500 (mixture)	[78]
PAN	Au@ZIF-8/TA/DT	155.5/0	97.8	— (SSE)	[79]
	SiO_2 NPs	144.2/0	—	3032.4 (mixture)	[80]
	—	159.5/—	98	3600 (mixture)	[81]
PI	Fe^{3+}-PA/OTMS	154.7/<20	>99	8400 (mixture)	[74]
	Al^3+-TA/PDMS	153.6/0	>99	6935 (mixture)	[82]
	PDMS/SiO_2 NPs	155.75/<20	98.81	4443.16 (mixture)	[83]
	PDMS/ZnO	120/0	99	3706 (mixture)	[84]
	PBZ/SiO_2 NPs	154/—	>99	4789 (mixture)	[85]
PI/CA	F-PBZ/SiO_2 NPs	162/0	>99	3106 (mixture)	[86]
PLA	PDA/AgNPs	158.6/0	98.4	2664.3 (mixture)	[87]
PLLA	PEO	143/<1	—	13600 (SFE)	[88]
F-PBZ	Al_2O_3/SiO_2	161/0	—	892.5 (SSE)	[89]

Abbreviations: WCA: water contact angle, OCA: oil contact angle, SSE: surfactant-stabilized emulsion, SFE: surfactant-free emulsion, NP: nanoparticle, PVDF: polyvinylidene fluoride, CNC: cellulose nanocrystal, P(MMA-*r*-FDMA): poly(methyl methacrylate-*random*-perfluorodecyl methacrylate), PAN: polyacrylonitrile, ZIF-8: zeolitic imidazolate framework-8, TA: tannic acid, DA: 1-dodecanethiol, PDMS: polydimethylsiloxane, PBZ: polybenzoxazine, CA: cellulose acetate, F-PBZ: fluorinated polybenzoxazine, PDA: polydopamine, PEO: polyethylene oxide, – refers to unavailable data in the reference.
Source: Adapted from Su [90].

membrane forming process control, inorganic particle deposition, electrospinning technology construction, template method, layer-by-layer self-assembly, and plasma pretreatment [91, 92].

Wu et al. fabricated a *Momordica-charantia*-like nanofibrous membrane (MCNM) with underwater superoleophobic performance through a sacrifice template strategy by the electrospinning solution of ZIF-8 and PAN particles [93]. The prepared MCNM showed excellent self-cleaning performance for crude oil underwater, avoiding the degradation of separation efficiency and flux caused by membrane contamination. Meanwhile, the separation efficiency of oil–water emulsion stabilized by various surfactants was higher than 99.6%, and the flux was up to $1580 \pm 30\,l\,m^{-2}\,h^{-1}$ completely driven by gravity. Polyvinyl alcohol hydrogel nanofibrous membrane

Table 6.2 Comparison of properties of hydrophilic and oleophobic membranes.

Polymer	Additive	WCA/OCA (°)	Separation efficiency (%)	Filtration flux ($l\,m^{-2}\,h^{-1}$)	References
PVDF	Ag/PVDF	0/132.2	—	1000	[94]
PAN	SiO_2	0/163	—	1000	[95]
	ZIF-8	0/160	99.6	1580	[93]
	GO	0/120	99	3500	[96]
	Fe	0/153	99.8	—	[97]
	AIBN	0/130	99.6	—	[98]
PLA	GA/SiO_2	0/<150	99.6	4200	[99]
PLLA	CH_2Cl_2	62.3/11	—	1950	[100]
PU	ACNTs/PDA	0/155	99.9	4195	[101]
PVA	GO	15/—	99.3	50.52	[102]
	SiO_2	0/161.8	96	1500	[103]
	Glu	0/150	99	10800	[104]
CA	GO/PDA/PEI	0/148.5	—	—	[105]
PEO/PPO	—	25/145	97.88	7115	[106]

Abbreviations:, AIBN: 2,2-azodiisobutyronitrile, GA: gallic acid, PU: polyurethane, ACNT: acidified carbon nanotube, PEI: polyetherimide, PPO: polypropylene oxide.
Source: Yan et al. [107].

(PVA-HNM) produced via electrospinning and vapor cross-linking process was capable of continuous separation of oil–water emulsions by using squeezing coalescence demulsification (SCD) [34]. During SCD process, the small oil droplets will agglomerate into larger droplets under the forced compression of the pore walls. The agglomerated oil droplets will then pass through the nanofibrous membrane and form an oil layer, which can be easily removed. PVA-HNM with selective wettability has a high separation flux (13969 $l\,m^{-2}\,h^{-1}$). A variety of polymeric materials has been prepared into hydrophilic and oleophobic nanofibrous membranes so far, and their filtration and separation properties are summarized in Table 6.2 [107]. Among the currently available hydrophilic polymers, PAN is the most widely used polymer for the preparation of superhydrophilic electrospun nanofibrous membranes.

6.4 Summary and Future Perspectives

Achieving efficient purification of oily wastewater emulsions is of great value for the protection of the ecological environment, the recycling of water resources and human health. With its extremely high specific surface area, porosity, and permeability, the electrospun nanofibrous membrane greatly enhances the wettability of the membrane surface, effectively breaking the limitation of low flux of membrane separation technology. Especially, the superiority of electrospun nanofibrous membrane in the field of oil–water separation is becoming more and more noticeable. The key to membrane oil–water separation lies in the construction of superhydrophobic and superoleophilic or superhydrophilic and superoleophobic structures based on the excellent properties of nanofibrous membrane, such as high porosity, high specific surface area, great permeability, and strong strength. This is essential to achieve the purpose of effective oil–water separation.

Currently, the research on superwetting materials for oil–water separation has made a series of progresses, not only in terms of how they can be prepared under laboratory conditions, but also toward scaling up production, and good results have been achieved. However, there are still some problems and much work to be done:

(1) In present, the investigations of researchers mainly focus on solution electrospinning, due to the limitations of the solution spinning process conditions, most researchers obtain superhydrophobic-superoleophilic (or superhydrophilic-superoleophobic) structure of electrospun nanofibrous membrane by post-processing such as solution impregnation to introduce functional groups and coating/doping with nanoparticles. These post-treatment processes are complex and the introduced functional groups or particles will inevitably fall off during the use of the membrane material, which not only reduces the performance of the membrane material, but also easily causes secondary pollution. Finding a simpler process to prepare nanofibrous oil–water separation membranes is significant.
(2) In an electrospun nanofibrous membrane for oil–water separation, the complete replacement of toxic solvents with safer toxic solvents is still at an early stage. This is considering that the separation efficiency and water flux of membranes produced using green alternatives are still inferior to those prepared with classical toxic solvents. Research efforts need to be devoted to the development of more efficient and less toxic strategies for the fabrication of polymeric electrospun nanofibrous membranes.
(3) Membranes with switchable wettability in specific environments are considered smart materials. They are selectively adapted to both water and oil environments (oil-in-water or water-in-oil) and are therefore capable of performing oil–water separation on demand. Although a lot of work has been done in this research field, the improvement of the separation efficiency and the sensitivity of the switch is still limited, which is an important direction for future advanced research.
(4) The stability of the superwetting membranes obtained at the present stage is good under laboratory testing conditions. However, when encountering solid particles, inorganic salts, and other organic molecules in the actual oily wastewater, the stability of the membrane and its impact on the separation effect need to be further explored.

References

1 ITOPF. Oil Tanker Spill Statistics (2021). International Tanker Owners Pollution Federation Ltd. https://www.itopf.org/knowledge-resources/data-statistics/statistics/.

2 Djomo, J.E., Dauta, A., Ferrier, V. et al. (2004). Toxic effects of some major polyaromatic hydrocarbons found in crude oil and aquatic sediments on *Scenedesmus subspicatus*. *Water Research* 38 (7): 1817–1821.

3 Shannon, M.A., Bohn, P.W., Elimelech, M. et al. (2008). Science and technology for water purification in the coming decades. *Nature* 452 (7185): 301–310.

4 Dalton, T. and Jin, D. (2010). Extent and frequency of vessel oil spills in US marine protected areas. *Marine Pollution Bulletin* 60 (11): 1939–1945.

5 Hassler, B. (2011). Accidental versus operational oil spills from shipping in the Baltic Sea: risk governance and management strategies. *Ambio* 40 (2): 170–178.

6 Shi, Z., Zhang, W., Zhang, F. et al. (2013). Ultrafast separation of emulsified oil/water mixtures by ultrathin free-standing single-walled carbon nanotube network films. *Advanced Materials* 25 (17): 2422–2427.

7 Nordvik, A.B., Simmons, J.L., Bitting, K.R. et al. (1996). Oil and water separation in marine oil spill clean-up operations. *Spill Science and Technology Bulletin* 3 (3): 107–122.

8 Ge, J., Fu, Q., Yu, J., and Ding, B. (2019). Electrospun nanofibers for oil–water separation. *Electrospinning: Nanofabrication and Applications* 391–417.

9 Reis, J.C. (1993). An overview of the environmental issues facing the upstream petroleum industry. In: *SPE Annual Technical Conference and Exhibition* OnePetro, Texas, USA, 57–62.

10 Wan, Z. and Chen, J. (2018). Human errors are behind most oil-tanker spills. *Nature* 560 (7717): 161–163.

11 Singh, B., Kumar, S., Kishore, B., and Narayanan, T.N. (2020). Magnetic scaffolds in oil spill applications. *Environmental Science: Water Research & Technology* 6 (3): 436–463.

12 Chavez N. (2020). Deepwater Horizon spill was about 30% bigger than previously thought, study says Cable News Network. https://www.cnn.com/2020/02/12/us/deepwater-horizon-oil-spill-reach-study/index.html.

13 Berenshtein, I., Paris, C.B., Perlin, N. et al. (2020). Invisible oil beyond the Deepwater Horizon satellite footprint. *Science Advances* 6 (7): eaaw8863.

14 Gao, W., Yin, X., Mi, T. et al. (2018). Microbial diversity and ecotoxicity of sediments 3 years after the Jiaozhou Bay oil spill. *AMB Express* 8 (1): 79.

15 Helton D. (1990). Mega Borg oil spill: NOAA Photo Library. https://en.wikipedia.org/wiki/Mega_Borg_oil_spill.

16 Mishra, S., Chauhan, G., Verma, S., and Singh, U. (2022). The emergence of nanotechnology in mitigating petroleum oil spills. *Marine Pollution Bulletin* 178: 113609.

17 Gohsep, L. (2007). *An Oiled Brown Pelican Near Grand*. Isle: Wikipedia https://en.wikipedia.org/wiki/Deepwater_Horizon_oil_spill.

18 Ethan, E. (2019). *How Do You Feel Oil Spills and Pollution are Affecting You-th and Your Environment?* Teens Now Talk. https://teensnowtalk.com/features/think-green-n-think-global/environment-report/feel-oil-spills-pollution-affecting-th-environment/.

19 Kim J-H. (2007). Shinduri beach: Members of the South Korean army shovel up the oil: Agence France-Presse. https://www.theguardian.com/environment/gallery/2007/dec/10/oilspills.wildlife.

20 Ge, J., Ye, Y.D., Yao, H.B. et al. (2014). Pumping through porous hydrophobic/oleophilic materials: an alternative technology for oil spill remediation. *Angewandte Chemie (International Ed. in English)* 53 (14): 3612–3616.

21 Zhang, W., Zhai, X., Xiang, T. et al. (2016). Superhydrophobic melamine sponge with excellent surface selectivity and fire retardancy for oil absorption. *Journal of Materials Science* 52 (1): 73–85.

22 Kujawinski, E.B., Kido Soule, M.C., Valentine, D.L. et al. (2011). Fate of dispersants associated with the deepwater horizon oil spill. *Environmental Science & Technology* 45 (4): 1298–1306.

23 Urakawa, H., Rajan, S., Feeney, M.E. et al. (2019). Ecological response of nitrification to oil spills and its impact on the nitrogen cycle. *Environmental Microbiology* 21 (1): 18–33.

24 Al-Shamrani, A.A., James, A., and Xiao, H. (2002). Destabilisation of oil-water emulsions and separation by dissolved air flotation. *Water Research* 36 (6): 1503–1512.

25 Ndimele, P.E., Saba, A.O., Ojo, D.O. et al. (2018). *Remediation of Crude Oil Spillage. The Political Ecology of Oil and Gas Activities in the Nigerian Aquatic Ecosystem*, 369–384. Academic Press.

26 French-McCay, D.P., Jayko, K., Li, Z. et al. (2021). Oil fate and mass balance for the Deepwater Horizon oil spill. *Marine Pollution Bulletin* 171: 112681.

27 Doshi, B., Sillanpaa, M., and Kalliola, S. (2018). A review of bio-based materials for oil spill treatment. *Water Research* 135: 262–277.

28 Wang, J. and Wang, H. (2019). Integrated device based on cauliflower-like nickel hydroxide particles-coated fabrics with inverse wettability for highly efficient oil/hot alkaline water separation. *Journal of Colloid and Interface Science* 534: 228–238.

29 Yu, B., Peng, Y., Luo, X. et al. (2023). Effects of particle wettability and water moisture on separation efficiency and drop size distribution of particle-laden droplets in oil. *Colloids and Surfaces A: Physicochemical and Engineering Aspects* 659.

30 Zhang, W., Shi, Z., Zhang, F. et al. (2013). Superhydrophobic and superoleophilic PVDF membranes for effective separation of water-in-oil emulsions with high flux. *Advanced Materials* 25 (14): 2071–2076.

31 Panizza, M., Michaud, P.A., Iniesta, J. et al. (2002). Electrochemical oxidation of phenol at boron-doped diamond electrode. Application to electro-organic synthesis and wastewater treatment. *Annali di Chimica* 92 (10): 995–1006.

32 Etchepare, R., Oliveira, H., Azevedo, A., and Rubio, J. (2017). Separation of emulsified crude oil in saline water by dissolved air flotation with micro and nanobubbles. *Separation and Purification Technology* 186: 326–332.

33 Zhang, Y., Wei, S., Liu, F. et al. (2009). Superhydrophobic nanoporous polymers as efficient adsorbents for organic compounds. *Nano Today* 4 (2): 135–142.

34 Zhu, X., Zhu, L., Li, H. et al. (2021). Enhancing oil-in-water emulsion separation performance of polyvinyl alcohol hydrogel nanofibrous membrane by squeezing coalescence demulsification. *Journal of Membrane Science* 630: 119324.

35 Motta, F.L., Stoyanov, S.R., and Soares, J.B.P. (2018). Application of solidifiers for oil spill containment: a review. *Chemosphere* 194: 837–846.

36 Saejung, C., Raksapon, D., and Chaiyarat, A. (2022). Synthesis and application of a magnetically recoverable biological nanocomposite material for waste cooking oil removal and commercial product recovery in wastewater treatment. *Journal of Cleaner Production* 363: 132425.

37 Pujol, R. and Chudoba, P. (2000). Technical solutions for upgrading high rate and medium loaded activated sludge plants for nutrient removal. *Water Science and Technology* 41 (9): 131–138.

38 Tong, K., Zhang, Y., Liu, G. et al. (2013). Treatment of heavy oil wastewater by a conventional activated sludge process coupled with an immobilized biological filter. *International Biodeterioration and Biodegradation* 84: 65–71.

39 Zhou, J., Du, E., He, Y. et al. (2020). Preparation of carbonized kapok fiber/reduced graphene oxide aerogel for oil-water separation. *Chemical Engineering & Technology* 43 (12): 2418–2427.

40 Xue, J., Wu, T., Dai, Y., and Xia, Y. (2019). Electrospinning and electrospun nanofibers: methods, materials, and applications. *Chemical Reviews* 119 (8): 5298–5415.

41 Deng, Y.F., Zhang, D., Zhang, N. et al. (2021). Electrospun stereocomplex polylactide porous fibers toward highly efficient oil/water separation. *Journal of Hazardous Materials* 407: 124787.

42 Wendorff, J.H., Agarwal, S., and Greiner, A. (2012). *Electrospinning: Materials, Processing, and Applications*. Wiley.

43 Baig, N., Salhi, B., Sajid, M., and Aljundi, I.H. (2022). Recent progress in microfiltration/ultrafiltration membranes for separation of oil and water emulsions. *Chemical Record* 22 (7): e202100320.

44 Miao, W., Tian, Y., and Jiang, L. (2022). Bioinspired superspreading surface: from essential mechanism to application. *Accounts of Chemical Research* 55 (11): 1467–1479.

45 Xiong, C., Quan, Z., Zhang, H. et al. (2020). Hierarchically tunable structure of polystyrene-based microfiber membranes for separation and selective adsorption of oil-water. *Applied Surface Science* 532: 147400.

46 Wu, J., Wang, N., Wang, L. et al. (2012). Electrospun porous structure fibrous film with high oil adsorption capacity. *ACS Applied Materials & Interfaces* 4 (6): 3207–3212.

47 Li, Y., Zhu, J., Cheng, H. et al. (2021). Developments of advanced electrospinning techniques: a critical review. *Advanced Materials Technologies* 6 (11): 2100410.

48 (a) Zhu, M., Han, J., Wang, F. et al. (2017). Electrospun nanofibers membranes for effective air filtration. *Macromolecular Materials and Engineering* 302 (1); (b) Selatile, M.K., Ray, S.S., Ojijo, V., and Sadiku, R. (2018). Recent developments in polymeric electrospun nanofibrous membranes for seawater desalination. *RSC Advances* 8 (66): 37915–37938.

49 Liao, Y., Wang, R., and Fane, A.G. (2013). Engineering superhydrophobic surface on poly(vinylidene fluoride) nanofiber membranes for direct contact membrane distillation. *Journal of Membrane Science* 440: 77–87.

50 Li, D. and Xia, Y. (2004). Direct fabrication of composite and ceramic hollow nanofibers by electrospinning. *Nano Letters* 4 (5): 933–938.

51 Subramanian, A., Vu, D., Larsen, G.F., and Lin, H.Y. (2005). Preparation and evaluation of the electrospun chitosan/PEO fibers for potential applications in cartilage tissue engineering. *Journal of Biomaterials Science, Polymer Edition* 16 (7): 861–873.

52 Bhattarai, N., Edmondson, D., Veiseh, O. et al. (2005). Electrospun chitosan-based nanofibers and their cellular compatibility. *Biomaterials* 26 (31): 6176–6184.

53 Kidoaki, S., Kwon, I.K., and Matsuda, T. (2005). Mesoscopic spatial designs of nano- and microfiber meshes for tissue-engineering matrix and scaffold based on newly devised multilayering and mixing electrospinning techniques. *Biomaterials* 26 (1): 37–46.

54 Wang, X., Ding, B., Yu, J., and Yang, J. (2011). Large-scale fabrication of two-dimensional spider-web-like gelatin nano-nets via electro-netting. *Colloids and Surfaces B: Biointerfaces* 86 (2): 345–352.

55 Yoon, K., Hsiao, B.S., and Chu, B. (2008). Functional nanofibers for environmental applications. *Journal of Materials Chemistry* 18 (44): 5326–5334.

56 Werber, J.R., Osuji, C.O., and Elimelech, M. (2016). Materials for next-generation desalination and water purification membranes. *Nature Reviews Materials* 1 (5): 1–15.

57 Winter, J., Barbeau, B., and Berube, P. (2017). Nanofiltration and tight ultrafiltration membranes for natural organic matter removal-contribution of fouling and concentration polarization to filtration resistance. *Membranes (Basel)* 7 (3): 34.

58 Jiang, S., Li, Y., and Ladewig, B.P. (2017). A review of reverse osmosis membrane fouling and control strategies. *Science of the Total Environment* 595: 567–583.

59 Feng, L., Li, S., Li, Y. et al. (2002). Super-hydrophobic surfaces: from natural to artificial. *Advanced Materials* 14 (24): 1857–1860.

60 Gao, X. and Jiang, L. (2004). Biophysics: water-repellent legs of water striders. *Nature* 432 (7013): 36.

61 Zhang, X., Shi, F., Niu, J. et al. (2008). Superhydrophobic surfaces: from structural control to functional application. *Journal of Materials Chemistry* 18 (6): 621–633.

62 Beedasy, V. and Smith, P.J. (2020). Printed electronics as prepared by Inkjet printing. *Materials (Basel)* 13 (3): 704.

63 Wang, X., Fu, C., Zhang, C. et al. (2022). A comprehensive review of wetting transition mechanism on the surfaces of microstructures from theory and testing methods. *Materials (Basel)* 15 (14): 4747.

64 Young, T. III, (1805). An essay on the cohesion of fluids. *Philosophical Transactions of the Royal Society of London* 95: 65–87.

65 Wenzel, R.N. (1936). Resistance of solid surfaces to wetting by water. *Industrial and Engineering Chemistry* 28 (8): 988–994.

66 Cassie, A.B.D. and Baxter, S. (1944). Wettability of porous surfaces. *Transactions of the Faraday Society* 40: 546–551.

67 Wei, Y., Qi, H., Gong, X., and Zhao, S. (2018). Specially wettable membranes for oil–water separation. *Advanced Materials Interfaces* 5 (23): 1800576.

68 Gao, L. and McCarthy, T.J. (2006). Contact angle hysteresis explained. *Langmuir* 22 (14): 6234–6237.

69 Qi, X., Song, W., Mao, Z. et al. (2013). Fabrication of a bionic needle with both super-hydrophobic and antibacterial properties. *Journal of Bionic Engineering* 10 (3): 377–382.

70 Feng, X.J. and Jiang, L. (2006). Design and creation of superwetting/antiwetting surfaces. *Advanced Materials* 18 (23): 3063–3078.

71 Su, B., Tian, Y., and Jiang, L. (2016). Bioinspired interfaces with superwettability: from materials to chemistry. *Journal of the American Chemical Society* 138 (6): 1727–1748.

72 Russell, T.P. (2002). Surface-responsive materials. *Science* 297 (5583): 964–967.

73 de Crevoisier, G., Fabre, P., Corpart, J.M., and Leibler, L. (1999). Switchable tackiness and wettability of a liquid crystalline polymer. *Science* 285 (5431): 1246–1249.

74 Ma, W., Zhang, M., Liu, Z. et al. (2018). Nature-inspired creation of a robust free-standing electrospun nanofibrous membrane for efficient oil–water separation. *Environmental Science: Nano* 5 (12): 2909–2920.

75 Zhang, J., Ge, J., Si, Y. et al. (2019). Taro leaf-inspired and superwettable nanonet-covered nanofibrous membranes for high-efficiency oil purification. *Nanoscale Horizons* 4 (5): 1174–1184.

76 Gao, J., Li, B., Wang, L. et al. (2018). Flexible membranes with a hierarchical nanofiber/microsphere structure for oil adsorption and oil/water separation. *Journal of Industrial and Engineering Chemistry* 68: 416–424.

77 Wang, D., Wei, L., Yang, Y., and Liu, H. (2018). Dietary supplementation with ketoacids protects against CKD-induced oxidative damage and mitochondrial dysfunction in skeletal muscle of 5/6 nephrectomised rats. *Skeletal Muscle* 8 (1): 18.

78 Sadeghi, I., Govinna, N., Cebe, P., and Asatekin, A. (2019). Superoleophilic, mechanically strong electrospun membranes for fast and efficient gravity-driven oil/water separation. *ACS Applied Polymer Materials* 1 (4): 765–776.

79 Zhang, Z., Yang, Y., Li, C., and Liu, R. (2019). Porous nanofibrous superhydrophobic membrane with embedded Au nanoparticles for the integration of oil/water separation and catalytic degradation. *Journal of Membrane Science* 582: 350–357.

80 Tai, M.H., Gao, P., Tan, B.Y. et al. (2014). Highly efficient and flexible electrospun carbon-silica nanofibrous membrane for ultrafast gravity-driven oil-water separation. *ACS Applied Materials & Interfaces* 6 (12): 9393–9401.

81 Feng, S., Luo, W., Wang, L. et al. (2019). Preparation and property of extremely stable superhydrophobic carbon fibers with core-shell structure. *Carbon* 150: 284–291.

82 Ma, W., Ding, Y., Zhang, M. et al. (2020). Nature-inspired chemistry toward hierarchical superhydrophobic, antibacterial and biocompatible nanofibrous membranes for effective UV-shielding, self-cleaning and oil-water separation. *Journal of Hazardous Materials* 384: 121476.

83 Zhang, M., Ma, W., Wu, S. et al. (2019). Electrospun frogspawn structured membrane for gravity-driven oil-water separation. *Journal of Colloid and Interface Science* 547: 136–144.

84 Cui, J., Li, F., Wang, Y. et al. (2020). Electrospun nanofiber membranes for wastewater treatment applications. *Separation and Purification Technology* 250: 117116.

85 Ma, W., Zhang, M., Liu, Z. et al. (2019). Fabrication of highly durable and robust superhydrophobic-superoleophilic nanofibrous membranes based on a fluorine-free system for efficient oil/water separation. *Journal of Membrane Science* 570–571: 303–313.

86 Ma, W., Guo, Z., Zhao, J. et al. (2017). Polyimide/cellulose acetate core/shell electrospun fibrous membranes for oil-water separation. *Separation and Purification Technology* 177: 71–85.

87 Liu, L. and Yuan, W. (2018). A hierarchical functionalized biodegradable PLA electrospun nanofibrous membrane with superhydrophobicity and antibacterial properties for oil/water separation. *New Journal of Chemistry* 42 (21): 17615–17624.

88 Zhao, J., Wang, W., Ye, C. et al. (2018). Gravity-driven ultrafast separation of water-in-oil emulsion by hierarchically porous electrospun poly(L-lactide) fabrics. *Journal of Membrane Science* 563: 762–767.

89 Huang, M., Si, Y., Tang, X. et al. (2013). Gravity driven separation of emulsified oil–water mixtures utilizing in situ polymerized superhydrophobic and superoleophilic nanofibrous membranes. *Journal of Materials Chemistry A* 1 (45): 14071–14074.

90 Su, R., Li, S., Wu, W. et al. (2021). Recent progress in electrospun nanofibrous membranes for oil/water separation. *Separation and Purification Technology* 256: 117790.

91 Rajab, F.H., Liu, Z., and Li, L. (2018). Production of stable superhydrophilic surfaces on 316L steel by simultaneous laser texturing and SiO_2 deposition. *Applied Surface Science* 427: 1135–1145.

92 Li, X., Hu, D., Huang, K., and Yang, C. (2014). Hierarchical rough surfaces formed by LBL self-assembly for oil–water separation. *Journal of Materials Chemistry A* 2 (30): 11830–11838.

93 Wu, M., Liu, W., Mu, P. et al. (2020). Sacrifice template strategy to the fabrication of a self-cleaning nanofibrous membrane for efficient crude oil-in-water emulsion separation with high flux. *ACS Applied Materials & Interfaces* 12 (47): 53484–53493.

94 You, X., Liao, Y., Tian, M. et al. (2020). Engineering highly effective nanofibrous membranes to demulsify surfactant-stabilized oil-in-water emulsions. *Journal of Membrane Science* 611: 118398.

95 Ge, J., Zhang, J., Wang, F. et al. (2017). Superhydrophilic and underwater superoleophobic nanofibrous membrane with hierarchical structured skin for effective oil-in-water emulsion separation. *Journal of Materials Chemistry A* 5 (2): 497–502.

96 Zhang, J., Pan, X., Xue, Q. et al. (2017). Antifouling hydrolyzed polyacrylonitrile/graphene oxide membrane with spindle-knotted structure for highly effective separation of oil-water emulsion. *Journal of Membrane Science* 532: 38–46.

97 Zhao, R., Tian, Y., Li, S. et al. (2019). An electrospun fiber based metal–organic framework composite membrane for fast, continuous, and simultaneous removal of insoluble and soluble contaminants from water. *Journal of Materials Chemistry A* 7 (39): 22559–22570.

98 Zang, L., Zheng, S., Wang, L. et al. (2020). Zwitterionic nanogels modified nanofibrous membrane for efficient oil/water separation. *Journal of Membrane Science* 612: 118379.

99 Zhang, Z.-M., Gan, Z.-Q., Bao, R.-Y. et al. (2020). Green and robust superhydrophilic electrospun stereocomplex polylactide membranes: multifunctional oil/water separation and self-cleaning. *Journal of Membrane Science* 593: 117420.

100 Zhang, D., Zhang, N., Ma, F.F. et al. (2018). One-step fabrication of functionalized poly(L-lactide) porous fibers by electrospinning and the adsorption/separation abilities. *Journal of Hazardous Materials* 360: 150–162.

101 Huang, X., Zhang, S., Xiao, W. et al. (2020). Flexible PDA@ACNTs decorated polymer nanofiber composite with superhydrophilicity and underwater superoleophobicity for efficient separation of oil-in-water emulsion. *Journal of Membrane Science* 614: 118500.

102 Ghaffar, A., Chen, C., Zhu, X., and Chen, B. (2019). Underwater superoleophobic PVA–GO nanofibrous membranes for emulsified oily water purification. *Environmental Science: Nano* 6 (12): 3723–3733.

103 Qing, W., Li, X., Wu, Y. et al. (2020). In situ silica growth for superhydrophilic-underwater superoleophobic Silica/PVA nanofibrous membrane for gravity-driven oil-in-water emulsion separation. *Journal of Membrane Science* 612: 118476.

104 Zhu, X., Zhu, L., Li, H. et al. (2021). Multifunctional charged hydrogel nanofibrous membranes for metal ions contained emulsified oily wastewater purification. *Journal of Membrane Science* 621: 118950.

105 Xiao, J., Lv, W., Song, Y., and Zheng, Q. (2018). Graphene/nanofiber aerogels: performance regulation towards multiple applications in dye adsorption and oil/water separation. *Chemical Engineering Journal* 338: 202–210.

106 Liang, Y., Kim, S., Yang, E., and Choi, H. (2020). Omni-directional protected nanofiber membranes by surface segregation of PDMS-Terminated triblock copolymer for high-efficiency oil/water emulsion separation. *ACS Applied Materials & Interfaces* 12 (22): 25324–25333.

107 Yan, X., Xiao, X., Au, C. et al. (2021). Electrospinning nanofibers and nanomembranes for oil/water separation. *Journal of Materials Chemistry A* 9 (38): 21659–21684.

7

Electrospun Nanofiber-based Evaporators for Interfacial Solar-driven Steam Generation

Huijie Liu and Xiaohong Qin

Donghua University, College of Textiles, Key Laboratory of Textile Science & Technology, Ministry of Education, 2999 Renmin North Road, Songjiang District, Shanghai 201620, China

7.1 Introduction

Fresh water is recognized as a global issue of strategic relevance, which is critical to the humankind's survival, as well as the economic development and societal progress [1]. So far, many people are suffering from fresh water shortage due to population growth, climate change, and environmental pollution in the last few decades [2]. Although water is one of the most abundant compounds, covering 71% of the Earth's surface, most of the seawater is not directly drinkable. Traditional desalination technologies such as reverse osmosis (RO) [3, 4] and multistage flash (MSF) [5] have high-energy consumption and expensive infrastructure. In recent years, interfacial solar-driven steam generation (ISSG) is growing to be an important technology to harvest solar energy and convert it to heat for water evaporation and storage of power [6–10].

The unique structure of electrospun nanofibers makes them exhibit many functional properties, which have shown broad application prospects in many fields [11], among which the application in the emerging field of ISSG is eye-opening. Electrospun nanofibers have high specific surface area and high porosity and permeability, which are considered ideal skeletons for water evaporators for sunlight absorption and fast water transport. Meanwhile, the nanofibers are easy to functionalize to meet the requirements of different water environments [12, 13]. In addition, the advantages of low cost, light weight, and excellent flexibility make them competitive water evaporation devices in practice. In fact, electrospun nanofiber-based evaporators have been used for desalination, wastewater purification, and electricity generation [9, 14–16], which may provide great potential opportunities in the field.

7.2 Interfacial Solar Steam Generation (ISSG) System

A typical ISSG system contains several components, including the solar absorbers, the substrates, bulk water reservoir, incident light, and the vapor [9]. The incident

Electrospinning: Fundamentals, Methods, and Applications, First Edition.
Edited by Liming Wang and Xiaohong Qin.

light was harvested by solar absorbers and converted into heat, heating water and producing the vapor at the water–air interface and ultimately on the substrate. Meanwhile, the water was pumped into the evaporation surface from the bulk water by the substrate. The generated heat raises the temperature of the water on the evaporation surface, which drives the evaporation process continuously with the unremitting supply of water to generate vapor.

In this chapter, we outline recent developments in electrospun nanofiber-based evaporators, including the material choice of various types of solar absorbers (e.g. carbonaceous materials, plasmonic nanoparticles, semiconductors, and polymers) and the preparation strategies, diverse potential applications, and some key challenges and opportunities that require further exploration.

7.3 The Photothermal Conversion Materials and Steam Generation Efficiency Calculation

7.3.1 Photothermal Materials

Materials with high light absorption over the whole solar spectrum and high-efficiency light-to-heat conversion is the first step to achieve a high-performance solar evaporation device. Three categories of solar-absorbing materials, namely carbon-based, plasmonic-based, and polymer-based absorbers, have been developed. Carbon materials are naturally black and are suitable for broadband solar absorption. Solar absorption in carbon-based materials involves the excitation of electrons and their subsequent relaxation. The optically excited electrons are quickly thermalized by electron–electron and electron–phonon scattering. Until now, several carbon-based solar absorbers can be found in this field such as graphite [17], hollow carbon sphere [18], reduced graphene oxide (RGO) [19], graphene oxide (GO) [20], carbon nanotubes (CNT) [21], and carbon dots [22]. However, the challenge with carbon-based absorbers for widespread applications is to decrease cost and to improve the stability of the system.

Plasmonic absorbers such as gold (Au), aluminum (Al), silver (Ag), palladium (Pd), and Au–Ag bimetallic have been used in ISSGs [23–25]. In some metallic nanomaterials, the localized surface plasmon resonance (LSPR) effect is exhibited, which is a resonant photon-induced coherent oscillation of charges occurring when the photon frequency matches the natural metal surface electrons frequency. Three sequential phenomena including near-field enhancement, hot electron generation, and photothermal conversion are induced by the LSPR effect. It should be noted that the light absorption ability, especially the absorption wavelength, highly depends on the size and shape of the metal particles. Therefore, the plasmonic nanoparticles of different scales should be immobilized on porous substrates to extend the solar spectrum because each plasmonic nanoparticles (NP) has a narrow absorb band of light.

Polymers also have been used as solar absorbers in ISSG due to its high broadband light absorption and have special features as bonding with water to facilitate the evaporation process. Several polymers, such as polypyrrole (PPy) [26, 27], polydopamine (PDA) [28], and poly(1,3, 5-hexahydro-1,3,5-triazines) [29], have been applied as solar absorbers for high-efficiency solar evaporation recently.

In addition, the metal oxides and chalcogenides have shown promising prospects due to the fine tunability in their absorption spectrum and large extinction coefficient in the NIR region. Up to now, titanium-based semiconductors (black TiO_2 [30], black TiOx [31], and narrow-bandgap Ti_2O_3 [17]), magnetic particles (Fe_3O_4, $MnFe_2O_4$, $ZnFe_2O_4$, and $CoFe_2O_4$) [32–34], Al-Ti-O [35], NiO [36], and MoO_{3-x} [37] have been reported and validated as effective photothermal evaporation materials.

7.3.2 Steam Generation Efficiency Calculation

Numerous studies have been conducted to develop various materials and structures for solar vaporization. Increasing solar absorptance and limiting thermal reflectance and emittance can be achieved through photothermal material selection, composition tuning, and surface engineering. Also, effective reduction of heat loss from conduction, convection, or radiation helps solar to thermal conversion. Several critical criteria are identified that can be employed for the judicious design and characterization of solar evaporation devices endowed with superior photothermal performance: solar absorption, light-to-heat conversion and heat-to-vapor process.

7.3.2.1 Efficient Solar Absorption

The range of absorption across the solar spectrum and the intensity of absorbance of each wavelength determine the light harvesting ability of the materials. Thus, solar absorbers are expected to have high absorbance and minimal transmittance and reflectance in the full solar spectrum range (300–2500 nm, AM 1.5). Solar absorptance is an important parameter to measure the ability of materials to absorb solar radiation, which is determined by the ratio of the total absorbed solar radiation to the incident radiation. If there is a light incidence θ, the solar absorptance of the device can be obtained by weighing the spectral absorptance of the photothermal materials with the spectral irradiance distribution of the standard solar spectrum (AM 1.5) and integrating over the range of wavelength, as shown in Eq. (7.1)

$$\alpha(\theta) = \frac{\int_{\lambda_{\min}}^{\lambda_{\max}} [1 - R(\theta, \lambda)]A(\theta, \lambda)\, \mathrm{d}\lambda}{\int_{\lambda_{\min}}^{\lambda_{\max}} A(\lambda)\, d\lambda} \tag{7.1}$$

where $\lambda_{\min}$ and $\lambda_{\max}$ are 0.3 and 2.5 μm, respectively. $A(\lambda)$ is the wavelength-dependent solar spectral irradiance. $R(\theta, \lambda)$ is the the total reflection at the corresponding wavelength λ.

7.3.2.2 Efficient Light-to-heat Conversion

The conversion of light to heat energy is crucial, which would require the material to convert the energy absorbed from the sun to heat energy effectively. A direct experimental method of determining the photothermal conversion efficiency is by measuring the temperature rise and, consequently, the generated heat energy by the incident sunlight. Typically, sunlight is passed through the water containing solar absorber materials, which convert the absorbed light to heat energy, resulting in an increase in the overall temperature of the water. Accounting for the heat losses by conduction and other mechanisms, the energy gained by the system (calculated by the increases in temperature and the specific heat capacity of water) is compared to the sunlight that is absorbed by the system to estimate the light-to-heat conversion efficiency.

7.3.2.3 Efficient Heat-to-Vapor Generation

In practice, some of the generated heat from the solar devices will be consumed through local water heating and losses to the container and surroundings by means of conduction, convection, and radiation. Therefore, there is another indicator of vaporization efficiency η, which is the sunlight-to-steam generation efficiency, to evaluate the photothermal performance. The efficiency η can be calculated using Eq. (7.2):

$$\eta = (\dot{m}h_{LV})/q_i \tag{7.2}$$

where $\dot{m}$ is the interfacial solar-driven evaporation rate of water under certain solar illumination ($\dot{m} = m_{light} - m_{dark}$), q_i is the incident power density of solar illumination in process of steam generation experiment, and h_{LV} was made up of the sensible heat and the enthalpy of vaporization calculated using Eq. (7.3):

$$h_{LV} = C\Delta T + \Delta h \tag{7.3}$$

where C is the specific heat capacity of water (4.18 J g^{-1} K^{-1}), ΔT is the temperature increase of water, and Δh is the enthalpy of vaporization on the relative temperature.

7.4 The Preparation of the Electrospun Nanofiber-Based Evaporators

7.4.1 Two-dimensional (2D) Photothermal Membrane

The two-dimensional (2D) electrospun nanofibrous membrane is usually used as the substrate to load the photothermal materials through mixed spinning [38–42], coating [43], and surface modification [44] to prepare the solar water evaporators.

Qin's group [38] constructed multiscale nanoarchitectured nanofiber/carbon nanotubes (NF/CNTs) networks by a needleless-electrospinning/spraying–netting strategy, enabling well-dispersed CNT networks (diameter ≈ 25 nm) welded on charged nanofibrous scaffolds (diameter >100 nm), layer by layer, as shown in Figure 7.1a. The NF/CNT can be used as high-performance solar vapor generators to desalinate seawater (3.56 L^{-2} d^{-1}). Coaxial electrospinning technique with AIEgen/oil solution as coaxial inner layer and poly(vinylidene fluoride-co-hexafluoropropylene) (PVDF-HFP) polymer solution as coaxial outer layer were also utilized in the fabrication of solar vapor generation devices to improve the photothermal conversion efficiency [41]. Song et al. [43] designed the evaporator with photocatalytic and photothermal performance by electrospun SiO_2

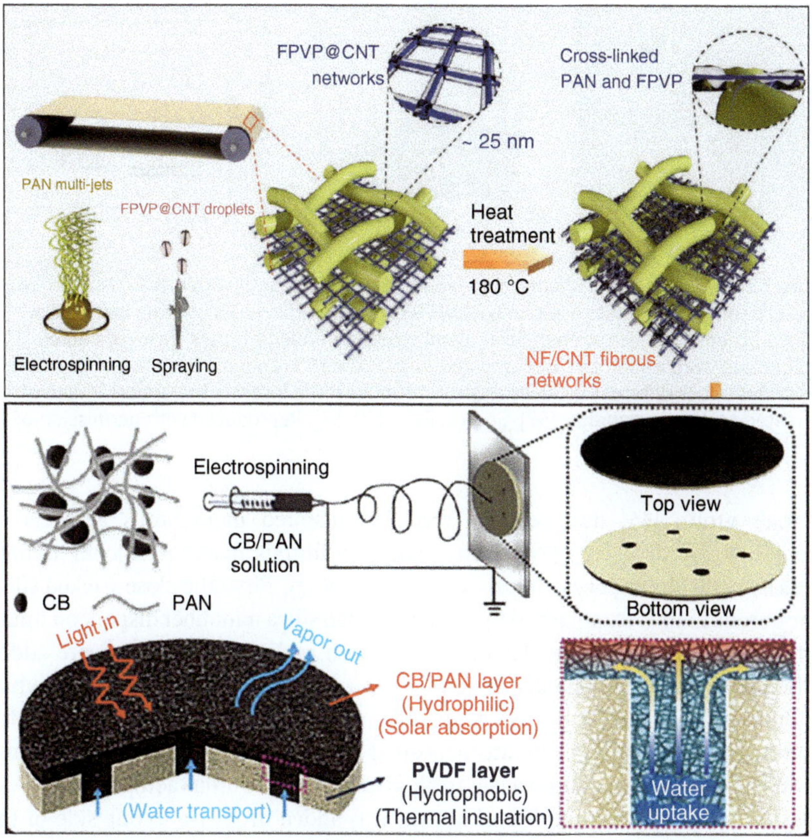

Figure 7.1 (a) The needless-electrospinning/spraying–netting and thermo-cross-linking process for the fabrication of NF/CNT fibrous network [38]. (b) Schematic illustration of the construction of the bilayer device using an electrospinning method [42]. Source: Refs. [38, 42]. Reproduced with permission of Wiley.

fiber as a template, followed by TiO_2 coating and magnesiothermic reduction. In order to further improve the energy efficiency, Chen et al. [44] developed a highly porous laser-induced graphene/polyimide (LIG/PI) photothermal membrane with a three-dimensional (3D) architecture through electrospinning and laser ablation techniques. The 3D structure of the LIG/PI membrane increases the evaporation area and reduces the energy loss caused by diffuse reflectance of light. Meanwhile, Janus-structured nanofibrous membranes are also widely used for solar steam generation by sequentially electrospinning polymers with different hydrophilic and hydrophobic properties (Figure 7.1b) [42].

7.4.2 3D Electrospun Nanofiber-Based Evaporators

There are mainly two types of 3D electrospun nanofiber-based evaporators: nanofiber-based aerogels [45–51] and the structural design of nanofibrous membranes [49, 52].

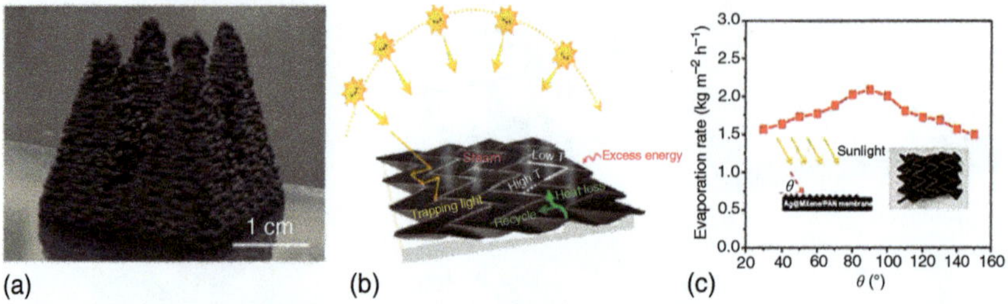

Figure 7.2 (a) 3D pyramid-shaped solar vapor generator (PSVG) with hierarchical porous structure combing co-electrospinning PAN/CNTs nanofibers and 3D printing technology [50]. (b) Schematic illustration of 3D origami structure under different incident angles of sunlight and evaporation rate of fabricated Ag@MXene/PAN nanofiber-based origami evaporator under different incident angles of sunlight; the insert is the optical image of fabricated origami evaporator [52]. Source: Refs. [50, 52]. Reproduced with permission of Elsevier.

Ding's group [46] designed the reed leaf-inspired nanofibrous aerogels by assembling the tightly packed electrospun nanofibrous mats into blocky nanofibrous aerogels through easy freeze-drying technology. First, the close-packed silica nanofibrous mats were smashed to an intertangled silica nanofiber dispersion under high-speed homogenization. Then, potato starch and glutaraldehyde were added to the dispersion. After the dispersion was frozen at −30 °C into ice cubes, white nanofibrous aerogels (W-NFAs) were obtained via freeze-drying. Subsequently, photothermal materials of PPy was deposited onto the surface of the W-NFAs, and the W-NFAs were freeze-dried again to obtain black nanofibrous aerogels (B-NFAs). Finally, the hydrolyzed solution of MTMS was added dropwise to one side of the B-NFAs to obtain R-NFAs with one-sided hydrophobicity. Overall, the architecture of the R-NFAs was similar to that of reed leaves and could be divided into three levels: vessels (150–200 μm), vessel walls (25–50 μm), and tetsubo-like nanofibers (<1 μm). The first two structure levels were inherited from the W-NFAs, and the last structure level was formed during the polymerization of the pyrrole. The well-designed hierarchical cellular structure can work stably in high-concentration brine (saturated, 26.3 wt%) under high-intensity light (up to 6 sun). A gas foaming technique was also used to stretch and expand the polyacrylonitrile (PAN) nanofibers to a self-floating foam, which enhanced the thermal insulation of the evaporators [49]. Liu et al. [50] designed and fabricated a 3D pyramid-shaped solar vapor generator (PSVG) by combining co-electrospinning PAN/CNTs nanofibers and the facile extrusion-based 3D printing technology (Figure 7.2a). The concave–convex shape and hierarchical porous structure achieve a high-level evaporation rate and has good self-cleaning property in high-concentration brine. Meanwhile, Liu et al. [52] utilized the flexibility and foldability of electrospun nanofiber to construct the evaporator of origami structure, which can enhance the broadband light absorption and achieve omnidirectional solar steam generation (Figure 7.2b).

7.5 Applications

The enormous progress that has been made in electrospun nanofiber-based solar-driven evaporators provides a great opportunity for boosting the practical application of sustainable energy and water toward a variety of fields. The transfer and conversion between various energies in solar evaporation systems enables the technology to convert unpurified water (such as seawater, wastewater, and polluted water) into clean water and solar energy into other forms of energy (such as thermal, chemical, electrical, and mechanical energy) that can be easily collected and stored for further use. In this section, we discuss potential applications of solar evaporation, including desalination, wastewater purification, and power generation.

7.5.1 Desalination

The fresh water supply becomes urgent for humans. Desalination is an effective way to obtain clean water from saline water. As compared to membrane technology, solar-driven desalination is less susceptible to the changes in component concentrations and low-quality water resource, though a highly polluted water feed may add complexity of the desalination. The current progress in developing electrospun nanofibers for solar-driven evaporation has greatly improved the conversion efficiency, so various systems ranging from solar absorber films and Janus structures to 3D devices have been employed for solar desalination.

Zhu et al. [53] fabricated flexible and washable carbon-nanotube-embedded PAN nonwoven fabrics by a simple electrospinning route. With a high concentration of simulated seawater as the model, the salts accumulated on the surface of fabrics can be effortlessly washed away through a plain hand-washing process. The washing process has a negligible influence on the morphology, photoabsorption, and evaporation performance of the fabric, demonstrating good durability.

Qi et al. [54] prepared the silicon dioxide/carboxylated multi-walled carbon nanotube/polyacrylonitrile (SiO_2/MWCNTs-COOH/PAN) fibrous membrane evaporator by electrospinning, which assembled the polystyrene (PS) foam and cotton yarns (CYs) used for sufficient water transportation. As a result, the interfacial water evaporator maintained a stable desalination rate of 1.22 kg m^{-2} h^{-1} for 20 cycles.

Janus absorbers have unique advantages in stable and efficient solar desalination [55–57]. Taking advantage of the unique structure of Janus, two functions of solar-steam generation and water pumping, are decoupled into different layers, with an upper hydrophobic layer for light absorption and steam generation, and a lower hydrophilic layer for pumping water. Therefore, salt can only be deposited in the hydrophilic layer and quickly be dissolved because of continuous water pumping. With the unique structural design achieved by scalable process, this flexible Janus absorber provides a stable and portable solar steam generator for direct solar desalination (1.3 kg m^{-2} h^{-1}, over 16 days) under 1 sun [55]. To further enhance the descaling capacity of evaporators, a self-descaling Janus evaporator (SJE) featuring a "moving interface" is developed for durable and efficient desalination of hypersaline water [56]. This evaporator consists of a Fe_3O_4-embedded poly(*N*-isopropylacrylamide) (PNIPAM) nanofibrous layer and a hydrophilic PAN nanofibrous layer constructed by sequential electrospinning. Integrating the photothermal conversion of Fe_3O_4 and the thermal responsiveness of PNIPAM, the SJE exhibits a reversible solar-induced wettability transition from asymmetric wettability (under sunlight) to hydrophilic wettability (at night). Under sunlight, the top surface and the evaporation interface (i.e. air–water interface) are separated to prevent salt accumulation on the evaporator, benefiting efficient solar steam generation. At night, the air–water interface moves to the upper surface, showing the self-descaling ability. For the stability of the electrospun hydrophilic/hydrophobic nanofibrous membrane in salty water, Qin's group fabricated the simply reconfigurable and entirely self-healing Janus evaporator [57]. The interface of the Janus membrane is stitched by the covalent bonds to directly improve the interface adhesion to improve the stability during the desalination process.

A lotus-inspired biomimetic evaporator (LBE) combines Janus electrospun membranes and thermal insulator was also fabricated to achieve rapid and stable long-term evaporation abilities without any accumulation of salt or solutes [51]. However, the nanofibrous evaporators prepared by electrospinning are usually packed into relatively dense membranes with unsatisfying connectivity and quantity of pores, which hinder the transportation of salt from the evaporation surface to bulk water. Most of the recently reported studies can only achieve salt resistance in low-salinity seawater (<3.5 wt%) and weak light intensity (1 sun), which is far behind the demands of industrial wastewater (10–20 wt%) and long-term evaporating solar evaporation system. Ding et al. [58] constructed CNTs@SiO_2 nanofibrous aerogels (CNFAs) with cellular architecture composed of vertically aligned vessels and porous vessel walls for the solar evaporator by assembling the inherent laminar electrospun nanofibrous membranes into blocky nanofibrous aerogels. Under the action of convection and diffusion promoted by this unique cellular architecture, the CNFAs exhibit an excellent salt transportation performance from high- to low-salinity brine along the vessels or through the pores on vessel walls. The excellent salt transportation performance can reduce the salinity of the evaporation interface and prevent salt crystallization even in high-salinity brine and intense light. With this unique cellular architecture, the CNFAs exhibit an excellent salt resistance with stable and efficient solar desalination even in 20 wt% brine and under 6-sun irradiation, better than most of the current reports. Moreover, by virtue of a high absorbance of 98%, the CNFAs exhibit an evaporation rate of $1.50\,kg\,m^{-2}\,h^{-1}$ under 1 sun irradiation without obvious decay even in extreme conditions or during long-term use (Figure 7.3).

7.5.2 Wastewater Purification

The electrospun nanofiber-based evaporators are easily functionalized to be suitable for all kinds of water sources. Like saline water desalination, similar solar devices

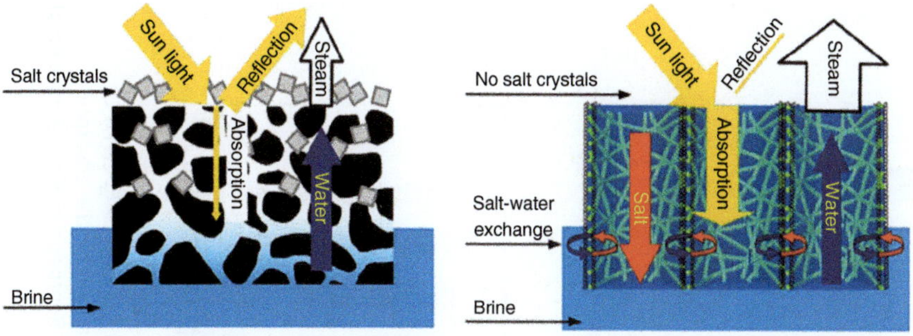

Figure 7.3 Comparison of water transportation, light absorption, evaporation, and salt crystallization in conventional solar evaporators and CNFAs [58]. Source: Dong et al. [58]. Reproduced with permission of Wiley.

are adopted for wastewater purification, including separation of emulsified oil/water mixtures [59], sterilization [44, 52] and degradation of dye molecules [60–63], and removal of heavy metal ions [64].

The reported [52] Ag@MXene/PAN nanofibrous evaporator can effectively catalyze reduction of the most common pollutants, 4-NP and 2-NA, and sterilize them due to the conjunction of Ag nanoparticles and MXene nanosheets, as shown in Figure 7.4a. Some contaminated water sources containing volatile organic components also face another challenge in that organic compounds tend to evaporate along with the collected water, leading to an impure water product. The mesoporous oxygen-vacancy-rich TiO_{2-x} nanofibrous membrane evaporator [43] can sustainably in situ intercept VOCs by providing more photocatalytic reactive sites for collision and longer tortuous channels for prolonging volatile organic compounds (VOCs) retention. When the m-TiO_{2-x} NFM is employed to purify practical river water in an evaporation prototype under real solar irradiation, complex volatile natural organic contaminants can be effectively intercepted, and the produced distilled water meets the drinking water standards. He et al. [59] designed a novel PDA-encapsulated carbon nanotube/polyurethane (PDA@CNT/PU) nanofibrous membrane-based solar steam generator with a hierarchical nanostructure by an ecofriendly one-step process, which exhibits superhydrophilicity, underwater superoleophobicity, and photothermal performance (Figure 7.4b). Li et al. [60] developed a dual-functional water purification device with a high evaporation rate and adsorption capacity by the fiber intercepting-particle structured ZIF-67 fabric (FIP-PZ) (Figure 7.4c). ZIF-67 can be considered as an aromatic compound that can interact with malachite green through π–π stacking interaction between them. The adsorption of malachite green by ZIF-67 was validated as an endothermic reaction due to the positive $\Delta H°$ in adsorption progress, which means the adsorption of malachite green can be enhanced by heating when evaporating. A compositive strategy to fabricate self-floatable catalytic evaporators was developed by Wen et al. [61] with both solar distillation and organic pollutant degradation, through Fe-MOF-loaded electrospinning and antibiofouling zwitterionic coating. Fe-based MOF particles proceeded photo-Fenton reaction to decompose VOC and nonvolatile pollutants in the

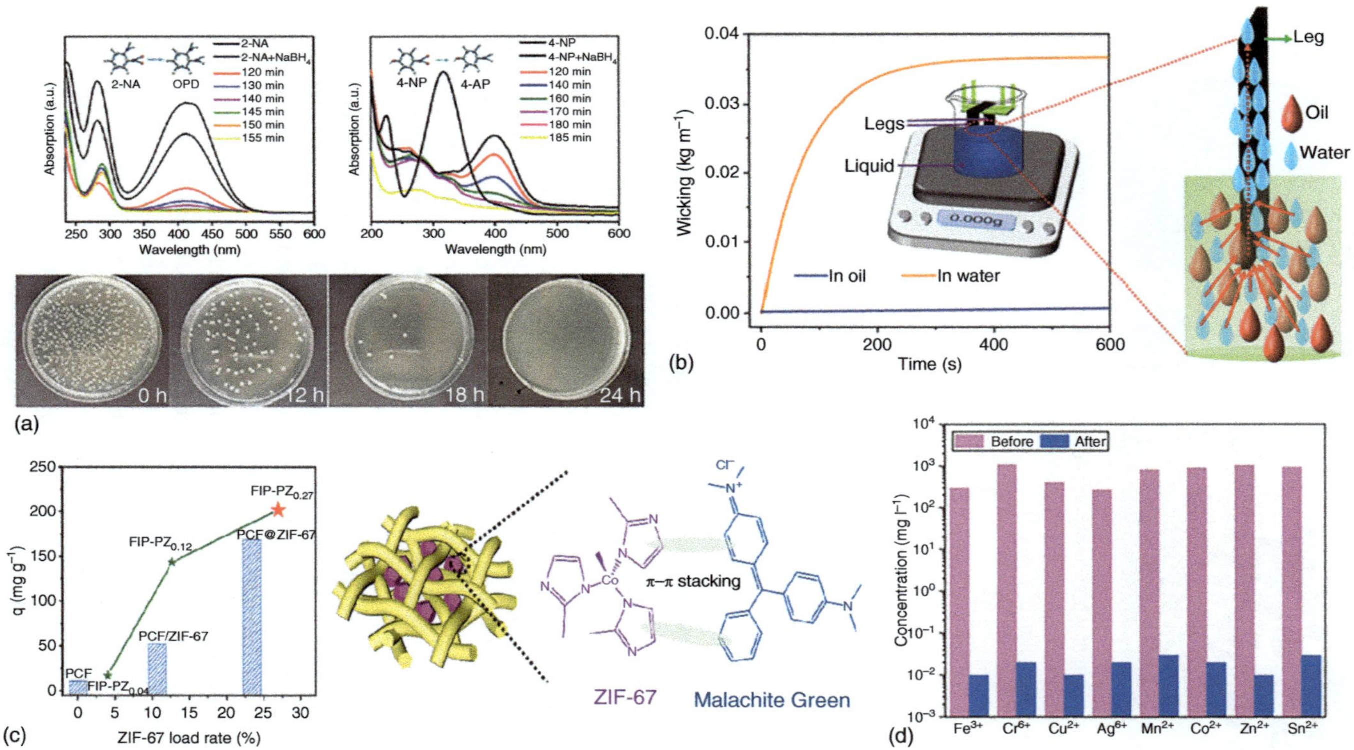

Figure 7.4 (a) The catalyzed reduction performance of 4-NP and 2-NA and sterilized of evaporators [52]. (b) Wicking ability of the PDA@CNT/PU nanofibrous membranes [59]. (c) MG adsorption capacity of different samples under 1-sun irradiation and the schematic diagram for the adsorption mechanism of MG by ZIF-67 [60]. (d) The heavy metal ion removal ability of the evaporators [64]. Source: Refs. [52, 60, 64]. Reproduced with permission of Elsevier. He et al. [59]. Reproduced with permission of RSC.

wastewater, while the Zwitterionic sulfobetaine (ZSB) hydrogel coating protected the fabric from various foulants, guaranteeing its high efficiency and long-term performance in complex conditions. The interfacial evaporation process can effectively remove various heavy metal ions exiting in wastewater, and the extracting condensed water meet the WHO drinking-water standard shown in Figure 7.4d [64].

7.5.3 Power Generation

Solar-induced energy generation through photovoltaic, photoelectrical, and photochemical processes has been widely investigated and shown excellent performance. Therefore, it is rational to solve water and energy scarcity issues at the same time by combining the solar vaporization with energy generation. Recently, efficient solar evaporation process, integrated with thermo-electric [65, 66], salinity-induced [67] or pyro-, piezo- [68] nanofiber-based device, has shown to be capable of fresh water production and electricity generation utilizing free sunlight. With the dual functions or concurrent demonstrations of fresh water production and electricity generation, the overall efficiency of photothermal conversion can be ameliorated.

Cai et al. [65] fabricated an assembled device containing a thermoelectric (TE) material wrapped by paper to facilitate water transport and a photothermal PLA/MXene@MoS_2 membrane attached on the TE material surface. When this device is exposed to solar irradiation, the temperature difference could form between the hot part (PLA/MXene@MoS_2 membrane) and the cold part of the TE material, which the voltage output is about 60 mV under 1 kW m^{-2}. The evaporator also showed great promise in voltage generation by turning on/off the light for five cycles, highlighting its excellent stability. A half-exposure of the dry membrane evaporator to solar illumination leads to a voltage of 1 mV due to the temperature difference at the two ends. [66].

The salinity gradient is naturally incurred between evaporation interfacial and bulk brines since the continuously rapid removal of water molecules from interfacial brine, hence providing the salinity gradient power. An evaporation-induced electricity generation in PA66/CB NF film evaporator is shown in Figure 7.5a. The electrical potential gradually augments with the evaporation time at the beginning thanks to the salinity gradient increasing between two ends of PA66/CB film in 3.5 wt% NaCl solution, and ultimately stabilized at 320 mV of open-circuit voltage and 0.18 μA of short-circuit current after 20 minutes when it reaches the equilibrium vaporization (Figure 7.5b and c) [67]. Meng et al. [68] provided an evaporation prototype based on the synergistic water activation of wave-triggered electricity and highly hydratable hydrogel, as shown in Figure 7.5d. The highly hydratable light-absorbing PVA hydrogel is customized with an oriented pore structure to enhance water transport. Under solar irradiation, this composite film evaporator can continuously convert wave energy reserved in the ocean into uninterrupted electrical energy, reactivating the water in the hydrogel that has been activated by highly hydrophilic polymer networks. The RGO in the hydrogel efficiently harvests solar energy and converts it into thermal energy to enable the water that has been dually activated to evaporate, giving rise to a high evaporation rate. A piezoelectric fibrous membrane is embedded into a

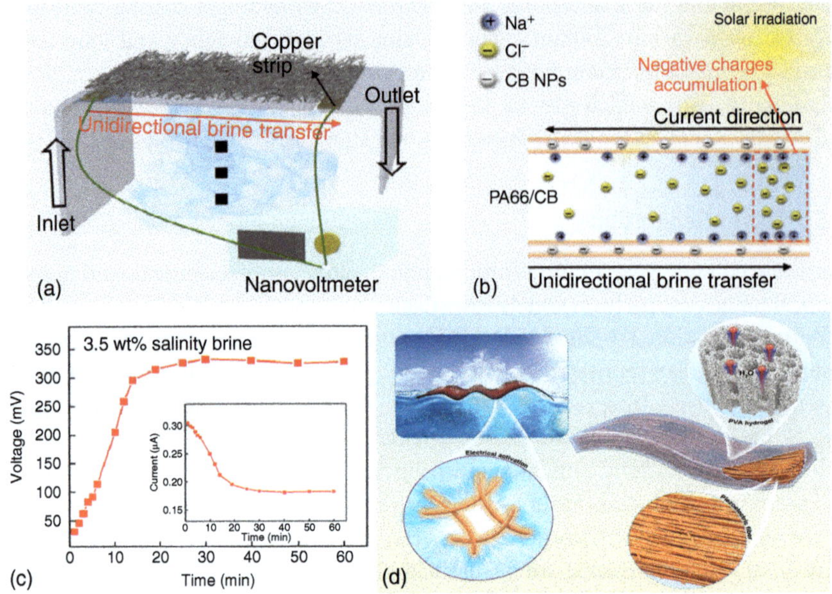

Figure 7.5 The schematic of the electricity generation device. (a) The evaporation induced electricity generation in PA66/CB NF film, (b) Schematic mechanism, and (c) Hydroelectric performance of integrated solar steam generator in 3.5 wt% salinity brine [67]. (d) Schematic illustration of the solar interfacial vapor generation system based on a piezoelectric composite film evaporator consisting of an anisotropic fibrous membrane and hydratable light-absorbing hydrogel for water purification [68]. Source: Refs. [67, 68]. Reproduced with permission of Wiley.

highly hydratable light-absorbing poly(vinyl alcohol) (PVA) hydrogel for synergistic water activation. The integrated evaporator is capable of continuously converting the wave energy into electrical energy, activating the water in the hydrogel to improve the evaporation performance.

This exciting progress of combining nanofiber-based evaporators with electricity generation made in the past few years provides this technology with greater prospects and potential for addressing challenges in global water and energy scarcity. We believe this kind of multifunctional device, given its sustainable nature and ability to simultaneously address fresh water and energy needs, will receive more attention in the foreseeable future. However, continuous efforts in material and structural design and interdisciplinary integration are still needed to further increase the energy conversion efficiency, as well as lower material and device costs toward commercial applications.

7.6 Conclusion and Future Perspective

In recent years, great achievements have been made in the past decades in the development of electrospun nanofiber-based evaporators. With the advantages of electrospinning technology and the innovative photothermal system designs, the united interfacial solar steam generation system performed effectively in clean water production, paving a potential way for highly efficient solar steam generation without the use of expensive optical concentrators. Hence, a handful of scalable and practical applications in seawater desalination, wastewater purification, and energy generation have been presented.

However, there are several important challenges that need to be confronted. First, the long-term stability of the devices when dealing with the various water sources, for example, seawater, wastewater, or sewage, has great significance. Materials engineering, surface coating, or structural design need to be integrated into the fabrication of the nanofiber-based evaporators. Second, a better understanding of the light transfer, water transport, and thermal diffusion behaviors and kinetics and the corresponding structure–property relationships is critical for future nanofiber-based solar evaporator designs and performance optimization. For example, water transport behavior is highly related to the nanofiber structure (e.g. porosity, pore size, and tortuosity) and surface properties (e.g. wettability). Thirdly, the manufacturing technology and cost of the developed materials and/or structures, which creates difficulty in scalable production for practical applications. Finally, interdisciplinary convergence of solar steam generation with other emerging fields, such as nanofluidics, thermoelectricity, rechargeable batteries, photo- and electro-catalysis, organic fuel production, and radiative cooling, will bring this technology to a broader prospect and future.

References

1 Alvarez, P.J.J., Chan, C.K., Elimelech, M. et al. (2018). Emerging opportunities for nanotechnology to enhance water security. *Nature Nanotechnology* 13 (8): 634–641.

2 Ni, G., Zandavi, S.H., Javid, S.M. et al. (2018). A salt-rejecting floating solar still for low-cost desalination. *Energy & Environmental Science* 11 (6): 1510–1519.

3 Verbeke, R., Gomez, V., and Vankelecom, I.F.J. (2017). Chlorine-resistance of reverse osmosis (RO) polyamide membranes. *Progress in Polymer Science* 72: 1–15.

4 Elimelech, M. and Phillip, W.A. (2011). The future of seawater desalination: energy, technology, and the environment. *Science* 333 (6043): 712–717.

5 Khawaji, A.D., Kutubkhanah, I.K., and Wie, J.M. (2008). Advances in seawater desalination technologies. *Desalination* 221 (1-3): 47–69.

6 Lewis, N.S. (2016). Research opportunities to advance solar energy utilization. *Science* 351: aad1920.

7 Chu, S., Cui, Y., and Liu, N. (2017). The path towards sustainable energy. *Nature Materials* 16: 16–22.

8 Crabtree, G.W. and Lewis, N.S. (2007). Solar energy conversion. *Physics Today* 60: 37–42.

9 Chen, C., Kuang, Y., and Hu, L. (2019). Challenges and opportunities for solar evaporation. *Joule* 3 (3): 683–718.

10 Zhang, P., Liao, Q., Yao, H. et al. (2019). Direct solar steam generation system for clean water production. *Energy Storage Materials* 18: 429–446.

11 Shi, S., Si, Y., Han, Y. et al. (2022). Recent progress in protective membranes fabricated via electrospinning: advanced materials, biomimetic structures, and functional applications. *Advanced Materials* 34: 2107938.

12 Chen, H.S., Huang, M.H., Liu, Y.B. et al. (2020). Functionalized electrospun nanofiber membranes for water treatment: a review. *Science of the Total Environment* 739: 139944.

13 Raza, Z.A., Munim, S.A., and Ayub, A. (2021). Recent developments in polysaccharide-based electrospun nanofibers for environmental applications. *Carbohydrate Research* 510: 108443.

14 Dao, V.D., Vu, N.H., and Yun, S. (2020). Recent advances and challenges for solar-driven water evaporation system toward applications. *Nano Energy* 68: 104324.16.

15 Liu, Y., Xiong, J., Li, A.L. et al. (2021). Plasmonic silver nanoparticle-decorated electrospun nanofiber membrane for interfacial solar vapor generation. *Textile Research Journal* 91 (21–22): 2624–2634.

16 Ghasemi, H., Ni, G., Marconnet, A.M. et al. (2014). Solar steam generation by heat localization. *Nature Communications* 5: 4449.

17 Zeng, Y., Yao, J., Horri, B.A. et al. (2011). Solar evaporation enhancement using floating light-absorbing magnetic particles. *Energy & Environmental Science* 4: 4074–4078.

18 Hu, X., Xu, W., Zhou, L. et al. (2016). Tailoring graphene oxide based aerogels for efficient solar steam generation under one sun. *Advanced Materials* 29: 1604031.

19 Zhu, L., Gao, M., Peh, C.K.N. et al. (2019). Recent progress in solar-driven interfacial water evaporation: advanced designs and applications. *Nano Energy* 57: 507–518.

20 Yang, P., Liu, K., Chen, Q. et al. (2017). Solardriven simultaneous steam production and electricity generation from salinity. *Energy & Environmental Science* 10: 1923–1927.

21 Zhu, L., Gao, M., Peh, C.C.K.N. et al. (2018). Self-contained monolithic carbon sponges for solar-driven interfacial water evaporation distillation and electricity generation. *Advanced Energy Materials* 8: 1702149.

22 Brongersma, M.L., Halas, N.J., and Nordlander, P. (2015). Plasmon-induced hot carrier science and technology. *Nature Nanotechnology* 10: 25–34.

23 Liu, T. and Li, Y. (2016). Photocatalysis: plasmonic solar desalination. *Nature Photonics* 10: 361–362.

24 Gao, M., Connor, P.K.N., and Ho, G.W. (2016). Plasmonic photothermic directed broadband sunlight harnessing for seawater catalysis and desalination. *Energy & Environmental Science* 9: 3151–3160.

25 Zhang, L., Tang, B., Wu, J. et al. (2015). Hydrophobic light-to-heat conversion membranes with self-healing ability for interfacial solar heating. *Advanced Materials* 27: 4889–4894.

26 Hao, D., Yang, Y., Xu, B. et al. (2018). Efficient solar water vapor generation enabled by water-absorbing polypyrrole coated cotton fabric with enhanced heat localization. *Applied Thermal Engineering* 141: 406–412.

27 Jiang, Q., Derami, H.G., Ghim, D. et al. (2017). Polydopaminefilled bacterial nanocellulose as a biodegradable interfacial photothermal evaporator for highly efficient solar steam generation. *Journal of Materials Chemistry A* 5: 18397–18402.

28 Chen, Q., Pei, Z., Xu, Y. et al. (2018). A durable monolithic polymer foam for efficient solar steam generation. *Chemical Science* 9: 623–628.

29 Zhu, G.L., Xu, J.J., Zhao, W.L. et al. (2016). Constructing black titania with unique nanocage structure for solar desalination. *ACS Applied Materials & Interfaces* 8 (46): 31716–31721.

30 Ye, M., Jia, J., Wu, Z. et al. (2017). Synthesis of black TiO_x nanoparticles by Mg reduction of TiO_2 nanocrystals and their application for solar water evaporation. *Advanced Energy Materials* 7: 1601811.

31 Wang, J., Li, Y., Deng, L. et al. (2017). High-performance photothermal conversion of narrow-bandgap Ti_2O_3 nanoparticles. *Advanced Materials* 29: 1603730.

32 Chen, R., Wu, Z., Zhang, T. et al. (2017). Magnetically recyclable self-assembled thin films for highly efficient water evaporation by interfacial solar heating. *RSC Advances* 7: 19849–19855.

33 Wang, X., Ou, G., Wang, N. et al. (2016). Graphene-based recyclable photo-absorbers for high-efficiency seawater desalination. *ACS Applied Materials & Interfaces* 8: 9194–9199.

34 Yi, L., Ci, S., Luo, S. et al. (2017). Scalable and low-cost synthesis of black amorphous Al-Ti-O nanostructure for high-efficient photothermal desalination. *Nano Energy* 41: 600–608.

35 Liu, H., Zhang, X., Hong, Z. et al. (2017). A bioinspired capillary-driven pump for solar vapor generation. *Nano Energy* 42: 115–121.

36 Ding, D., Huang, W., Song, C. et al. (2017). Non-stoichiometric MoO_{3-x} quantum dots as a light-harvesting material for interfacial water evaporation. *Chemical Communications* 53: 6744–6747.

37 Chala, T.F., Wu, C.M., Chou, M.H. et al. (2018). Melt electrospun reduced tungsten oxide/polylactic acid fiber membranes as a photothermal material for light-driven interfacial water evaporation. *ACS Applied Materials & Interfaces* 10 (34): 28955–28962.

38 Xiong, J., Li, A.L., Liu, Y. et al. (2022). Multi-scale nanoarchitectured fibrous networks for high-performance, self-sterilization, and recyclable face masks. *Small* 18: 2105570.

39 Zhang, R., Zhou, Y., Xiang, B. et al. (2021). Scalable carbon black enhanced nanofiber network films for high-efficiency solar steam generation. *Advanced Materials Interfaces* 8: 2101160.

40 Ding, Q., Guan, C., Li, H. et al. (2020). Solar-driven interfacial evaporation based on double-layer polylactic acid fibrous membranes loading Chinese ink nanoparticles. *Solar Energy* 195: 636–643.

41 Li, H., Wen, H., Zhang, Z. et al. (2020). Reverse thinking of the aggregation-induced emission principle: amplifying molecular motions to boost photothermal efficiency of nanofibers. *Angew.* 59 (46): 20371–20375.

42 Gao, T., Li, Y., Chen, C. et al. (2019). Architecting a floatable, durable, and scalable steam generator: hydrophobic/hydrophilic bifunctional structure for solar evaporation enhancement. *Small Methods* 3: 1800176.

43 Song, C.J., Qi, D.P., Han, Y. et al. (2020). Volatile-organic-compound-intercepting solar distillation enabled by a photothermal/photocatalytic nanofibrous membrane with dual-scale pores. *Environmental Science & Technology* 54 (14): 9025–9033.

44 Chen, Z.C., Li, Q., and Chen, X.M. (2020). Porous graphene/polyimide membrane with a three-dimensional architecture for rapid and efficient solar desalination via interfacial evaporation. *ACS Sustainable Chemistry & Engineering* 8 (36): 13850–13858.

45 Ren, Y., Lian, R., Liu, Z. et al. (2022). CNT/polyimide fiber-based 3D photothermal aerogel for high-efficiency and long-lasting seawater desalination. *Desalination* 535: 115836.

46 Dong, X.Y., Si, Y., Chen, C.J. et al. (2021). Reed leaves inspired silica nanofibrous aerogels with parallel-arranged vessels for salt-resistant solar desalination. *ACS Nano* 15 (7): 12256–12266.

47 Jia, J. and Wang, C. (2019). A facile restructuring of 3D high water absorption aerogels from methoxy polyethylene glycol-polycaprolactone (mPEG-PCL) nanofibers. *Materials Science and Engineering* 94: 965–975.

48 Li, H.X., Wen, H.F., Li, J. et al. (2020). Doping AIE photothermal molecule into all-fiber aerogel with self-pumping water function for efficiency solar steam generation. *ACS Applied Materials & Interfaces* 12 (23): 26033–26040.

49 Liang, T., Wang, C., Li, B. et al. (2022). Ultralight electrospun fiber foam with tunable lamellar macropores for efficient interfacial evaporation. *Journal of Environmental Chemical Engineering* 10 (3): 107522.

50 Liu, Y., Luo, B., Liu, H.J. et al. (2022). 3D printed electrospun nanofiber-based pyramid-shaped solar vapor generator with hierarchical porous structure for efficient desalination. *Chemical Engineering Journal* 452 (3): 139402.

51 Zhang, R., Xiang, B., Wang, Y. et al. (2022). A lotus-inspired 3D biomimetic design toward an advanced solar steam evaporator with ultrahigh efficiency and remarkable stability. *Materials Horizons* 9 (4): 1232–1242.

52 Liu, H.J., Liu, Y., Wang, L.M. et al. (2021). Nanofiber based origami evaporator for multifunctional and omnidirectional solar steam generation. *Carbon* 177: 199–206.

53 Zhu, B., Kou, H., Liu, Z. et al. (2019). Flexible and washable CNT-embedded PAN nonwoven fabrics for solar-enabled evaporation and desalination of seawater. *ACS Applied Materials & Interfaces* 11 (38): 35005–35014.

54 Qi, Q., Wang, W., Wang, Y. et al. (2020). Robust light-driven interfacial water evaporator by electrospinning SiO_2/MWCNTs-COOH/PAN photothermal fiber membrane. *Separation and Purification Technology* 239: 116595.

55 Xu, W., Hu, X., Zhuang, S. et al. (2018). Flexible and salt resistant janus absorbers by electrospinning for stable and efficient solar desalination. *Advanced Energy Materials* 8: 1–7.

56 Li, H.N., Yang, H.C., Zhu, C.Y. et al. (2022). A self-descaling Janus nanofibrous evaporator enabled by a "moving interface" for durable solar-driven desalination of hypersaline water. *Journal of Materials Chemistry A* 10: 20856.

57 Liu, H., Gu, J., Liu, Y. et al. (2022). Reconfiguration and self-healing integrated Janus electrospinning nanofiber membranes for durable seawater desalination. *Nano Research* 16: 489–495.

58 Dong, X., Cao, L., Si, Y. et al. (2020). Cellular structured CNTs@ SiO_2 nanofibrous aerogels with vertically aligned vessels for salt-resistant solar desalination. *Advanced Materials* 32 (34): 1908269.

59 He, M.T., Liu, H.J., Wang, L.M. et al. (2021). One-step fabrication of a stretchable and anti-oil-fouling nanofiber membrane for solar steam generation. *Materials Chemistry Frontiers* 5 (9): 3673–3680.

60 Li, A.L., Xiong, J., Liu, Y. et al. (2022). Fiber-intercepting-particle structured MOF membranes for simultaneous solar vapor generation and organic pollutant adsorption. *Chemical Engineering Journal* 428: 131365.

61 Wen, C., Yang, J., Guo, H. et al. (2022). Zwitterionic functionalized catalytic evaporator enables simultaneous solar distillation and organic pollutant degradation. *Applied Energy* 321: 119372.

62 Yin, X., Zhang, Y., Xu, X. et al. (2021). Bilayer fiber membrane electrospun from MOF derived Co_3S_4 and PAN for solar steam generation induced sea water desalination. *Journal of Solid State Chemistry* 303: 122423.

63 Wang, R., Deng, J.H., Wu, P. et al. (2022). Sandwich-type absorber for synergistically enhanced solar water evaporation and photocatalysis. *Journal of Environmental Chemical Engineering* 10 (4): 108173.

64 Liu, Y., Liu, H.J., Xiong, J. et al. (2022). Bioinspired design of electrospun nanofiber based aerogel for efficient and cost-effective solar vapor generation. *Chemical Engineering Journal* 427: 131539.

65 Cai, C., Wang, Y., Wei, Z. et al. (2021). Biomimetic 3D membranes with MXene heterostructures for superior solar steam generation, water treatment, and electricity generation. *Solar RRL* 5 (11): 2100593.

66 Wu, D., Liang, J., Zhang, D. et al. (2020). Solar evaporation and electricity generation of porous carbonaceous membrane prepared by electrospinning and carbonization. *Solar Energy Materials & Solar Cells* 215: 110591.

67 Mao, J., Li, J.J., Xie, A.Q. et al. (2022). High-performance integrated solar steam generator for synergetic freshwater production, salt harvesting, and electricity generation. *Solar RRL* 6 (9): 2200330.

68 Meng, S., Tang, C.Y., Yang, J. et al. (2022). A wave-driven piezoelectrical film for interfacial steam generation: beyond the limitation of hydrogel. *Advancement of Science* 9 (33): 2204187.

8

Electrospinning Waterproof and Breathable Membrane

Lu Zhang, Lei Zhang, and Li Liu

Donghua University, College of Textiles, Key Laboratory of Textile Science & Technology, Ministry of Education, No.2999, Renmin North Road, Songjiang District, Shanghai 201620, China

8.1 Introduction

A waterproof and breathable membrane refers to a class of film materials that can prevent the passage of liquid water molecules under certain external pressures, but allow the transport of gaseous water molecules [1]. The pore size plays a role in it. When the pore diameter of the microporous fabric is between the size of water vapor molecules and water droplets, the external liquid droplets will not pass through the fabric, and only water vapor can be diffused through the fabric to the outside world, thus giving the fabric waterproof and breathable performance [2].

The original waterproof and breathable fabric was designed by the British Shirley Institute. Ventile fabric, breathable fabric, with large pores between yarns under dry conditions, provides pore channels for the transfer of water vapor, then it turns waterproof as the occurrence of moisture absorption swelling of cotton fibers causes the reduction of voids among fibers and yarns [2]. Gore-Tex was first invented by American R. W. Gore in 1969. It is a laminated waterproof and breathable fabric further developed on the basis of biaxially stretched polytetrafluoroethylene (PTFE) microporous film, which has a large number of micropores in its structure that are 20,000 times smaller than water droplets but 700 times larger than water vapor molecules, giving it waterproof and breathable properties [3].

In 2001, Gibson et al. [4] first investigated the moisture permeability of electrospun nanofibrous membranes by comparing the moisture resistance of polybenzimidazole fibrous membranes, commercial PTFE microporous membranes, and polyacrylonitrile fibrous membranes prepared by the electrostatic spinning method. The study showed that the evaporative resistance of electrostatically spun fibrous membranes was similar to, and even lower than, that of PTFE microporous membranes. It is thus known that the fibrous membranes fabricated by electrostatic spinning [5] can exhibit excellent moisture permeability properties similar to those of PTFE microporous membranes, but the good hydrophilicity of polyacrylonitrile itself leads to electrostatically spun fibrous membranes not

Electrospinning: Fundamentals, Methods, and Applications, First Edition.
Edited by Liming Wang and Xiaohong Qin.

having water resistance. Although this study did not show the waterproof and breathable properties of electrospun fibrous mat, at least it demonstrated that the unique microporous structure of electrospun fibrous mat can contribute a lot to its excellent moisture permeability. It proposed a new strategy to make use of electrospinning technology to design and fabricate waterproof and breathable fabrics. Since then, the preparation of waterproof and breathable membranes using electrostatic spinning technology has attracted the attention of a large number of researchers.

8.2 Waterproof and Breathable Theory

8.2.1 Waterproof Mechanism

Liquid water, in effect, is a condensed state of its molecules due to the hydrogen bonds between them. Therefore, as an accumulation of water molecules, the liquid state always presents in the form of rain, snow, dew, and frost with different physical sizes. Waterproofness refers to the ability of the fabric to resist the penetration of liquid water from the external environment [6]. Liquid water permeates through the functional membrane to the inside in two ways: wetting and penetrating. First, after the external water comes into contact with the membrane, the liquid water is transferred through the pores of the membrane to the surface and the inside of the functional membrane due to the capillary effect, which makes the membrane wet. This is a spontaneous process that is closely related to surface roughness and surface energy. Second, upon being subjected to external pressure, besides surface roughness and surface energy, the synergistic effect of outer pressure, pore size, porosity of the mat, and kinetic energy of water droplets contribute to the water transportation [7]. Therefore, there are two principles to prevent the fabric from being wetted and penetrated by liquid water.

8.2.1.1 Wetting Theory

The properties of functional film materials, such as water repellency and absorption, are influenced by the wetting properties of the functional film. According to Young's equation

$$\cos\theta = \frac{\gamma_{SV} - \gamma_{SL}}{\gamma_{LV}}$$

where γ_{SL} is the solid–liquid surface free energy, γ_{SV} is the gas–solid interface free energy, γ_{LV} is the liquid-surface free energy, and θ is the contact angle.

According to Young's equation, the greater the surface energy of the solid, the easier it is to be wet, and vice versa; thus, a solid with low surface energy has a waterproof performance [8]. Wettability means that if the liquid does not spread smoothly on the surface of the solid, it is instead retained in the shape of a water droplet on the surface of the solid. Solid, liquid, and gas at the junction of the three phases will show an angle formed by the tangent line between the solid surface and the edge of the liquid, that is, the contact angle θ. This angle is usually used to quantitatively describe the ability to be wet. The larger the more hydrophobic. As shown

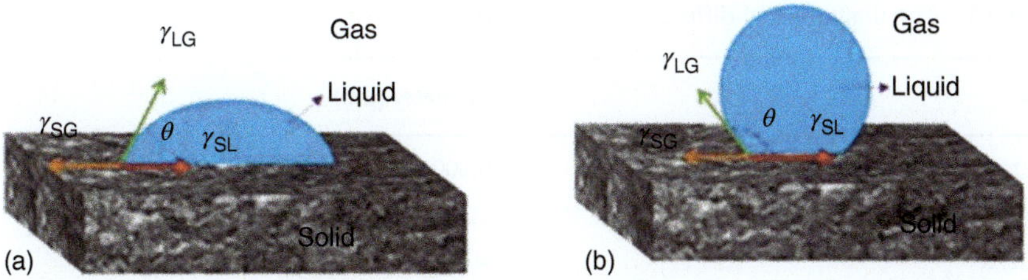

Figure 8.1 Schematic diagram of contact angle θ in Young's equation. (a) Schematic of contact angle for hydrophilic surface, $\theta < 90°$; (b) Schematic of contact angle for hydrophobic surface, $\theta > 90°$. Source: Tehrani-Bagha [9]. Reproduced with permission of Elsevier Science.

in Figure 8.1, $\theta < 90°$ and $\theta > 90°$ represent hydrophilic and hydrophobic surfaces, respectively. A hydrophilic surface allows liquid to wet the surface spontaneously even without external aid.

8.2.1.2 Penetration Theory

The water repellency of functional membranes is not only because water droplets cannot be infiltrated on the surface but also because the continuity of the nonporous membrane or coating itself and the surface tension prevent the water droplets from passing through [10]. Therefore, the liquid must have external pressure to pass through the functional membrane to form the capillary effect, so we can use this applied pressure to express the waterproofness of the functional membrane. Studies have shown that the water resistance of functional membranes depends mainly on the contact angle of the membrane surface and the pore size of the micropores in the material. For hydrophilic membranes, this additional pressure is an ability to enhance liquid water permeation through the membrane; that is, when rainwater tries to penetrate the membrane, it weakens the waterproof performance of the membrane; for hydrophobic membranes, the additional pressure is a force that weakens liquid water permeation through the membrane [11]. Therefore, the capillary effect of functional membranes indicates that hydrophilic surfaces are less waterproof, while hydrophobic functional membranes exhibit relatively good waterproofness.

8.2.2 Breathable Mechanism

The mechanism of moisture permeability of waterproof and permeable fabrics can be divided into two major aspects: one is the microporous diffusion mechanism, and the other is the "adsorption–diffusion–desorption" process of polymeric hydrophilic groups [12].

8.2.2.1 "Adsorption–Diffusion–Desorption" Mechanism of Polymer Hydrophilic Groups

The "adsorption–diffusion–desorption" mechanism of polymer hydrophilic groups has nothing to do with the internal structure of the waterproof and breathable membrane, but is related to the molecular structure of the polymer raw material, which is applicable to some materials carrying a large number of hydrophilic groups.

Table 8.1 The diameters of different types of water.

Type	Diameter (μm)
Water vapor molecule	0.0004
Fog	20–200
Drizzle	400
Light rain	900
Moderate rain	2000
Heavy rain	3000–4000
Rainstorm	6000–10000

Because the membrane contains a large number of hydrophilic groups, a humidity gradient is formed on both sides of the membrane, and hydrophilic functional groups form hydrogen bonds with water molecules while water vapor molecules are transferred along the molecular chain gaps, and water vapor molecules are transferred from the side of high humidity to the side of low humidity, forming the process of "adsorption–diffusion–desorption" [13].

8.2.2.2 Micropore Diffusion Mechanism

Micropore diffusion mechanism is realized by using the diameter difference between microporous membrane and water droplets, water vapor, or air [14]. Generally, the pore size of microporous membrane materials is less than 2 μm. Table 8.1 lists the diameters of various types of water . It can be seen that except for water vapor and air molecules, the diameter of the remaining liquid water molecules is much larger than the pore diameter of the membrane material, so that water vapor can diffuse freely through these micropores, while the water droplets cannot penetrate through the film.

8.3 Classification of Waterproof and Breathable Membranes

Since the late 1970s, waterproof and breathable membranes have been widely concerned, and various types of waterproof and breathable membranes have emerged in an endless stream. Usually, they can be divided into hydrophilic nonporous membranes and hydrophobic microporous membranes according to their wettability and pore structure characteristics [15].

8.3.1 Hydrophilic Nonporous Membrane

Waterproof and permeable films without micropores are generally hydrophilic and permeable films without permeable micropores on the film, and their permeability is achieved through hydrophilic chain segments on the macromolecular chain [16].

Hydrophilic nonporous films are a class of dense films prepared by hydrophilic polymers, which have the advantages of high mechanical strength, good chemical stability, good processing performance, environmental protection, and nontoxicity, and are usually composed of alternating hydrophilic and hydrophobic segments. Due to the presence of hydrophilic chain segments, hydrogen bonding force is generated between polar groups in the polymer and water molecules. While the thermal movement of hydrophilic chain segments provides intermolecular pores, so that moisture is absorbed on the side with higher humidity through "adsorption–diffusion–desorption" and diffused and migrated to the side with lower humidity along the humidity gradient in the direction of membrane thickness, thus achieving the effect of moisture permeability. Whereas the presence of hydrophobic chain segments and the membrane structure itself do not have the property of pores, a hydrophilic nonporous membrane can well prevent the passage of liquid water, as shown in Figure 8.2. The diameter of many common bacteria and viruses is between 20 and 300 nm, which is much larger than the diameter of water molecules (0.4 nm), so they cannot pass through the film, making the film have a certain protective effect. For example, Polytechnic (Qualitex) film produced by American Polytech International Inc [18]. Polyurethane (PU) macromolecule diol and isocyanate connected to form a flexible long-chain soft segment. The soft segment can produce hydrogen bonds with moisture to transfer water vapor. Its molecular structure of small-molecule polyol and isocyanate are connected to form a rigid short-chain hard segment to prevent liquid penetration, and due to the interaction between the rigid chain segments, PU hydrophilic nonporous film has high tensile strength, elongation and tear strength, excellent wear resistance, oil resistance, and corrosion resistance. However, thermoplastic polyurethane (TPU) hydrophilic film has a high water pressure resistance due to its nonporous structure, but is limited to the low-end market due to poor moisture permeability and discomfort caused by easy wetting with water.

The moisture permeability of hydrophilic nonporous film mainly depends on the chemical composition and molecular structure of raw materials, film thickness, and

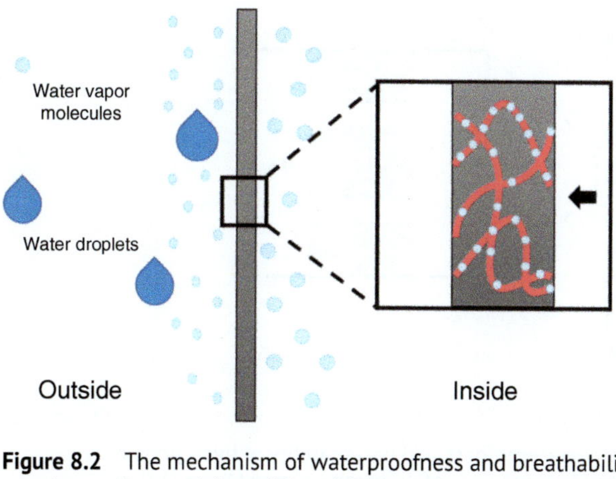

Figure 8.2 The mechanism of waterproofness and breathability of hydrophilic nonporous membrane. Source: Adapted from Mukhopadhyay and Midha [17].

temperature/moisture gradient on both sides. Generally speaking, the thinner the film and the larger the moisture gradient, the better the moisture permeability of the film.

8.3.2 Hydrophobic Microporous Membrane

Hydrophobic microporous membranes are made of hydrophobic polymers with a large number of interconnected micron-sized pores, and their waterproof and moisture permeability functions depend mainly on the screening effect of their micropores. PTFE, polyvinylidene fluoride (PVDF), and polystyrene are common raw materials having advantages of high toughness, high corrosion resistance, and good aging resistance.

These membrane materials have various types of microporous structures, such as three-dimensional mesh, sponge, and slit [19]. Their diameters are usually less than 2 μm, while the diameter of liquid water droplets is usually >100 μm, the diameter of water vapor molecules is about 0.4 nm, and the diameter of air molecules is smaller than that of water molecules. Although the diameter of common bacteria and viruses is <1 μm, most of them use water as the medium, so they cannot pass through the film, as shown in Figure 8.3.

The water and moisture permeability of hydrophobic microporous film mainly depends on the diameter and number of micropores. The moisture permeability and air permeability of microporous film are proportional to the pore diameter and number, and the water blocking capacity is inversely proportional to the pore diameter [20].

In summary, hydrophobic microporous membrane has better waterproof and moisture permeability performance than hydrophilic nonporous membrane; moisture can be freely diffused through the micropores and rapidly discharged, and interconnected microporous structure makes the film both breathable and waterproof at the same time. Although PTFE microporous film has better waterproof

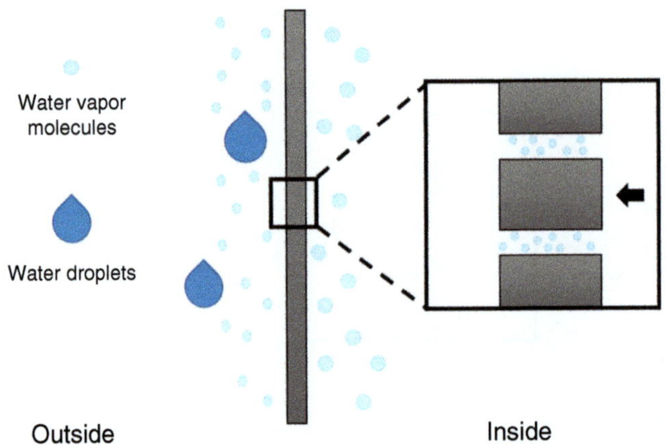

Figure 8.3 The mechanism of waterproofness and breathability of hydrophobic microporous membrane. Source: Adapted from Mukhopadhyay and Midha [17].

and moisture permeability performance, it is expensive, monopolized by foreign countries, and has unstable product quality. Therefore, electrostatic spinning technology, as a new preparation technology for microporous membrane, is the main direction of future research on waterproof and moisture-permeable functional membranes [21].

8.4 Methods of Fabricating Waterproof and Breathable Membrane

Waterproof and breathable membranes are widely used in various fields because of their functionality and practical value, but different fields have different requirements for membrane performance and different membrane structures; therefore, different processes are adopted for different raw materials, and their main preparation methods include melt extrusion, biaxial stretching, phase separation, flash evaporation, and electrostatic spinning [22].

Various approaches to fabricate waterproof and breathable membrane have been developed. This section mainly describes the characteristics of these methods and highlights the waterproof and breathable fabrics by electrospinning.

8.4.1 Biaxial Stretching

The bidirectional stretching method is to control the ambient temperature above the glass transition temperature and below the melting point of polymeric film materials, and then perform the transverse/longitudinal bidirectional stretching on the cast sheet and thick film, and then maintain the tension state for proper heat setting or cooling treatment, which is one of the microporous film forming methods that have attracted wide attention in recent years [23]. PTFE is very stable and insoluble in almost any solvent, so PTFE microporous films are often fabricated by biaxial stretching method.

8.4.2 Melt Extrusion

Melt extrusion is a common method to prepare hydrophilic nonporous film. After adding thermoplastic polymer raw materials to the feed port of the extruder, it is pushed to the extrusion port with the rotation of the screw [24], during which the polymer melts under the dual action of its own frictional heat and external heat source heating, and is extruded in the extrusion device to obtain polymer film by stretching.

The melt extrusion method is divided into two types: blow extrusion and flat film extrusion. The flat film extrusion method can precisely control the film thickness and produce uniform film thickness because it can control the film stretching rate in both warp and weft directions [25], but this method can only process polymers with high melt strength and good calendering characteristics to avoid film tearing during the stretching process [26]. The blown extrusion method has higher production efficiency, lower equipment costs, a more controllable blowing ratio, and a wider range of film thicknesses compared to the flat film extrusion method, but the thickness uniformity is poor [27].

8.4.3 Phase Separation

Phase separation method is one of the common methods for microporous membrane preparation by inducing phase separation of homogeneous solutions of polymers and their solvents to form two phases (high polymer concentration phase and low polymer concentration phase) [28], where the low polymer concentration phase is uniformly distributed in the high polymer concentration phase and the solvent continuously generates microporosity when the high polymer concentration phase solidifies into a membrane by taking advantage of the boiling point difference. Therefore, the phase separation method is further divided into thermogenic phase separation and non-solventogenic phase separation according to the way of inducing phase separation [29].

8.4.4 Flash Method

The flash evaporation method is a new spinning technology researched by DuPont in the early 1990s [30]. It is a dry spinning technology, also known as the instantaneous solvent volatilization method. The flash evaporation method usually uses high-density polyethylene as the polymer raw material, and dichloromethane and dichloroethylene as solvents. Through the rapid expansion and rapid volatilization of the solution, the polymer solution is sprayed at the spinneret at extremely high pressure and speed. The fibers are stretched into very fine fibers to form a fiber web on the receiving plate. At the same time, the fibers are separated from each other by an electrostatic method to maintain a single-fiber state [31].

8.4.5 Electrospinning

Electrostatic spinning is a new method for preparation of nanofibers. The conventional electrostatically spun nanofiber diameter can be as low as 50 nm, with a pore size distribution range of 0.2–5 μm. They are very suitable for waterproof and breathable field and this is a new preparation method of hydrophobic microporous membrane. Electrostatic spinning technology is mainly with the help of high-voltage electrostatic field polymer solution or melt charged with deformation, at the end of the nozzle to form a dangling Taylor cone-shaped droplets. When the charge repulsion on the surface of the droplet exceeds its surface tension, in the droplet surface will be high-speed jet polymer micro-jet; these jets in a short distance through the electric field force of high-speed stretching, solvent volatilization and curing, and are finally deposited on the acceptor plate to obtain hydrophobic nanofibrous film with small fiber diameter, small pore size, and high porosity [32]. Electrostatic spinning method is simple because the manufacturing process has low cost, continuous production, and can directly produce composite fabrics as the receiving substrate, and has gradually become one of the important ways to produce waterproof and breathable materials [33]. Melt electrostatic spinning compared to solvent electrostatic spinning, the process is simpler, but its processing to obtain a coarser diameter of the fiber, and can be spun polymer raw material restrictions, high equipment requirements, high energy consumption in production, therefore, the current electrostatic spinning waterproof and breathable nanofibrous membrane is mainly used solution electrostatic spinning method [34]. According to the difference in the raw materials used and the preparation method, the existing research on electrostatic spinning waterproof and breathable nanofibrous membrane can be divided into two main categories: direct spinning method and post-processing modification method.

8.4.5.1 Direct Spinning

The direct spinning method refers to the direct preparation of nanofibrous membranes with water and moisture permeability using only electrostatic spinning technology without processing such as post-treatment. In 2006, Lee et al. [35] fabricated polypropylene nanofibrous membranes by melt electrostatic spinning and found that polypropylene nanofibrous membranes have excellent permeability resistance to liquids with high surface energy and poor permeability resistance to liquids with lower surface energy; however, in 2007, Kang et al. [36] prepared a nanofibrous waterproof and breathable membrane from PU. By adjusting parameters such as the diameter of fibers in the fibrous membrane, the moisture permeability flux of the PU nanofibrous membrane was ideal, but its resistance to water permeation pressure was poor and still could not meet the requirements of waterproof and moisture permeability. Later, researchers found through a study of PTFE microporous membrane materials, microporous membrane materials with excellent water and moisture permeability of two important prerequisites: a large number of small size connected pores and hydrophobic surface properties, so the hydrophobic properties of polymer raw materials become the first choice for the preparation of waterproof and permeable membrane. According to the choice of raw materials with the way these membranes can be divided into hydrophobic polymer single-component nanofibrous waterproof and permeable membrane and doping modified nanofibrous waterproof and breathable membranes [37].

8.4.5.2 Post-Treatment Modification

The post-treatment method refers to the modification of the surface or internal structure of the fibrous membrane prepared by electrostatic spinning method through coating and other means to improve the water and moisture permeability of the fibrous membrane. Although the direct spinning method can produce nanofibrous membrane with excellent waterproof and moisture permeability properties, the electrostatic spinning method of fibrous membrane is composed of random fiber accumulation, the relative slip between the fibers makes it easy to deform the fibrous membrane, and different application scenarios will also use some hydrophilic polymers (such as polyacrylonitrile and polyvinyl alcohol), so doping modification is still difficult to effectively improve its waterproof performance. In this case, it is necessary to use post-treatment modification method to prepare waterproof and breathable membrane [38, 39]. The composite fibrous membrane thus obtained has both a porous fibrous structure and a modified hydrophobic surface, resulting in good waterproof and moisture permeability properties. The main post-treatment modification methods include impregnation coating method and scratch coating method [40].

8.5 Applications of Waterproof and Breathable Membrane

Waterproof and breathable fibrous membrane not only can windproof, waterproof, and breathable, but also can balance the pressure on both sides of the membrane, and electrostatic spinning technology has the characteristics of simple and fast, and low cost, and the fiber made shows the advantages of small pore size, high porosity, and large specific surface area. In this way, electrostatic spinning waterproof and breathable materials have a wide range of application prospects such as in protective clothing, filter materials, medical and health fields, and other fields.

8.5.1 Clothing

Waterproof and breathable membranes were initially used in the field of clothing, not only for aesthetic purposes, but also for wearable comfort, which is reflected in the moisture permeability and waterproof performance of clothing. The first use of waterproof and breathable membrane materials to improve the comfort of clothing is the US [41]. Army Special Forces field uniforms. Gore-Tex waterproof and breathable fabric introduced by a US company was first used in the manufacture of US Army uniforms. This uniform has excellent waterproof, windproof and perspiration function, greatly improving the comfort of US soldiers in the field activities. Electrostatic spinning nanofiber material has a microstructure of porous channels, so the waterproof and moisture-permeable material prepared by it has the characteristics of high porosity and has certain advantages in regulating thermal comfort. Therefore, the electrostatic spinning waterproof and moisture permeable functional clothing fabric is pushed to various fields.

Firefighting clothing is a special functional clothing to protect the personal safety of frontline rescue personnel. Its role is to prevent various forms of heat intrusion into the inside of the clothing and to protect the human body from harm. It is usually a multilayer structure from the inside and outside in the order of thermal insulation layer, vapor barrier layer, and flame retardant outer layer [42], as shown in Figure 8.4. In which, the vapor barrier layer is by waterproof moisture-permeable membrane and the bottom fabric is composite, whose role is to prevent external high-temperature firefighting water penetration into the inner layer of the clothing from scalding the human body but also to allow the timely exclusion of the human body sweat vapor to ensure its wearing comfort. So the membrane waterproof moisture permeable function is a key factor affecting the safety of individual firefighters and the comfort of dressing heat and moisture. Zhu et al.

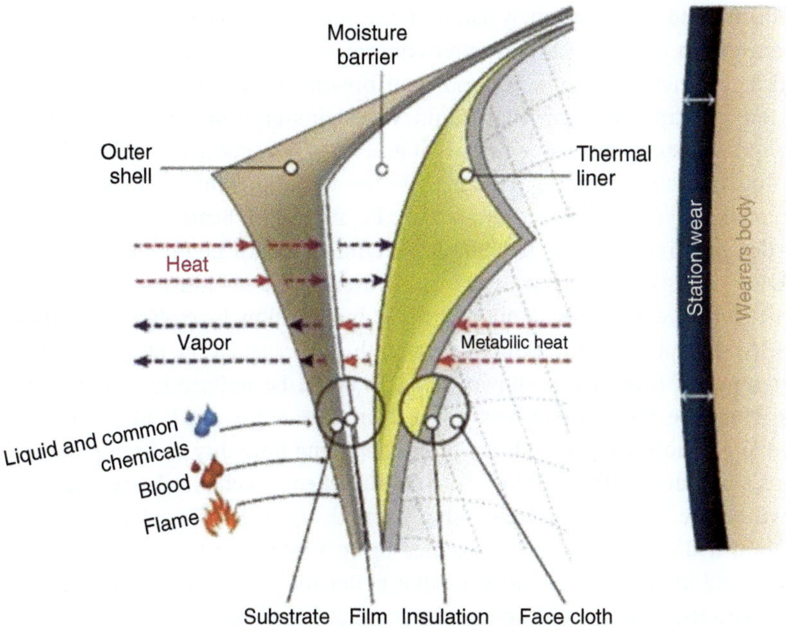

Figure 8.4 Typical structure of the firefighting jacket with waterproof and breathable membranes. Source: Ferreira et al. [43]. Reproduced with permission of Elsevier Ltd.

[44] established orthogonal experiments and constructed hydrophilic–hydrophobic composite fibrous membranes by using electrostatic spinning technology, which can directly laminate fabrics as the receiving substrate, with hydrophilic thermoplastic membranes as the substrate and hydrophobic polymers as the spinning materials, in order to obtain an optimized scheme for preparing well-shaped fibrous membranes. The composite membrane with microporous membrane as the support layer and hydrophilic nonporous membrane as the selective separation layer has hydrophilic–hydrophobic bipolarity, with a contact angle of 100° for the hydrophobic layer and 30° for the hydrophilic layer, a hydrostatic pressure resistance of 6240.3 Pa, and a moisture permeability of 3515.9 $g\,m^{-2}$–24 h. The starting decomposition temperature of the composite membrane is 393.4 °C, which is expected to become a protective fabric for firefighting suits. Mao et al. [45] also prepared a composite membrane with quartz nano (SNF) fibrous membrane by electrostatic spinning technology, and used it as the support layer and impregnated and coated new polyurethane membrane (FPU) to prepare a composite membrane with high water and moisture permeability performance.

In recent years, with the improvement of people's health awareness, outdoor activities have become more and more popular, and the requirements for protective sports clothing have gradually increased, such as ski clothing, mountaineering clothing, and rushing clothes. In outdoor activities, activists tend to produce a lot of sweat due to the large amount of exercise. To ensure the comfort of the personnel themselves, clothing with good moisture permeability is needed. In the event of storms and other extremely bad weather, you need to have good waterproof performance from the raincoat and other equipment to take protective measures. The clothing fabricated with waterproof and breathable materials can not only prevent the penetration of external liquid water or raindrops, but also protect against wind and snow, discharge the sweat and moisture produced by the human body during strenuous exercise, and regulate the thermal comfort of the human body. Hu et al. [21] synthesized thermoplastic PU films with heat–wet response by polymer design using electrostatic spinning technology with highly inclusive raw materials and adjustable product structure, and then prepared industrially available heat–wet-responsive THSPUs all-around smart hybrid films by co-blending silica aerogel (SA) particles, which exhibited good thermal insulation properties due to the presence of SAs. It was shown that SAs could be uniformly dispersed in THSPUs when the SAs content was lower than 2.0%. This smart film achieves typical adaptive moisture permeability under changing humidity and temperature conditions. In addition, the THSPUs film has excellent mechanical strength and strong hydrostatic pressure, with breaking strength up to 34 MPa, elongation up to 700%, and hydrostatic pressure resistance up to 17 800 mmH_2O. The prepared hybrid film and its hot-pressed fabric exhibit better thermal insulation properties compared with the well-known commercial product Gore-Tex.

8.5.2 Construction

In terms of building materials, outdoor rain and indoor moisture are important factors affecting building safety. When building a house, in order to avoid the harm of water on the structure of the house, it is usually necessary to apply waterproof protection treatment on some areas of the building, for example, using asphalt, lime slurry, or paint for coating. Although the use of the abovementioned materials for coating can prevent rainwater from penetrating the wall, water vapor inside the wall is equally impermeable and will still cause problems such as mold and mildew [46]. Therefore, the use of waterproof and breathable membrane materials for building wall protection has become a common practice in the modern construction field. As shown in Figure 8.5, waterproof and breathable membranes are often used for the decorative layer of exterior walls and the interior insulation layer of buildings [47]. Most polymer materials are easy to burn. As China's fire-retardant waterproofing and breathable membrane is not clearly defined, the fire-retardant research on waterproofing and breathable membrane is rarely reported, resulting in a large number of polymer flammable waterproofing and breathable membrane flooded in the market, laying a serious hidden danger. The US National Institute of Technology and Standards research results show that flame-retardant materials can make the fire response time 15 times more, the grassroots emergency response capacity is greatly enhanced, so the waterproof breathable membrane flame-retardant modification is very necessary, which can control the burning of materials in the early stages, or even self-extinguishing, to win valuable time for disaster relief. Lu et al. [48] developed and prepared a waterproof and breathable membrane with flame-retardant property. The material can be used in various curtain wall exterior

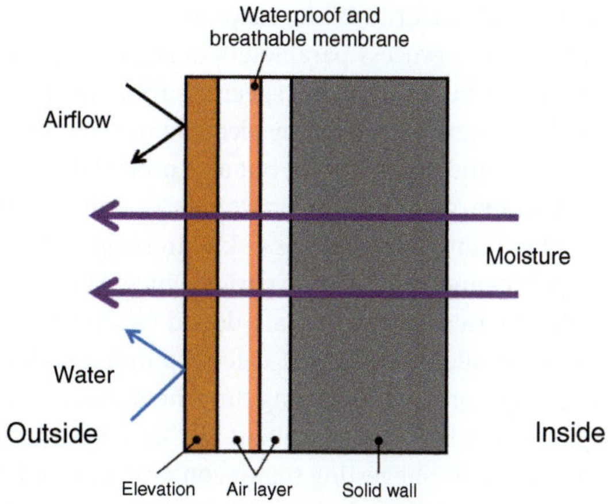

Figure 8.5 The diagram of waterproof and breathable membranes in the construction field. Source: Lu Zhang.

insulation systems of glass wool insulation board as well as rock wool insulation board exterior surfaces. Ding et al. [49] used electrostatic spinning technology with PU as the main raw material to prepare a flame-retardant, waterproof, and breathable membrane. Through the feature of easy adjustment of electrostatic spinning, suitable flame retardants were screened, and the addition amounts of PU, LiCl, and fluorinated hydrophobic agent (FPU) were added to determine the optimal spinning formulation of the waterproof and permeable membrane, so as to produce a flame-retardant, waterproof, and permeable membrane that meets the application standards in all aspects of performance.

8.5.3 Healthcare

Waterproof and breathable functional films in the medical and healthcare field mainly include medical protective clothing and sheets, surgical towels, wound dressings, disinfection equipment, wrapping materials, and other medical materials [50]. During the new epidemic, medical personnel lived in an environment with more bacteria and viruses, and because of the epidemic prevention requirements, hospitals and isolation areas were not allowed to use central air conditioning. Medical personnel who use multilayer protective equipment will sweat a lot, and in order to avoid or reduce the infection of medical personnel, but also to enhance the comfort of medical personnel wearing, medical protective clothing with waterproof and breathable function is particularly suitable. Tsai et al. [51] prepared PU fibrous membranes using electrostatic spinning process, which were experimentally proven to have excellent moisture permeability and breathability to ensure comfort in hot and stuffy environments. Zhu et al. [52] fabricated polyaniline/polystyrene electrospun composite fibrous membranes with intentional superhydrophobic properties and acid and alkali resistance, which can be effectively used in protective clothing.

Wound dressing materials also need to be waterproof, otherwise body fluids in the vicinity of the wound will not be distributed, causing wound ulceration and infection. The pore diameter of electrostatically spun nanofibrous membrane ranges from hundreds of nanometers to several microns, which can effectively prevent the invasion of bacteria. While its process is relatively simple and can be adjusted by adjusting the process parameters to achieve the fiber diameter and fiber pore size, and it also has a super high specific surface area. Therefore, the waterproof and permeable material prepared by electrostatic spinning can not only meet the high permeability and breathability, but also protect the wound from external infection, and has become one of the ideal materials for trauma dressing. Among them, electrospun PU fiber film is often used as a skin dressing because of its excellent bacterial barrier properties and breathability performance. Mani et al [53]. developed a PU nanofiber scaffold based on zinc nitrate doped ($AgNO_3$) through the unique product structure controllable feature of electrospinning technology. The superior mechanical properties, prolonged clotting time, hemolysis, and cytocompatibility studies of this material confirmed its excellent low hemolysis index and enhanced fibroblast proliferation rate, suggesting that it could be a candidate material for wound dressings. Supaphol et al. [54] designed an electrospun gelatin fibrous mat containing silver nanoparticles (average diameter ~13 nm) and cross-linked by glutaraldehyde vapor to improve the substrate's stability in aqueous medium for stability. The material has good antibacterial properties against some common bacteria found in burn wounds, such as *Staphylococcus aureus* and *Escherichia coli*. Using it on burn wounds can produce significant therapeutic effects. In recent years, with the continuous progress of electrostatic spinning technology, handheld electrostatic spinning equipment has been developed and applied, which can spray the dressing material directly on the injured area of the skin with the assistance of electrostatic field, greatly enhancing the therapeutic effect.

In addition, the medical material used to wrap the scalpel for sterilization, because of its waterproof and breathable function, not only blocks the entry of dust in the air, but also allows the disinfectant to enter from the outside to sterilize the scalpel on the inside.

8.5.4 Electronic and Electrical

In the field of electronic appliances, waterproof and breathable materials can effectively resist the effects of water, salt, and other corrosive liquids while allowing water vapor to pass freely. So the materials can prevent condensation to the greatest extent possible, reduce the stress on the housing seal by quickly achieving pressure balance inside and outside the housing, and make the products easy to maintain. For sensors, sensitivity is an important index, and sensitivity is proportional to the surface area of the membrane per unit mass, and since electrostatically spun nanofibers have a large specific surface area and can be used for sensors to effectively enhance their sensitivity, electrostatically spun waterproof and breathable membranes can be applied to the development and use of sensors. Pedicini et al. [55] added carbon black to the electrostatic spinning solution of polycarbonate and other polymers to obtain microfibrous mats whose color changes with temperature and can be used in temperature sensors. In addition, electrostatically spun waterproof and moisture-permeable films can also be applied to sensors, electric toothbrushes, vacuum cleaners, video cameras, and other electronic appliances.

8.5.5 Others

In addition to applications in apparel, medical, construction, electronics, and other fields, electrostatically spun waterproof and breathable functional fabrics also play a valuable role in aerospace and aquatic transportation. For example, in the manufacture of spacecraft, in order to better regulate the relative humidity inside the spacecraft, researchers use functional materials with high-efficiency moisture absorption to transform the absorbed water vapor into liquid water and store it. The material used for spacecraft protection is a waterproof and breathable membrane, which not only ensures the effective transmission of water vapor, but also prevents the penetration of liquid water [56].

8.6 Summary

Waterproof and breathable function, essentially, reflect the selection preference of fabric to the states of water molecules. The targets of waterproof and breathable properties of the fabric are, respectively, liquid water and gaseous water molecules, depending on the accumulation degree of water molecules. Electrospinning is such a technology that allows fabrics to adjust this kind of selection preference at the same or similar dimension with water droplets or vapor by applying functional materials and controlling the structure/morphology. A lot of work has been conducted in laboratories to offer additional functionalities to waterproof and breathable electrospun fibrous mat, but the poor mechanical properties still limit its real applications. Nowadays, we still highly rely on the use of finishing agents to produce commercial waterproof and breathable fabrics, even though there is requirement of application environment and service life for them. In the future, to achieve longer service life or permanent waterproof and breathable fabrics, we could consider highly improving the mechanical properties while still maintaining the outstanding waterproof and breathable properties.

References

1 Roey, M.V. (1992). Water-resistant breathable fabrics. *Journal of Industrial Textiles* 22 (1): 20–31.

2 Ahn, H., Park, C., and Chung, S. (2011). Waterproof and breathable properties of nanoweb applied clothing. *Textile Research Journal* 81 (14): 1438–1447.

3 Zhang, T., Xu, Z., Zhao, J. et al. (2020). Design the SBS elastomer electrospun fibermat/polyester composite textiles: morphology effect on waterproof-breathable performance. *Macromolecular Materials and Engineering* 305 (12).

4 Gandhimathi, C., Sundarrajan, S., Matsuura, T. et al. (2020). Fabrication and characterization of high flux poly (vinylidene fluoride) electrospun nanofibrous membrane using amphiphilic polyethylene-block-poly (ethylene glycol) copolymer. *Journal of Applied Polymer Science* 138 (17): 50296.

5 Chen, C., Zhao, Y., and Liu, W. (2013). Electrospun polyethylene glycol/cellulose acetate phase change fibers with core-sheath structure for thermal energy storage. *Renewable Energy* 60: 222–225.

6 Raza, A., Li, Y., Sheng, J. et al. (2014). Protective clothing based on electrospun nanofibrous membranes. In: *Electrospun Nanofibers for Energy and Environmental Applications*, 355–369. Berlin Heidellberg: Springer.

7 Kim, E.Y., Lee, J.H., Lee, D.J. et al. (2013). Synthesis and properties of highly hydrophilic waterborne polyurethane-ureas containing various hardener content for waterproof breathable fabrics. *Journal of Applied Polymer Science* 129 (4): 1745–1751.

8 Tian, X., Verho, T., and Ras, R.H.A. (2016). Moving superhydrophobic surfaces toward real-world applications. *Science* 352 (6282): 142–143.

9 Tehrani-Bagha, A.R. (2019). Waterproof breathable layers – a review. *Advances in Colloid and Interface Science* 268 (10): 114–135.

10 Conley, J.A., Guckert, J.R., Marin, R.A. et al. (2013). Breathable waterproof fabrics with a dyed and welded microporous layer. US Patent 8, 470,722.

11 Gretton, J.C., Brook, D.B., Dyson, H.M. et al. (1998). Moisture vapor transport through waterproof breathable fabrics and clothing systems under a temperature gradient. *Textile Research Journal* 68 (12): 936–941.

12 Hale, W.R., Dohrer, K.K., Tant, M.R. et al. (2001). A diffusion model for water vapor transmission through microporous polyethylene/$CaCo_3$ films. *Colloids and Surfaces, A: Physicochemical and Engineering Aspects* 187 (01): 483–491.

13 Ruckman, J.E. (1997). An analysis of simultaneous heat and water vapour transfer through waterproof breathable fabrics. *Journal of Industrial Textiles* 26 (4): 293–307.

14 Zhao, F., Wang, X., and Lin, J. (2011). Nanoparticle decorated fibrous silica membranes exhibiting biomimetic superhydrophobicity and highly flexible properties. *RSC Advances* 1 (8): 1482–1488.

15 Holmes, D.A. (2000). Waterproof breathable fabrics. In: *Handbook of Technical Textiles*, 282–315.

16 Jaiswal, P.K., Binder, K., and Puri, S. (2012). Formation of metastable structures by phase separation triggered by initial composition gradients in thin films. *The Journal of Chemical Physics* 137 (6): 064704.

17 Mukhopadhyay, A. and Midha, V.K. (2008). A review on designing the waterproof breathable fabrics part I: fundamental principles and designing aspects of breathable fabrics. *Journal of Industrial Textiles* 37 (3): 225–262.

18 Park, S.-H., Lee, H.B., Yeon, S.M. et al. (2016). Flexible and stretchable piezoelectric sensor with thickness-tunable configuration of electrospun nanofiber mat and elastomeric substrates. *ACS Applied Materials & Interfaces* 8 (37): 24773.

19 Jeong, W.Y. and An, S.K. (2002). A study on physical and mechanical properties of breathable waterproof fabrics manufactured with PTFE membrane-fabric composite. *Journal of the Korean Society of Clothing and Textiles* 26 (12): 1685–1693.

20 Liao, Y., Lon, C.-H., Tian, M. et al. (2018). Progress in electrospun polymeric nanofibrous membranes for water treatment: fabrication, modification and applications. *Progress in Polymer Science* 77: 69–94.

21 Gu, J., Gu, H., Cao, J. et al. (2018). Robust hydrophobic polyurethane fibrous membranes with tunable porous structure for waterproof and breathable application. *Applied Surface Science* 439: 589–597.

22 Zhu, F., Xin, Q., Feng, Q. et al. (2015). Novel poly (vinylidene fluoride)/ thermoplastic polyester elastomer composite membrane prepared by the electrospinning of nanofibers onto a dense membrane substrate for protective textiles. *Journal of Applied Polymer Science* 132 (26): 6986–6994.

23 Tsunekawa, T., Higashioji, T., Hosokawa, H. et al. (2012). Development and industrialization of a biaxially stretched nano-alloyed film. *Polymer Journal* 44 (12): 1170–1178.

24 Rastogi, V.K. and Samyn, P. (2015). Bio-based coatings for paper applications. *Coatings* 5 (4): 887–930.

25 Yu, X., Wu, X., Si, Y. et al. (2019). Waterproof and breathable electrospun nanofibrous membranes. *Macromolecular Rapid Communications* 40 (8): 1800931.

26 Mukhopadhyay, A. and Midha, V.K. (2016). Waterproof breathable fabrics. *Handbook of Technical Textiles: Technical Textile Applications* 2: 27.

27 Wang, H., Zhou, Y., He, M. et al. (2015). Effects of soft segments on the waterproof of anionic waterborne polyurethane. *Colloid and Polymer Science* 293 (3): 875–881.

28 Zheng, L., Wang, J., Wu, Z. et al. (2016). Preparation of interconnected biomimetic poly (vinylidene fluoride-co-chlorotrifluoroethylene) hydrophobic membrane by tuning the two-stage phase inversion process. *ACS Applied Materials & Interfaces* 8 (47): 32604–32615.

29 He, Y., Sun, S., Han, N. et al. (2020). Thermal energy regulated and thermochromic composite film with temperature-sensitive "breathable" stomata. *Journal of Materials Science* 55 (27): 12921–12939.

30 Jiayue, W.B.W.J.S. and Wenxiang, H. (2002). Progress on flash-spun polyethylene nonwovens. *Chemical Industry Times* 9: 003.

31 Zhu, W., Zhao, J., Wang, X. et al. (2019). Facile fabrication of fluorine-free breathable poly (methylhydrosiloxane)/polyurethane fibrous membranes with enhanced water-resistant capability. *Journal of Colloid and Interface Science* 556: 541–548.

32 Ramakrishnan, R., Gimbun, J., Samsuri, F. et al. (2016). Needleless electrospinning technology–an entrepreneurial perspective. *Indian Journal of Science and Technology* 9 (15): 1–11.

33 Hu, J., Wang, X., Ding, B. et al. (2011). One-step electro-spinning/netting technique for controllably preparing polyurethane nano-fiber/net. *Macromolecular Rapid Communications* 32 (21): 1729–1734.

34 He, F., Fan, J., and Chan, L. (2014). Preparation and characterization of electrospun poly (vinylidene fluoride) /poly (methyl methacrylate) membrane. *High Performance Polymers* 26 (7): 817–825.

35 Lee, S. and Kay Obendorf, S. (2006). Developing protective textile materials as barriers to liquid penetration using melt-electrospinning. *Journal of Applied Polymer Science* 102 (4): 3430–3437.

36 Park, C.H., Kang, Y.K., Kim, J., and Kang, T.J. (2007). Application of electrospun polyurethane web to breathable water-proof fabrics. *Fibers and Polymers* 8 (5): 564–570.

37 Chang, M.J., Chai, X.J., Cui, W.N. et al. (2018). Facile fabrication of electrospun silica nanofibrous membrane with hydrophobic, oleophilic and breathable performances. *Fibers and Polymers* 19 (4): 760–766.

38 Zhu, W., Zhao, J., Wang, X. et al. (2019). Facile fabrication of fluorine-free breathable poly (methylhydrosiloxane)/polyurethane fibrous membranes with enhanced waterresistant capability. *Journal of Colloid and Interface Science* 556: 541–548.

39 Sheng, J., Zhang, M., Luo, W. et al. (2016). Thermally induced chemical cross-linking reinforced fluorinated polyurethane/polyacrylonitrile/polyvinyl butyral nanofibers for waterproof-breathable application. *RSC Advances* 6 (35): 29629–29637.

40 Zhang, M., Sheng, J., Yin, X. et al. (2017). Polyvinyl butyral modified polyvinylidene fluoride breathable–waterproof nanofibrous membranes with enhanced mechanical performance. *Macromolecular Materials and Engineering* 302 (8).

41 Zhao, J., Wang, X., Liu, L. et al. (2018). Human skin-like, robust waterproof, and highly breathable fibrous membranes with short perfluorobutyl chains for eco friendly protective textiles. *ACS Applied Materials & Interfaces* 10 (36): 30887–30894.

42 Özek, H.Z. (2018). Development of waterproof breathable coatings and laminates. In: *Waterproof and Water Repellent Textiles and Clothing*, 25–72.

43 Ferreira, A.D.B.L., Nóvoa, P.R.O., and Marques, A.T. (2016). Multifunctional material systems: a state-of-the-art review. *Composite Structures* 151: 3–35.

44 Yang, W., Li, R., Fang, C. et al. (2019). Surface modification of polyamide nanofiber membranes by polyurethane to simultaneously improve their mechanical strength and hydrophobicity for breathable and waterproof applications. *Progress in Organic Coatings* 131: 67–72.

45 Mao, X., Chen, Y.C., Si, Y. et al. (2013). Novel fluorinated polyurethane decorated electrospun silica nanofibrous membranes exhibiting robust waterproof and breathable performances. *RSC Advances* 20: 7562–7569.

46 Liu, L., Li, H., Lazzaretto, A. et al. (2017). The development history and prospects of biomass-based insulation materials for buildings. *Renewable and Sustainable Energy Reviews* 69: 912–932.

47 Katti, D.S., Robinson, K.W., Ko, F.K. et al. (2004). Bioresorbable nanofiber-based systems for wound healing and drug delivery: optimization of fabrication parameters. *Journal of Biomedical Materials Research* 70B (2): 286–296.

48 Zhao, J., Li, Y., Sheng, J. et al. (2017). Environmentally friendly and breathable fluorinated polyurethane fibrous membranes exhibiting robust waterproof performance. *ACS Applied Materials & Interfaces* 9 (34): 29302–29310.

49 Amini, G., Samiee, S., Gharehaghaji, A.A. et al. (2016). Fabrication of polyurethane and Nylon 66 hybrid electrospun nanofiber layer for waterproof clothing applications. *Advances in Polymer Technology* 35 (4): 419–427.

50 Xia, D.L., Chen, Y.P., Wang, Y.F. et al. (2016). Fabrication of waterproof, breathable composite liquid dressing and its application in diabetic skin ulcer repair. *Advances in Skin & Wound Care* 29 (11): 499–508.

51 Ge, J., Si, Y., Fu, F. et al. (2013). Amphiphobic fluorinated polyurethane composite microfibrous membranes with robust waterproof and breathable performances. *RSC Advances* 3 (7): 2248–2255.

52 Yuan, W., Zhou, N., Shi, L., and Zhang, K.Q. (2015). Structural coloration of colloidal fiber by photonic band gap and resonant Mie scattering. *ACS Applied Materials & Interfaces* 7 (25): 14064–14071.

53 Schreuder-Gibson, H., Gibson, P., and Senecal, K. (2002). Protective textile materials based on electrospun nanofibers. *Advanced Materials* 34 (3): 44–55.

54 Beck, W.C. and Collette, T.S. (1952). False faith in the surgeon's gown and surgical drape. *American Journal of Surgery* 83 (2): 125–126.

55 Akindoyo, J.O., Beg, M.D.H., Ghazali, S. et al. (2016). Polyurethane types, synthesis and applications-a review. *RSC Advances* 6 (115): 114453–114482.

56 Ding, B., Kim, J., Miyazaki, Y., and Shiratori, S. (2004). Electrospun nanofibrous membranes coated quartzcrystal microbalance as gas sensor for NH_3 detection. *Sensors and Actuators B: Chemical* 101 (3): 373–380.

9

Preparation and Application of Electrospun Nanofibers in Heat Insulation

Mantang He and Xiaohong Qin

Donghua University, College of Textiles, Key Laboratory of Textile Science & Technology, Ministry of Education, No. 2999 Renmin North Road, Songjiang District, Shanghai 201620, China

9.1 Introduction

In recent years, the resulting temperature rise, sea level rise, and extreme weather and climate phenomena have intensified, posing a huge threat to the human survival and ecological environment with the global warming [1–4]. One of the main causes of global warming is deforestation and energy consumption. With the intensification of the global energy crisis, reducing heat loss in the process of production, transmission, and utilization, improving energy efficiency and reducing energy waste have become urgent problems to be solved. In order to achieve harmonious coexistence with nature and sustainable development of resources, building insulation, pipeline insulation, thermal battery insulation, and other energy-saving insulation materials have great development potential [5, 6].

Heat insulation material refers to the material with heat insulation ability, can hinder the transmission of heat flow, usually has the characteristics of light weight, loose, high porosity, and low thermal conductivity [7]. Therefore, it is widely used in industries such as thermal equipment and pipeline, construction, aerospace, and other fields. There are many types of insulation materials with different properties. According to the composition, they can be divided into mechanical heat insulation materials and non-mechanical heat insulation materials. According to the application temperature range, they can be divided into low temperature heat insulation materials, heat insulation materials, and high-temperature heat insulation materials. According to the density, they can be divided into general light heat insulation materials, commonly used light heat insulation materials, and super light heat insulation materials; According to the structure, they can be divided into granular heat insulation materials, fibrous heat insulation materials, microporous heat insulation materials, and layered heat insulation materials [8, 9]. Electrospun nanofibers possess a high specific surface area and controllable pore structure, which are often used as heat insulation materials [10]. Furthermore, aerogel refers to a gel composed of

Electrospinning: Fundamentals, Methods, and Applications, First Edition.
Edited by Liming Wang and Xiaohong Qin.

microporous solids in which the gas is the dispersed phase and the nanofiber aerogel can be synthesized from nanofibrous membrane by freeze-drying processes. More importantly, the high specific surface area and high thermal stability of nanofibers make the prepared ceramic aerogels more stable at high temperatures [11, 12].

In this chapter, the heat insulation mechanism and conductivity of high-temperature heat insulation materials are described. The high-temperature insulation material is mainly divided into two-dimensional (2D) nanofibrous membrane and three-dimensional (3D) nanofiber-based aerogel according to the structure. The research status of 2D electrospun nanofibrous membrane, 3D electrospun nanofiber-based aerogel insulation materials, and application of electrospun nanofiber-based insulation materials is emphatically introduced, and the development of high-temperature insulation materials is prospected.

9.2 Heat Transfer Mechanisms in Nanofiber-Based Insulation Materials

When sunlight shines on the surface of materials, there are mainly three forms: reflection, absorption, and penetration. The principle of heat insulation is to use the property of material itself or the material of finishing technology after the process, when heat radiation irradiates the material surface, a small part passes through fabric, and the vast part is reflected and absorbed by fabric itself or finishing coating on the surface, thus achieving the effect of heat insulation [13]. Based on these three phenomena, the current heat insulation principle is mainly barrier type, reflection type, and radiation type, among which the reflection type is the easiest to implement and the most studied. The key problem of thermal insulation material is the research on heat transfer inside thermal insulation material. This process is not fully understood because of the limitations of existing research methods. The heat transfer of insulating materials is very complex, including (i) the solid conducts heat and is connected to the thermal contact resistance at the fiber contact edge; (ii) gaseous heat conduction; (iii) natural convection between fibers; (iv) heat radiation between heating media; and (v) water transport in pores is related to water adsorption and desorption [14].

The heat insulation mechanism of 2D nanofibrous membrane and 3D nanofiber-based aerogel is shown in Figure 9.1. For dry insulation materials (without moisture), the total thermal conductivity can be calculated using the following formula:

$$H_{tot} = H_{sc} + H_{gc} + H_c + H_r \tag{9.1}$$

where H_{sc} is the solid thermal conductivity (W m^{-1} K^{-1}), H_{gc} is the gas molecule's thermal conductivity, H_c is the gas flow thermal conductivity, and H_r is the radiative thermal conductivity. The heat transfer performance of thermal insulation materials is mainly related to such parameters as heat conduction, heat radiation, heat convection, and water transport.

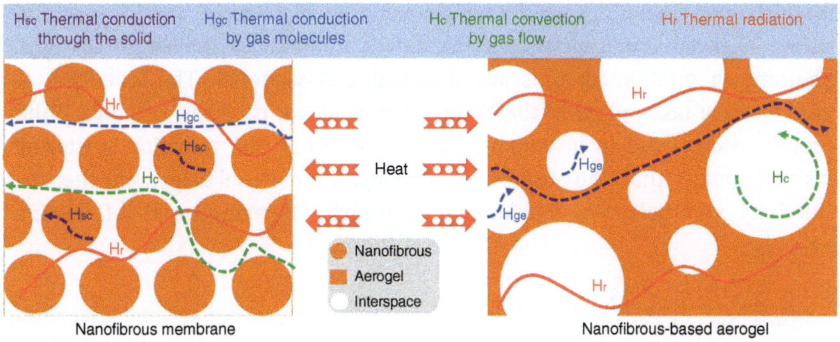

Figure 9.1 Comparison of heat conduction mechanism between nanofibrous-based membranes and aerogels.

9.2.1 Heat Conduction

The heat conduction of insulating materials is mainly divided into solid and gaseous aggregates. Generally speaking, heat conduction follows Fourier's law:

$$H_k = -k_{ef} \nabla T \tag{9.2}$$

where k_{ef} is the effective thermal conductivity (W m^{-1} K^{-1}), H_k is the heat flux, and ∇T is the temperature gradient.

The magnitude of solid heat conduction is related to the propagation of elastic waves. This energy is transmitted through quantum energy called phonons. Heat conduction is caused by the interaction between phonons or between phonons and electrons, crystal defects, and grain (or fiber) boundaries. The heat transfer rate of solids is affected by the ratio of the mean free path of phonons to the heat transfer area. With the increase of phonon velocity and volume density, the heat flux density of phonons through the solid state increases with the decrease of phonon ratio. Often in other heat transfer mechanisms, solid heat transfer is usually ignored, mainly because the volume fraction of solids is very low, the thermal conductivity is generally very low. The thermal conductivity of solid increases slightly with the increase of insulation density.

Gas heat conduction is often negligible, especially at low pressure. In low-density insulating materials, gas conduction will occur with an increase in pressure. This heat transfer mechanism is important for opaque insulating materials.

9.2.2 Thermal Radiation

Thermal radiation follows the long-distance heat transfer mechanism. It is mainly affected by the direction of wave propagation and wavelength. The contribution of radiation to the overall heat transfer equilibrium depends on density, temperature, solid transparency, solid refractive index, and structural elements such as the shape, size, and spatial orientation of fibers, particles, or pores.

With the increase of density, the contribution of radiation decreases and the conductivity of solid increases. Thermal radiation comes from external or internal sources, including insulating elements such as gases, liquids, or solid particles. Thermal radiation travels through a material at the speed of light, as photons. In motion, photons are absorbed and scattered. The scattering of photons usually takes place at the boundaries of fibers, particles, or pores. Solid absorption occurs due to its maximum density. When the radiation scattering intensity is greater than the absorption intensity, the material's apparent thermal conductivity will decrease. Radiative heat transfer depends on the following factors:

$$H_k = \frac{\sigma\left(T_1{}^4 - T_2{}^4\right)}{\frac{1}{\varepsilon_1} + \frac{1}{\varepsilon_2} + \frac{3\tau}{4} - 1} \tag{9.3}$$

where T_1 and T_2 are the ambient wall temperatures in Kelvin units, τ is the optical thickness, ε_1 and ε_2 are the emittance, and σ is the Stefan–Boltzmann constant in 5.67×10^{-8} W m^{-2} K^{-4}.

The radiant heat flux is also related to the intensity of monochromatic radiation $e\Omega$, and it is affected by the time, location, and direction of radiation propagation. It can be calculated by the radiative transport equation, and will also change due to the absorption, emission, and scattering processes.

9.2.3 Heat Convection

Convective heat transfer is the result of gas motion phenomenon caused by gas pressure difference. The pressure difference comes from the density difference, which is caused by the temperature difference inside the insulation. The convective heat transfer is related to the Rayleigh number, and the calculation formula is as follows (for the perforated unit):

$$H_a = \frac{g\beta\Delta T\delta K(\rho_g c_g)}{u_g k} \tag{9.4}$$

where g is the apparent gravity (m s^{-2}), β is the coefficient of the thermal expansion of gas in K^{-1}, ΔT is the temperature difference in kelvins, δ is the thickness of the material, K is the insulation transmission coefficient, ρ_g is the density of gas, c_g is the gas specific heat in J kg^{-1} K^{-1}, μ_g is the dynamic viscosity of gas, and k is the coefficient of thermal conductivity (W m^{-1} K^{-1}).

The permeability of insulating layer increases with an increase in porosity and pore diameter. For high-porosity and large-porosity insulation, convection heat transfer must be included. It refers to $H_a < 40$.

9.2.4 Water Transport

Water vapor may occur due to the influence of moist air or during the manufacturing processes. The difference in water concentration caused by temperature difference leads to the diffusion of water vapor and liquid inside the insulation. Due to changes in temperature and humidity, adsorption and desorption can occur.

9.3 2D Electrospun Nanofibrous Membrane for Heat Insulation

Thermal-conductive and insulating polymer materials are widely used in light-emitting diodes (LEDs), integrated electronics, energy storage and conversion systems, military weapons, and the aerospace industry due to their appropriate thermal management capabilities [15, 16]. With the rapid development of electrical systems and electronic devices, traditional polymer composites cannot meet the high requirements of heat management. For example, the metal-oxide-semiconductor field-effect transistor (MOSFET) can generate more heat due to its miniaturization and the higher power density of power or power adapter modules. To this end, the thermal interface material between the MOSFET and the radiator should have sufficient heat transfer capacity and electrical insulation level, otherwise high temperature may shorten the life of the module. Polymer materials have excellent electrical insulation performance, flexibility, and design freedom. However, low intrinsic thermal conductivity does not provide much benefit to the applicability of its thermal management. Therefore, the advantages of composite materials with high thermal conductivity combined with polymers and fillers are considered an ideal solution. In addition, filler-induced thermal conductivity enhancement should preserve as much flexibility, workability, lightness, and electrical insulation property of the composite as possible. A novel material processing and manufacturing method for electrospinning copper sulfide/polyvinylpyrrolidone (CuS/PVP) nanowires was reported to improve the heat shielding efficiency. Electrospinning makes it possible to use well-dispersed CuS nanoparticles for coating thermal shielding membranes, which is impossible in typical solution coating processes. CuS/PVP nanowires deposited on flexible polymer films in this manner can be fluorinated to reduce electrostatic repulsion at the interface. CuS/PVP nanowirings showed excellent near-infrared shielding efficiency (87.15%) by monitoring in situ temperature changes of the shielding box [17]. By combining the membrane directly with a model car and exposing it to sunlight, it proved to be about twice as effective as a commercial tungsten oxide membrane. In general, the synthesis of nanomaterials, surface treatment, and integration technology can produce highly flexible and reliable thermal shielding membranes.

Currently, the industrialization that hinders membrane distillation is constrained by ideal high-performance membranes. In recent years, engineered electrospun nanofibrous membranes (ENMs), such as double- or multilayer ENMs, have shown better permeability performance than conventional membranes in membrane distillation (MD). Due to the delamination and performance degradation of the membrane, the performance is still unsatisfactory after long-term operation. On this basis, inspired by nested structures with interleaved branches, An et al. [18] prepared an interleaved lamellar fiber complex (iHFC) membrane composed of interconnected polyvinylidene perhexafluoropropylene (PH) nanofibers and polyethylene terephthalate (PET) microfibers through convective electrospinning. Hydrophobic PH nanofibers enhance moisture resistance and high salt repulsion. PET microfibers can significantly reduce the impact of mass transfer resistance, thus improving the heat insulation performance. The results show that the PH/PET ratio of the optimized IHFC membrane is 1.5/0.8, a film thickness of 80 μm, and a membrane flux of 65 LMH under the condition of a temperature difference of 40 °C, with stable performance up to 60 hours. In addition, the interwoven structure gives iHFC membranes high mechanical strength and good long-term stability. The strategies explored in this paper and the unique membranes developed will pave the way for the industrialization of MD membranes. Moreover, one of the important components of multilayer insulation is a high-quality reflective screen. However, the metal foil currently used as a thermal reflection screen is prone to oxidation in an oxidizing environment. Nano-yttrium oxide has excellent stability in oxidation environment, mainly because of its high reflectivity in near-infrared (NIR) region. The material also shows potential as a high-temperature reflecting screen. Xie et al. [19] spun the synthetic precursor of polyyttrium acetylacetone (PAY) into precursor fibers through electrostatic spinning, and then subjected it to heat treatment, and successfully prepared high near-infrared reflection Y_2O_3 nanofibrous membranes. A new multi-organic metering precursor PAY with 42% Y_2O_3 was synthesized by adjusting the ratio of acetylacetone to $YCl_3{\cdot}6H_2O$. The crystallization, microstructure, and near-infrared reflectivity of Y_2O_3 nanofiber film were analyzed, and the tensile strength of Y_2O_3 nanofiber film was evaluated. The Y_2O_3 nanofiber films with good elasticity, perfect morphology, and high crystallinity have a strength of 1.6 MPa and a near-infrared reflectivity of 95%. High-temperature thermal reflection experiments show that Y_2O_3 nanofiber films can reflect thermal radiation at 1500 °C. They can be continuously used as reflecting screens of high-temperature insulating material in high-temperature oxidation environment. Hollow SiO_2/TiO_2 microspheres (HSTS) were used as porous agents and infrared shielding agents. A novel silica composite nanofiber (SiO_2-NF) with closed cell structure and excellent heat insulation performance was designed and prepared by electrospinning technology [20]. At the same time, the material has the characteristics of excellent flexibility, good tensile strength, low density, good thermal stability, and so on, which has a good thermal insulation application prospect. This study not only offers a new strategy for the preparation of porous nanocomposite fibers, but also provides a brand new idea for the multifunctional development of porous materials.

9.4 3D Electrospun Nanofiber-Based Aerogels for Heat Insulation

The thermal insulation of nanofibrous membrane can be effectively realized by blending thermal insulation material with polymer and electrospinning. However, due to the limitation of 2D fiber, the thermal insulation effect of nanofibers cannot be maximized. Therefore, many researchers began to consider the preparation of 2D fiber membrane into 3D aerogel to further improve its thermal insulation performance. The structure of traditional inorganic aerogel has the fragility problem, which is mainly due to the weak interaction force between the network nodes formed by the sol–gel process. In general, with the decrease of fiber diameter, the flexibility and specific surface area of the fiber can be significantly improved. The nucleation of sol particles and the construction of gel networks are affected by the size, aspect ratio, and surface activity of nanofibers. Nanofibrous aerogels can be divided into two categories according to their preparation methods: nondirectional freeze-drying and directional freeze-drying aerogels.

9.4.1 Nondirectional Freeze-Drying Aerogel

Nondirectional freeze-drying aerogels are a method for rapid preparation of aerogels, and the prepared aerogels are distributed with different pores in various directions, which are conducive to heat insulation. Si et al. [21] reported a new strategy that enables the creation of fibrous, isotropically bound elastic reconstruction (FIBER) with layered cell structures and superelastic NFA by combining electrospinning nanofibers and fiber cryoforming techniques (Figure 9.2a). Our method enables the mass assembly of intrinsically layered electrospun nanofibers into elastic bulk aerogels with adjustable density and ideal shape. The NFA density of the prepared FIBER is $0.12 > mg\,cm^{-3}$, and can quickly recover from deformation, effectively absorb energy, and has versatility in terms of insulation, sound absorption, emulsion separation, and elastic response conductance. The successful synthesis of this excellent material may provide new insights into the design and development of multifunctional NFAs for multiple applications. Li et al. [22] reported an aerogel-type microwave absorber composed of multidimensional organic and inorganic components (Figure. 9.2b). Polyacrylonitrile fiber and polybenzoxazine film were used as skeleton and cross-linking agents to form a 3Dframe, carbon nanotubes were connected to form a conductive network, and Fe_3O_4 nanoparticles were uniformly dispersed in the aerogel. What's more, aerogels are ultralight, ultrathin (1.5 mm), and highly absorbent (reflection loss −59.85 dB). Moreover, the specific reflection loss value is much better than that of current magnetoelectric hybrid materials with similar components. In addition, aerogels have high hydrophobicity and good heat insulation, with self-cleaning, infrared stealth, heat insulation, and other attractive properties, and many properties can even be comparable to commercial products. The excellent multifunctional properties depend on the cell structure of aerogels, the assembly of multidimensional nanomaterials, and the cooperation of organic–inorganic components [23]. At the same time, this study provides a new way to design next-generation absorbent materials with wide application potential. Silica aerogels have the characteristics of low thermal conductivity and good heat resistance, and are a very attractive thermal insulation materials. In addition, a scalable strategy has been reported to be the development of ultralight, ultra-flexible, and washable micro/nanofibrous sponge (MNFS) made of high-modulus polyethylene terephthalate microfibers due to its rigid and flexible coupling structure. It is bridged with flexible polyacrylonitrile nanofibers through a strong binding structure, while the in situ doping of fluoropolymer makes the micro/nanofibers have good moisture resistance [25]. The MNFS has ultrahigh elasticity and compressive fatigue resistance (5.7% per

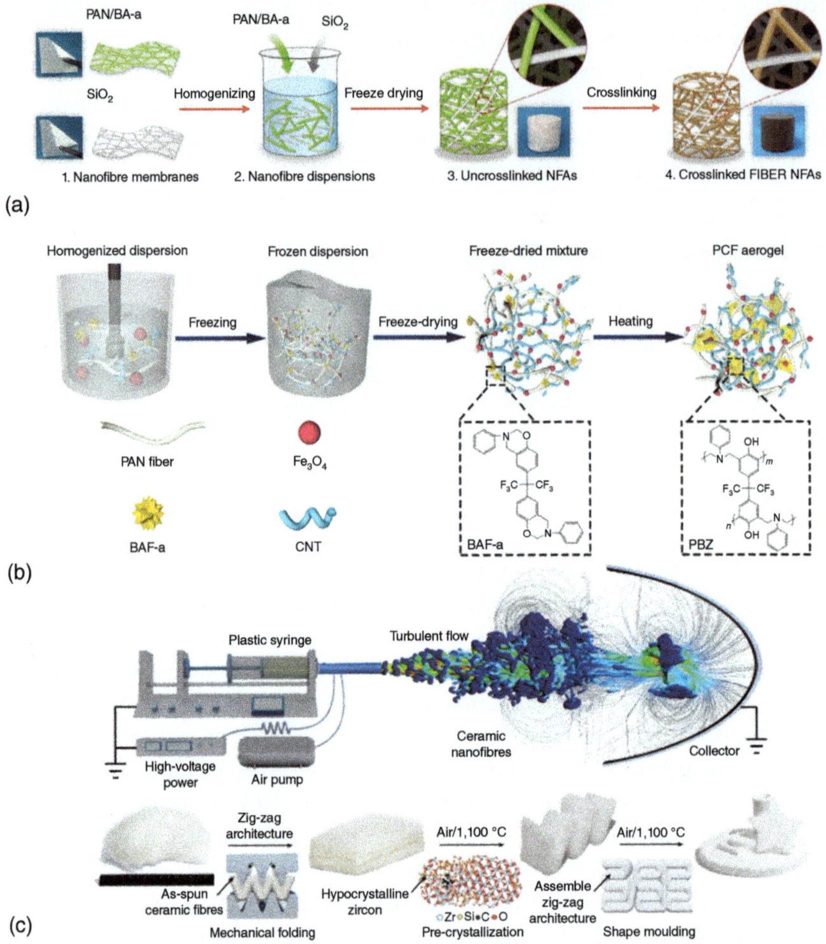

Figure 9.2 (a) Schematic diagram of fibrous isotropically bonded elastic reconstructed nanofibrous aerogels (FIBER NFA) preparation. Source: Si et al. [21] / with permission of Springer Nature. (b) Synthesis process of organic–inorganic mixed aerogel. Source: Li et al. [22] / with permission of John Wiley & Sons. (c) Schematic diagram of zircon nanofibrous aerogels (ZAGs) prepared by electrospinning assisted by turbulence. Source: Guo et al. [24] / with permission of Springer Nature.

1000 residual strain), low temperature superelasticity (up to −196 °C), and excellent washable superelasticity. At the same time, the attractive structure of high porosity, high flexural, and small porosity gives MNFS ultralight performance (7.5 mg cm^{-3}) and effective insulation (28.51 mW m^{-1} K^{-1}). In addition, MNFS has remarkable dirt resistance, robustness, and long service life.

Inorganic nanofibers can often be added to inorganic aerogel networks to enhance the structural stability. In general, inorganic fibers and particles could be added into aerogel networks by step-by-step assembly (Figure 9.2c) [24]. 3D zircon fiber matrix was prepared by electrospinning in air turbulent flow field. Zircon nanofiber aerogels have a zigzag structure, excellent thermo-mechanical stability, and ultralow thermal conductivity at high temperature. It features near-zero Poisson's ratio (3.3×10^{-4}) and near-zero coefficient of thermal expansion (1.2×10^{-7}), which ensure excellent structural flexibility and thermo-mechanical properties. Subjected to severe thermal shock, they exhibit ultrahigh thermal stability and ultralow strength degradation (less than 1%), as well as high-temperature resistance (up to 1300 °C). By wrapping the residual carbon material in the zircon fibers that make up the crystals, thermal radiation is able to transfer heat and realize one of the lowest high-temperature thermal conductivity to date in ceramic aerogels, reaching $104\,\text{mW}\,\text{m}^{-1}\,\text{K}^{-1}$ at 1000 °C. The combination of thermal machinery and insulated property provides an attractive material system with robust insulation in extreme conditions. Si et al. [26] use PAN nanofibers to enhance SiO_2 aerogels Cross-linked composite aerogels have good elasticity and can be compressed 1000 times at 60% ε.

9.4.2 Directional Freeze-Drying Aerogel

Compared with nondirected lyophilized aerogels, directed lyophilized aerogels have obvious anisotropy. They exhibit thermal insulation in two different directions and have directional heat transfer capabilities. For example, an air inhalation effect induction (ASEI) strategy for the preparation of superadiabatic SiC aerogels (STISA) has been reported. ASEI strategy can adjust the directional flow of SiO_2 gas by air inhalation effect, and induce the directional growth and assembly of SiC nanowires to obtain the oriented layered structure. The sintering time was shortened by 90%. The compression and elastic properties of STISA are significantly improved by using ASEI strategy to form directional layered structure. In addition, the layered structure gives STISA an ultralow thermal conductivity of $0.019\,\text{W}\,\text{m}^{-1}\,\text{K}^{-1}$. ASEI strategy provides a new idea for structural design of superadiabatic advanced ceramic aerogels [27]. Dou et al. [28] reported the layered cell structure of a silica nanofiber aerogel, in which electrospun SiO_2 nanofibers (SNFs) and SiO_2 nanofibers (SNAs) were used as the matrix, and SiO_2 sol was used as a high-temperature nanogel. This pathway leads to the assembly of essentially randomly deposited SNFs into fibrocyte structures, with SNAs evenly distributed on fibroblast walls. The unique layered cell structure of ceramic nanofiber aerogels gives them an ultralow density, negative Poisson's ratio, ultralow thermal conductivity ($23.27\,\text{mW}\,\text{m}^{-1}\,\text{K}^{-1}$), temperature-invariant superelasticity from −196 to 1100 °C, and comprehensive properties of large-scale editable shapes. These advantages make aerogels ideal for industrial, aerospace, and even extreme environments. The core–shell structure of SiC/SiO_2 nanowire aerogel constructed from SiC nanowires and SiO_2 molds is presented. After annealing under environmental conditions, adjacent nanowires fuse together through the oxide layer [28]. This fusion and the anisotropic microstructure of the nanowires result in high compressibility. In addition, inorganic sol nanoparticles can be used as cross-linking agents to form stable cross-linking structure between nanofibers and nanoparticles. In addition, directional freeze-drying provides a directional structure for the growth of ice crystals. The aerogels prepared by directional freeze-drying have good microstructure and mechanical properties. The core–shell structure of SiC/SiO_2 nanowire aerogel was formed by directional

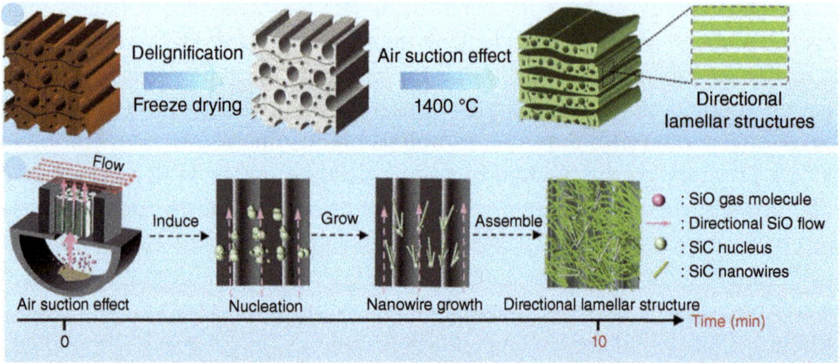

Figure 9.3 Manufacturing process diagram of STISA (top); the growth mechanism of STISA (downward). Application of Spinning Electrospun Nanofiber-Based Insulation Materials. Source: Yan et al. [27] / with permission of John Wiley & Sons.

freezing casting and subsequent heating. Compared with inorganic nanofibers, polymer nanofibers have excellent flexibility, high aspect ratio, and abundant active groups, which can provide high flexibility and versatility for inorganic aerogels (Figure 9.3).

As shown in Figure 9.4, three polymerization strategies are presented to summarize the most promising thermal insulation scenarios for different aerogels [14]. The first polymerization strategy is to use the sol–gel process of polymer nanofibers embedded in the wet gel network composed of inorganic sol nanoparticles. The most typical inorganic sol nanoparticles are SiO_2. A second strategy is to combine inorganic nanofibers into a network of polymer nanofibers. The third strategy is to combine inorganic aerogel particles with nonwoven fabric to obtain aerogel blanket and realize the industrial-scale production and application of aerogel products [29]. Of course, all of these strategies require subsequent drying processes to produce the final aerogel material, such as vacuum drying, CO_2 supercritical drying, or ambient pressure drying. The work done so far indicates that the hybrid polymerization of polymer fibers and inorganic nanomaterials provides a multifunctional platform with controllable microstructure for a variety of thermal insulation applications, adequate mechanical strength and flexibility, as well as the ability to customize. But we also notice that the application of various aerogel materials in the field of heat preservation far exceeds what we have summarized. Therefore, we will discuss the advantages and disadvantages of aerogel materials obtained through these three strategies, as well as some surprising new trends beyond these strategies.

At present, aerogel felt is mainly used in oil and gas pipeline insulation. High-temperature steam, natural gas, oil, and fluid medium pipelines are important equipment components in chemical industry, oil refining, thermal power, and other fields. Aerogel as pipeline insulation material can prevent high temperature or cold environment in the pipeline, a lot of heat loss, to ensure the quality of products. To date, some of the world's largest petrochemical companies, such as ExxonMobil, Shell, and PetroChina, have used aerogel blankets to insulate their pipelines. In the future, however, the market for aerogel blankets used in construction is expected

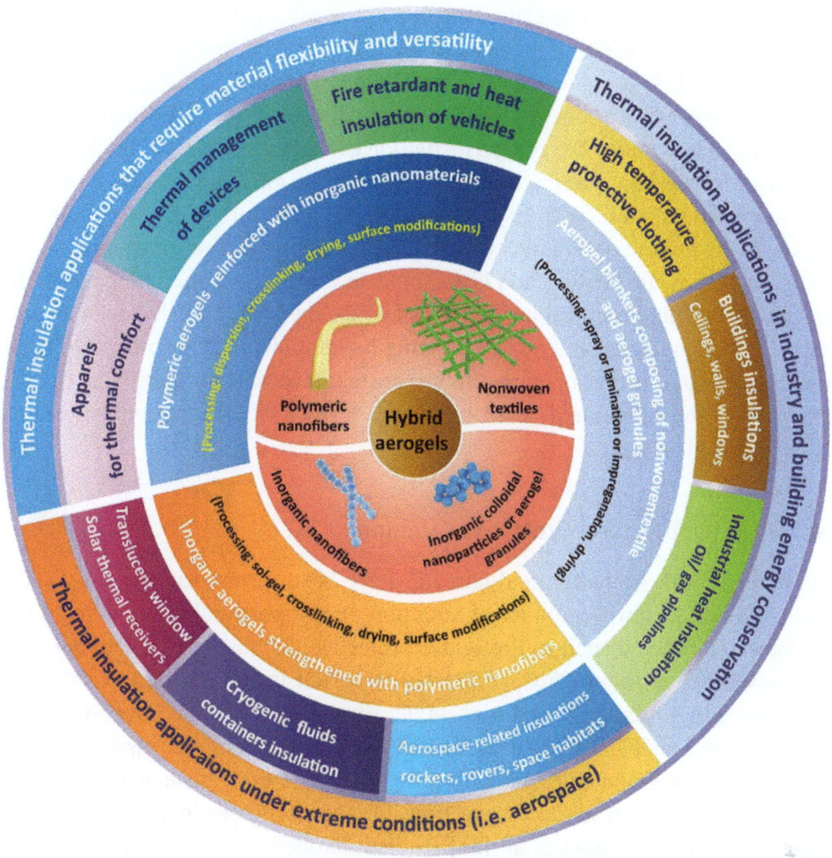

Figure 9.4 Three aggregation strategies for aerogel mixtures and a summary of the most promising thermal insulation application scenarios for different aerogels. Source: Liu et al. [14] / with permission of John Wiley & Sons.

to grow. In addition, the aerogel blanket has important applications in automobile fire prevention, ski field fire prevention, and other fields, such as clothing, outdoor clothing, and military applications.

9.4.3 Insulation for Buildings and Constructions

For high-latitude countries in central and northern Europe, large amounts of energy consumption and carbon dioxide emissions are mainly due to building heating. According to statistics, more than one-third of household energy used for indoor heating. Especially the old buildings' thermal insulation performance is generally poor, resulting in high energy consumption and poor thermal comfort. Therefore, the insulation inside and outside the buildings is very important for these countries. Spaceloft aerogel blanket is a commercial product produced by Aspen Aerogel for building and clothing insulation. It has a very low density of $0.15\,g\,cm^{-1}$ and a very low thermal conductivity of only $14\,mW\,m^{-1}\,K^{-1}$. The maximum temperature of this blanket is 200 °C. As they claim, the blanket is soft, hydrophobic, breathable, and can be used in a wide range of services, such as walls (indoor and outdoor), floors, ceilings, skylights and pitched roofs, terraces and balconies, gutters and arches, windows, window openings, points, repeating and linear thermal bridges, pipes, and hot water pipes. The aerogel insulation used to retrofit the old brick house, as well as a thermal image of the wooden wall, with the top nail insulated by a thin layer of aerogel insulation. Ibrahim et al. [30], the proposed aerogel-based composite system is applied to the concrete modification of exterior and interior walls.

Up to now, many universities and institutions in Europe are conducting research on aerogel in building renovation. Aerogel's low thermal conductivity, open steam diffusion, hydrophobicity, and good fire rating make it ideal for historic buildings. In addition, aerogel with its open porous structure is favorable for traditional buildings because of the diffusion of steam, which can further avoid indoor moisture and possible bacterial growth.

9.4.4 High-Temperature–Protective Clothing

So far, aerogel nonwovens for cold climates have been commercialized very successfully. There has also been a lot of attention on the potential use of aerogel in heat-protective clothing such as firefighter protective clothing (FPC). The normal body temperature is about 37 °C. °C started to be felt in human skin at 44 years of age. The skin starts to feel pain at 44 °C, receives first-degree burn at 48 °C, and receives irreversible second-degree burn at 55 °C. Human skin tissue will be instantly destroyed when come in contact to 72 °C. The time gap between feeling pain and receiving burn is the escape time for the firefighter to withdraw from a dangerous situation. Shaid et al. [31] studied aerogel nonwovens as protective properties for FPC reinforcing materials and flocculating materials. SiO_2 aerogel nonwovens can be purchased online from Buygeogel.com. The fabric weighs $285\,g\,m^{-1}$ and has a thermal conductivity of approximately $23\,mW\,m^{-1}\,K^{-1}$. They found that aerogel nonwovens for firefighter protective clothing (FPCS) could provide firefighters with an escape time of more than one minute, while for commercial FPCS using existing thermal linings and reinforcing materials, it could only provide a five-minute escape. Therefore, it is fully proved that the use of aerogel can significantly improve the protective performance of FPC.

In addition to thermal liners for fire suits, the use of aerogel blankets in space suit insulation, cold weather clothing, multifunctional protective and thermal comfort clothing, and military applications (gloves, insoles, jackets, pants, etc.) is highlighted.

9.4.5 Insulation for Ski Resorts

Aerogel is also commonly used to combat the cold in public sports facilities such as ski resorts. In addition to their low thermal conductivity, aerogels should be airtight, lightweight, easy to install and maintain, durable, resistant to bacteria and fungi, and resistant to poor temperature. Pilipenko et al. [32] reported experience using seamless polyethylene casings in Arctic fixed and mobile home modules, and seamless plastic casings in insulation systems based on the "hot blanket principle used in ski resorts. The comparison of polyethylene foam with other thermal insulation materials under extreme conditions shows that the use of polyethylene foam is limited by the unstable adhesion to metal surfaces due to the breathability of fiber products and sprayed polyurethane foam. Polyethylene can be used to insulate stationary or mobile home modules designed for polar environments and seasonal snow protection in ski resorts.

9.5 Conclusion

Electrospinning nanofibers can be combined with a variety of organic and inorganic materials by blending and other means, showing many unique advantages in the field of thermal insulation applications, such as low thermal conductivity and strong adhesion. What's more, while nanofibrous membranes are widely used for thermal insulation due to their thickness, nanofibers can also be processed into aerogels, which also improve thermal insulation. Currently, most of the composite aerogel networks are based on electrostatic forces, Johannes Diderik van der Waals forces, and hydrogen bonding between them. Of course, there is a long way to go before cheap aerogels can be used on a large scale. It is expected that with the development of technology and the improvement of global energy-saving awareness, the demand for environment-friendly, degradable ultralow thermal conductivity new insulation materials will continue to grow. I can see that, the growth in demand will focus on energy efficiency in industrial pipelines, safety protection for new energy vehicles, fireproofing of rail transit trains and vehicle bodies, heat flow management in the semiconductor industry, special cold suits, high-temperature-protective suits, space shuttle insulation, and various building insulation materials.

References

1 Zhang, E., Zhang, W., Lv, T. et al. (2021). Insulating and robust ceramic nanorod aerogels with high-temperature resistance over 1400 °C. *ACS Applied Materials & Interfaces* 13: 20548.

2 Zhao, X., Yang, F., Wang, Z. et al. (2020). Mechanically strong and thermally insulating polyimide aerogels by homogeneity reinforcement of electrospun nanofibers. *Composites Part B: Engineering* 182.

3 Wang, D., Peng, H., Yu, B. et al. (2020). Biomimetic structural cellulose nanofiber aerogels with exceptional mechanical, flame-retardant and thermal-insulating properties. *Chemical Engineering Journal* 389: 124449.

4 Li, J., Lu, Z., Xie, F. et al. (2021). Highly compressible, heat-insulating and self-extinguishing cellulose nanofiber/aramid nanofiber nanocomposite foams. *Carbohydrate Polymers* 261: 117837.

5 Zhang, X., Wang, B., Wu, N. et al. (2021). Multi-phase SiZrOC nanofibers with outstanding flexibility and stability for thermal insulation up to 1400 °C. *Chemical Engineering Journal* 410, 128304.

6 Cheng, X., Liu, Y.T., Si, Y. et al. (2022). Direct synthesis of highly stretchable ceramic nanofibrous aerogels via 3D reaction electrospinning. *Nature Communications* 13: 2637.

7 Zhang, X., Cheng, X., Si, Y. et al. (2022). All-ceramic and elastic aerogels with nanofibrous-granular binary synergistic structure for thermal superinsulation. *ACS Nano* 16: 5487–5495.

8 Guo, P., Su, L., Peng, K. et al. (2022). Additive manufacturing of resilient SiC nanowire aerogels. *ACS Nano* 16: 6625–6633.

9 Kizildag, N. (2021). Smart composite nanofiber mats with thermal management functionality. *Scientific Reports* 11: 4256.

10 He, M., Liu, H., Wang, L. et al. (2021). One-step fabrication of a stretchable and anti-oil-fouling nanofiber membrane for solar steam generation. *Materials Chemistry Frontiers* 5: 3673–3680.

11 Hu, F., Wu, S., and Sun, Y. (2019). Hollow-structured materials for thermal insulation. *Advanced Materials* 31: e1801001.

12 Tychanicz-Kwiecień, M., Wilk, J., and Gil, P. (2019). Review of high-temperature thermal insulation materials. *Journal of Thermophysics and Heat Transfer* 33: 271.

13 Zhao, R., Guo, H., Yi, X. et al. (2020). Research on thermal insulation properties of plant fiber composite building material: a review. *International Journal of Thermophysics* 41, 1–48.

14 Liu, Q., Yan, K., Chen, J. et al. (2021). Recent advances in novel aerogels through the hybrid aggregation of inorganic nanomaterials and polymeric fibers for thermal insulation. *Aggregate* 2: e30.

15 Yuan, K., Li, H., Jin, X. et al. (2022). Electrospun flexible calcium zirconate fiber membrane with excellent thermal stability and alkali resistance. *Ceramics International* 48: 12408.

16 Chen, J., Huang, X., Sun, B., and Jiang, P. (2019). Highly thermally conductive yet electrically insulating polymer/boron nitride nanosheets nanocomposite films for improved thermal management capability. *ACS Nano* 13: 337–345.

17 Kwon, Y.T., Ryu, S.H., Shin, J.W. et al. (2019). Electrospun CuS/PVP nanowires and superior near-infrared filtration efficiency for thermal shielding applications. *ACS Applied Materials & Interfaces* 11: 6575–6580.

18 An, X., Bai, Y., Xu, G. et al. (2020). Fabrication of interweaving hierarchical fibrous composite (iHFC) membranes for high-flux and robust direct contact membrane distillation. *Desalination* 477: 114264.

19 Xie, Y., Wang, L., Liu, B. et al. (2018). Flexible, controllable, and high-strength near-infrared reflective Y_2O_3 nanofiber membrane by electrospinning a polyacetylacetone-yttrium precursor. *Materials & Design* 160: 918.

20 Zhang, B., Tong, Z., Pang, Y. et al. (2022). Design and electrospun closed cell structured SiO_2 nanocomposite fiber by hollow SiO_2/TiO_2 spheres for thermal insulation. *Composites Science and Technology* 218: 109152.

21 Si, Y., Yu, J., Tang, X. et al. (2014). Ultralight nanofibre-assembled cellular aerogels with superelasticity and multifunctionality. *Nature Communications* 5: 5802.

22 Li, Y., Liu, X., Nie, X. et al. (2019). Multifunctional organic–inorganic hybrid aerogel for self-cleaning, heat-insulating, and highly efficient microwave absorbing material. *Advanced Functional Materials* 29: 1807624.

23 Hu, Y., Yang, G., Zhou, J. et al. (2022). Proton donor-regulated mechanically robust aramid nanofiber aerogel membranes for high-temperature thermal insulation. *ACS Nano* 16: 5984–5993.

24 Guo, J., Fu, S., Deng, Y. et al. (2022). Hypocrystalline ceramic aerogels for thermal insulation at extreme conditions. *Nature* 606: 909.

25 Wu, H., Cai, H., Zhang, S. et al. (2022). Ultralight, superelastic, and washable nanofibrous sponges with rigid-flexible coupling architecture enable reusable warmth retention. *Nano Letters* 22: 830.

26 Si, Y., Fu, Q., Wang, X. et al. (2015). Superelastic and superhydrophobic nanofiber-assembled cellular aerogels for effective separation of oil/water emulsions. *ACS Nano* 9: 3791.

27 Yan, M., Zhang, H., Fu, Y. et al. (2022). Implementing an air suction effect induction strategy to create super thermally insulating and superelastic SiC aerogels. *Small* 18: e2201039.

28 Dou, L., Zhang, X., Cheng, X. et al. (2019). Hierarchical cellular structured ceramic nanofibrous aerogels with temperature-invariant superelasticity for thermal insulation. *ACS Applied Materials & Interfaces* 11: 29056–29064.

29 Berglund, L., Nissila, T., Sivaraman, D. et al. (2021). Seaweed-derived alginate-cellulose nanofiber aerogel for insulation applications. *ACS Applied Materials & Interfaces* 13: 34899–34909.

30 Ibrahim, M., Sayegh, H., Bianco, L., and Wurtz, E. (2019). Hygrothermal performance of novel internal and external super-insulating systems: in-situ experimental study and 1D/2D numerical modeling. *Applied Thermal Engineering* 150: 1306.

31 Shaid, A., Wang, L., Padhye, R., and Bhuyian, M.A.R. (2018). Aerogel nonwoven as reinforcement and batting material for firefighter's protective clothing: a comparative study. *Journal of Sol-Gel Science and Technology* 87: 95.

32 Pilipenko, A., Ter-Zakaryan, K., Bobrova, E., and Zhukov, A. (2019). Insulation systems for extreme conditions. *Materials Today: Proceedings* 19: 2475.

10

Research Progress on Sound Absorption of Electrospun Fibrous Materials

Jinyu He and Xinxin Li

Donghua University, College of Textiles, Key Laboratory of Textile Science & Technology, Ministry of Education, 2999 Renmin North Road, Songjiang District, Shanghai, 201620, China

10.1 Introduction

With the rapid development of global industrialization, especially in urban construction and electromechanical systems, noise pollution is regarded as one of the four significant pollutions and poses considerable threats to human health. The health threats include physical, physiological, emotional, the efficiency of a human body's mental or physical operation, undesirable economic effects, and privacy invasion. Noise can also cause a decline in hearing and disruption of the physiological system. Therefore, a worldwide noise [1–3] control is quite necessary by using sound-absorbing materials or acoustic insulation properly. For sound-absorbing materials, the sound energy will be converted into thermal energy and dissipated due to the viscous and thermal effects. However, acoustic insulation usually forms an acoustic barrier to block the transmission of sound waves from entering or leaving [4].

According to the sound absorption mechanisms, sound-absorbing materials can be classified as the resonant absorber and the porous absorber. The resonant absorber is the equivalent parallel connection of multiple Helmholtz resonators, where the sound waves are consumed by transforming the resonator into a vibration. In addition, it can be divided into a single resonator, a perforated, and the microperforated panel in combination with an air-back cavity and membrane absorbent. This kind of material exhibits a good sound absorption performance at low frequencies, but the sound absorption bandwidth is narrow and shows poor processability [5]. The porous absorber, including fiber, foam, and granular acoustic materials, possesses numerous internal apertures that perform different sound absorbability varied with frequency. It is an adiabatic process when sound waves incident into the pores of porous materials at high frequencies, leading to heat loss due to friction. However, at low frequencies, sound energy is consumed by heat exchange which is an isothermal process [6].

The present sound-absorbing materials meet a bottleneck that low-frequency sound waves can easily bypass obstacles and are hard to be absorbed. More

Electrospinning: Fundamentals, Methods, and Applications, First Edition.
Edited by Liming Wang and Xiaohong Qin.

importantly, traditional acoustic materials are commonly combined with other materials or set rigid backing; therefore, these bulk structures greatly limited the application and industrial production [7]. To tackle the abovementioned problems, electrospinning, as a technique for creating fibers with a diameter ranging from micro to nano, has attracted much attention in the field of sound absorption [8]. Benefiting from the reasonably designed microstructure of electrospun nanofibers with lightweight, high surface area, small pore size, and high porosity, the resultant nanofibrous membranes promote the collision between the sound waves and materials, even the sound waves and atmosphere, thus the acoustic is damped and converted into heat [9].

The fibers made of electrospinning present outstanding acoustic performance and other good mechanical properties such as impressive air permeability, superhydrophobic, ultralight, and high strength [10–13]. Additionally, the simplicity and processability make this type of acoustic material more economical than conventional materials. Accordingly, electrospun nanofibers are promising for sound absorption in traffic, construction, military, and aerospace [8, 14]. Meantime, regulating electrospinning parameters rationally such as electrospinning temperature, humidity, time, and the concentration of electrospinning solution could produce sound-absorbing nanofibrous materials with excellent integrated performance for broader applications.

Due to the range of advantages of electrospinning and acoustic nanofibrous materials, we focus on the sound absorption of electrospun fibrous materials in this review. First, the mechanism of sound absorption and classification of sound-absorbing materials are discussed. Then, we summarize recent research on electrospun fibrous materials and electrospinning parameters to realize better sound absorption performance. Finally, we outline the existing deficiencies of sound-absorbing electrospun materials and shed new light on future development.

10.2 Mechanism of Sound Absorption

Nanofibers prepared by an electrical spinning process have been recognized as highly efficient sound absorbers due to their superior absorption of lower-frequency sound, which is attributable to the high specific surface area, high porosity, and nanoscale fiber diameter [9]. Moreover, electrospun nanofibers belong to porous sound-absorbing materials, which possess many internal pores for sound waves to enter and convert acoustic energy into thermal energy by friction [7].

Figure 10.1a can be applied to explain the sound absorption process for porous materials. When sound waves strike into the material, the sound energy is presented in three ways, some reflect at the surface, some through the tortuous path of the internal pores resulting in absorption, and others pass through the material leading to noise [16]. Sound-absorbing materials typically focus on the process of dissipation and conversion of sound energy. Much has been done to investigate achieving high-efficiency sound consumption, like designing a gradient structure by multistep electrospinning for absorption of broadband sound waves which provides a new idea

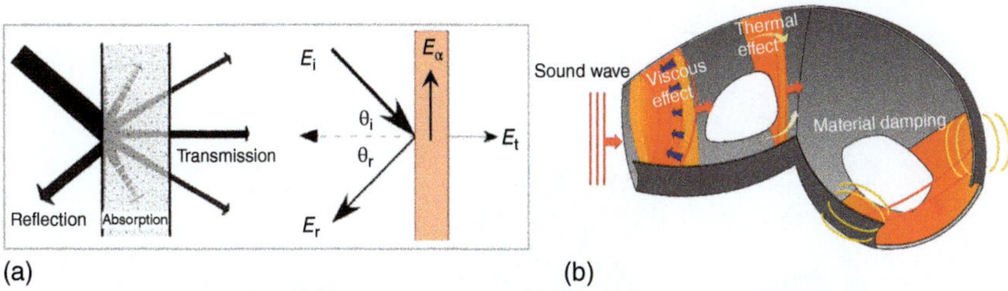

Figure 10.1 (a) Schematic illustration of the sound absorption process for porous materials [15]. (b) Illustration of common sound energy dissipation mechanisms in a porous sound-absorbing material [15]. Source: Leitao Cao et al. [15] / with permission of Elsevier.

to enhance sound absorption band [17] or add silver nanoparticles on nanofibrous membranes to increase conversion of acoustic energy and therefore reduce sound transmission at low frequencies [18].

For porous materials, as demonstrated in Figure 10.1b, sound waves will convert to heat mainly by viscous effect, thermal effect, and material damping. Sound waves that enter the pores inside will excite the back-and-forth movement of the air within the material; this vibration of the fluid can convert acoustic energy into thermal energy by viscous effect. Meantime, friction also exists between the fluid and solid; therefore, sound waves and porous material walls have continuous relative motion, with alternating compression and expansion, finally producing heat and dissipating sound waves by thermal effect. In addition, the acoustic pressure wave will be damped due to the flexible motion of the solid structure consequently facilitating the dissipation of the acoustic vibration [7] [19].

Compared to conventional porous materials, electrospun nanofibers are more suitable to be applied to obtain good absorbing capability for full frequencies because they have both resonant and porous absorption characteristics [20]. It will induce a strong resonance when the frequency of the material coincides with the sound wave, occurring along the transmission loss and the amplitude, as well as vibration speed, reaching their maximum so that sound energy consumption reaches the peak [9, 21]. Along with the sound waves vibrating the resonant nanofibrous system, acoustic energy at the resonant frequency is partially converted to kinetic energy, and other frequencies accumulated in the materials, dissipation by heat ultimately [9]. In general, electrospun nanofibers combine small pore size, high porosity, and good acoustic property, which have been considered a good candidate for new acoustic products.

10.3 Classification of Sound-Absorbing Materials

Porous materials have been widely utilized as noise reduction products due to their massive porosity and effective airflow resistance, which can be divided into the fiber, foam, and granular acoustic materials [22]. Some types of porous materials for sound absorption are shown in Figure 10.2. In this chapter, inorganic fibrous

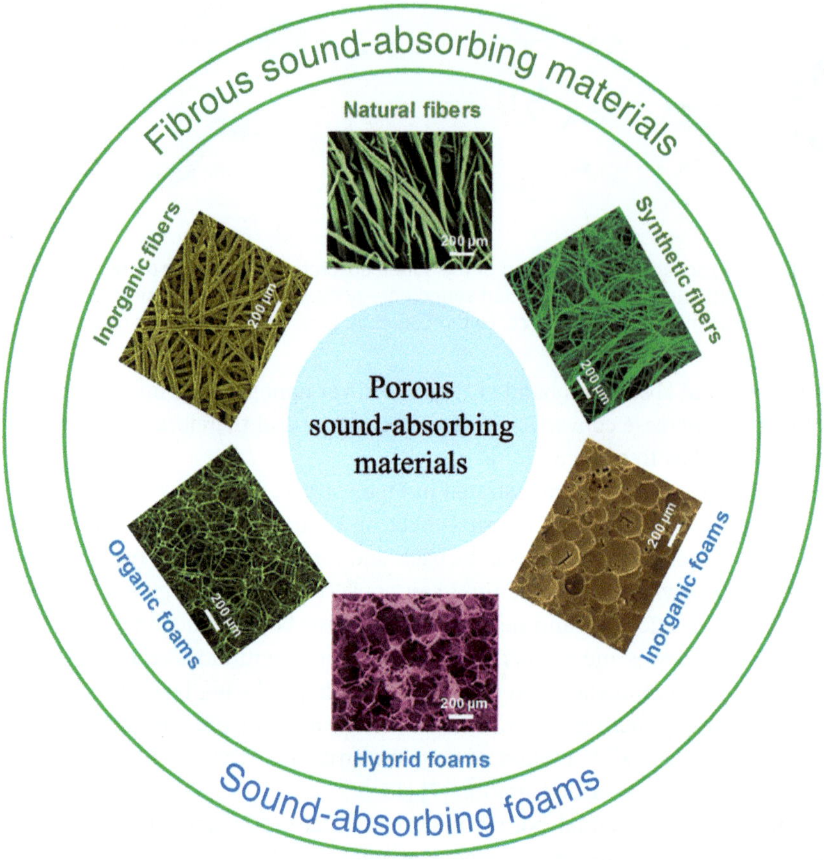

Figure 10.2 Porous materials for sound absorption. Source: Leitao Cao et al. [15] / Elsevier.

materials, natural fibrous materials, and nanofibrous materials have been reviewed for fabricating sound absorbers.

Among inorganic fibrous materials, metal fibers possess stable performance, high-temperature resistance, and high strength; thus, they can be used in noise control in aircraft engine liners at high temperatures [23]. The stainless steel fiber porous material prepared by the loose sintering method has an internal gradient structure in the interior of the material, with better sound absorption properties than compacting sintering and positioning sintering [24]. Moreover, sound absorption performance will be obviously enhanced with the decrease of fiber diameters and increase of sample thickness and porosity [25]. Apart from metallic fibrous sound-absorbing materials, glass fibers have advantages in sound absorption insulation and are very important for aircrafts, trains, and so on. The previous research shows that the incorporation of polystyrene (PS) solution into glass fiber fabric epoxy composite leads to some voids at the cross section of the composite which provides a higher sound absorption coefficient and improvement in thermal insulation [26]. However, glass fibrous materials suffer from the disadvantages of fragileness and the harmfulness of the fragments which need improvement.

Natural fibrous materials, which are commonly utilized in textile acoustic applications, are harmless and have comprehensive sound absorption performance, good designability, softness, and a relative loss cost [27]. The natural fiber sound-absorbing materials reported include vegetative fibers such as flex, cotton, kenaf, wood, hemp, coconut, straw, and cane, as well as animal fibers such as sheep wool and silk. [28]. It is indicated that the low-frequency sound absorption coefficient of materials can be enhanced by increasing the cavity length and material density, but there is little effect even a reverse effect for the high-frequency sound absorption property [22]. It urgently needs to be solved for the poor sound absorption of the natural fibrous material [29]. There is evidence indicating that controlling the densities of the pineapple leaf fibers and introducing the air gap behind the samples can reach a sound absorption coefficient of 0.9 on average above 1 kHz [30]. Furthermore, ecofriendly sound-absorbing composite materials have been created by compression bonding technique using bamboo charcoal and sugarcane bagasse fibers, which can absorb the sound resistance of more than 70% and have adequate moisture resistance at extreme conditions [31]. All in all, natural fibrous materials exhibit the advantages of being biodegradable, abundantly available, and having fewer human health hazards during processing and handling, but enhancing the sound absorption performance at a low-frequency range is still a research hotspot [15].

As a facile and straightforward technique, electrospinning can fabricate nanofibrous materials that have the superiorities of excellent filtration efficiency and high air permeability. When the nanofibrous is exposed to the incident acoustic wave, the air movement and friction inside the nanopores and scattering from the fibers can well consume the acoustic energy. Altering the materials or adding other matter can affect the sound performance and change the physical and chemical properties at the same time [32]. Therefore, choosing correcting materials plays a crucial role in making electrospun nanofibers. The present review of nanofibrous materials focuses mainly on polyvinyl alcohol (PVA), which is a water-soluble polymer that can be mixed in any proportion, and it can film easily with well-mechanical properties, thus PVA nanofibrous membranes have become one of the major candidates for electrospun acoustic materials. According to fabricating a series of PVA membranes loaded with different concentrations of ZrC and TiO_2 nanoparticles, the sound absorption coefficient reached the best at the frequency range from 500 to 1500 Hz With 3 wt% TiO_2 [33]. Moreover, Selvaraj et al. [34] proposed a novel approach that blended PVA with protein hydrolysate which was prepared from the fleshing waste through acid hydrolysis, producing nanofibers by optimizing the electrospinning parameters and with good sound absorbing potential in the lower frequencies. Shen et al. [35] used PVA as the raw material and flexible PVA microperforated membranes were fabricated through a punching process. The resultant absorber showed that the flexible microperforated material had favorable sound absorption performance both in low and high frequency.

Considering the role of piezoelectric effects on acoustical behavior, piezoelectric electrospun nanofibrous membranes have been widely utilized as acoustic materials which can promote the conversion of acoustic energy into electric energy

[32]. Take polyvinylidene fluoride (PVDF) as an example. This kind of polymer has the β phase which is responsible for the piezoelectric properties, with light, strong, and conformable, and can be cut into various shapes and sizes, resulting in widespread applications of energy acquisition and novel sound absorption [36]. Chang et al. showed PVDF electrospun membranes exhibit high surface area providing a large number of contact sites with the sound waves and they are crucial in converting sound energy to electric potential and absorbing sound waves in a low-frequency region. Besides, the carbon nanotubes (CNTs) addition improved piezoelectricity and shifted sound absorption toward low-frequency regions because the PVDF/CNTs composites not only combine the piezoelectric properties of PVDF and conducting properties of CNTs but also possess the high levels of β-phase formation in the electrospun nanofibers. Moreover, electrospun PVDF/acoustic foam has the potential to be a smart material using shunt damping in piezoelectric PVDF, therefore, achieving superior absorption performance for low-frequency sound waves [37]. There also have been studies that add graphene (Gp) [38], silver nanoparticles (AgNPs) [18], and polymethyl methacrylate (PMMA) [39] to present acoustic–electric conversion characteristics so that exhibit a great enhancement in sound absorption at low frequencies.

Recently, polyacrylonitrile (PAN) has been electrospun for noise reduction owing to its high viscoelasticity, easy processing, good thermal performance, and relative low cost. Shao et al. [40] indicated that regulating the spinning ratio of PS coarse fiber (1053 nm) and PAN/CNT fine fiber can cause an obvious three-dimensional structure with a specific surface area of 103.16 m^2/g and a pore size of 2.25 μm, exhibiting excellent sound absorption abilities. Additionally, thermoplastic polyurethane (TPU) and thermoplastic polyester elastomer (TPEE) combines the flexibility of rubbers and the strength and processability of thermoplastics which were composed of alternating hard and soft segments. It has been indicated that compared with TPU and TPEE fiber membranes, the PAN membrane has good sound absorption behaviors in the low and medium frequency range of 100–2500 Hz [41]. Apart from the above-listed polymers, polyvinyl chloride (PVC) [42], poly(lactic acid) (PLA) [43], and prepared polyurethane (PU) [44] are all promising electrospun sound-absorbing materials and have better acoustic performances when composite with other materials.

10.4 Electrospun Fibrous Materials for Sound Absorption

In the field of acoustics, electrospun nanofibrous materials have been designed to improve sound quality and prevent unpleasant reverberation inside material pores by absorbing sound energy [45]. In practical applications, the structure configurations of electrospun nanofibrous materials mainly include membranes, composites, aerogel, multilayered structures, and other surface modification nanofibrous materials (Figure 10.3) [32].

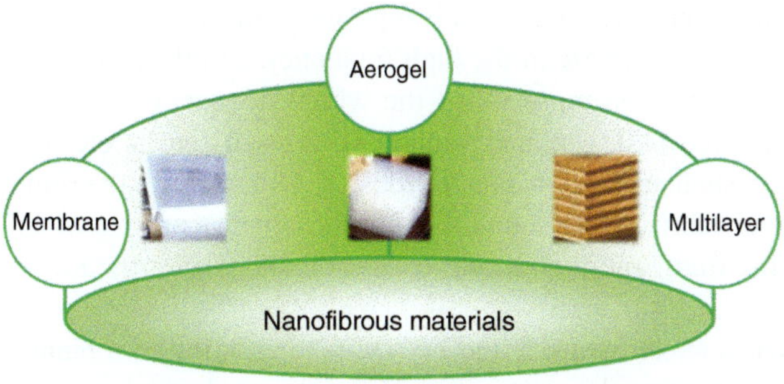

Figure 10.3 Schematic illustration of nanofiber-based noise reduction structures and materials. Source: Xiaoning Tang et al. [32] / with permission of Taylor & Francis.

10.4.1 Electrospun Nanofibrous Membrane for Sound Absorption

Due to the high porosity and extremely low weight of electrospun nanofibrous membranes, they are attractive for sound-absorbing materials. Shen et al. [35] fabricated nanofibrous membranes by electrospinning using PVA, which is environmentally friendly raw material, and combined nanofiber, porous material, and perforated panel to make flexible PVA microperforated membranes through the punching process. The macroscopic photo of the microperforated samples presented that the upper sound absorption value increased gradually with increasing perforation rate and the width of each sound absorption band also broadened. Moreover, as the diameter increased, the resonance absorption peak value of the flexible membrane first increased and then decreased slightly and was overall consistent with different perforation diameters within the range of 100–2500 Hz. Liu et al. [46] prepared nanofibrous membranes by spiral vane electrospinning with different contents of PVA and polyethylene oxide (PEO). The results demonstrated greatly improved sound absorption properties by changing the morphological characteristics and construction of PVA. In order to enhance the sound absorption behavior of acoustic felts, electrospun PAN nanofibrous membranes were used to coat them, and the resultant samples indicate that having a higher amount of PAN nanofibers tended to demonstrate better sound absorbency [47].

10.4.2 Nanocomposite Materials for Sound Absorption

Electrospun nanofibers are often used to combine with kins of sound-absorbing materials for a high sound absorption coefficient in all frequency ranges while keeping the materials' minimum thickness and lightweight. In brief, they can bind with conventional acoustic materials, such as nonwovens, foam, and even knitted spacer fabrics to increase acoustic absorption. Meanwhile, integrating nanofibrous materials with CNTs and graphene is also an effective way to decrease the cell size and increase the tortuous paths of the material, thus increasing the acoustic absorption properties [5].

Employing a thin PAN nanofibrous membrane on the surface can effectively improve the sound absorption coefficients of traditional materials. It caused an increase when PAN nanofibrous membrane was composited with BASF foam at a large frequency, unless a slight decrease in the high frequency, and the acoustical performance of the composite was improved in the whole frequency range for polypropylene (PP) nonwoven [48]. Hajimohammadi et al. used coaxial electrospinning to combine core–shell and hollow nanofibrous membranes with the front or back side of a nonwoven fabric, which found that it is a resonance process when the membranes are in the front, and the absorption peak shifts toward the lower frequencies [49].

Inspired by the layered microstructure found on owl feather surfaces, a bionic sound absorber was made of a thin nanofibrous membrane backed with a substrate melamine foam layer (NMSMF), and the nanoscale poly(vinylidene fluoride-co-hexafluoropropylene)(PVDF-HFP) fibrous membrane is considered as the upper canopy layer. This research identifies a cut-off frequency at which the sound absorption coefficient suddenly increases and sheds new insight into designing novel semipermeable sound absorbers possessing a cut-off effect for engineering applications [50]. Compared to the pristine PU foam, the sound insulation properties of foam that was combined with electrospun nylon-6 nanofibers were greatly enhanced, which can broaden the application of electrospun nanofibers for sound pollution control [51].

As for spacer warp-knitted fabrics, the advantages of high capacity air trap, large and double-faced structures, which make the acoustic wave decrease by the collision with the surface structure and the connecting yarns between the two surfaces, the SACs (sound absorption coefficient) of all samples are increased by nanofiber enhancement at all frequencies [52]. It has been investigated that by tuning deposition amount and surface coating arrangement of PAN nanofibers combined with spacer knitted fabrics, the sound absorption coefficient can value up to 0.7 in the low and medium frequency ranges with no weight and thickness, therefore, providing a novel approach to achieve lightweight textile acoustic structure [53].

It has been reported that the addition of CNTS into the electrospun membranes can increase acoustic absorption. A study shows that the addition of CNT further improves the piezoelectric properties, and the ES PVDF/CNT membranes with the highest piezoelectricity thus shift the sound absorption to a lower-frequency region [36]. Shao et al. designed and developed composite nanofibrous materials with fluffy structures, as well as blended structures of coarse and fine fibers through electrospinning technology, and the nanofibrous membrane materials doped with CNT exhibited excellent sound absorption abilities, which can be applied to areas with serious air pollution and high noise pollution [40].

It was found that the use of graphene oxide as a reinforcing phase in nanofibers can increase the sound absorption coefficient of the materials. Liu et al. fabricated blend films based on PEO doped with different concentrations of graphene oxide. The result showed that the fibers were intertwined in a loop and formed a network, and the area density and surface roughness of the nanofibrous membrane were reduced, leading to an increase in sound absorption properties [54]. Furthermore, adding graphene further improved the piezoelectric properties through interfacial polarization. Electrospun nanofibrous membranes exhibited an increase in surface area;

consequently, their contact with sound waves was increased, which enhanced the sound-energy absorption in the middle-frequency region through the friction and vibration of the internal nanofiber [38].

10.4.3 Nanofibrous Aerogel for Sound Absorption

Nowadays, sound absorption made of aerogel has attracted attention due to its open solids with high porosity and low density. Low-density nanofibrous aerogels enable various applications but are often hindered by their fragile bendability and poor tensile strength. Leitao et al. designed nanofibrous aerogels with robust bendability and superelastic properties, which have been achieved by fabricating bamboo lashing-like structures through a freeze-drying method. The resultant hierarchical porous structures and hydrophobicity of the materials provide them with ultralight properties and efficient sound absorption capability (noise reduction coefficient of 0.41) [55]. There is a novel strategy to create fibrous, isotopically bonded elastic reconstructed materials with a hierarchical cellular structure and superelasticity by combining electrospun nanofibers and the fibrous freeze-shaping technique. The sound absorption of this kind of material is better than that of commercial nonwoven at the frequency at the whole sound band [56]. Moreover, nanofibrous aerogel which has a hierarchical maze-like microstructure presents excellent sound absorption performance with an NRC (noise reduction coefficient) of 0.58 as well as robust mechanical properties. This kind of aerogel used the freeze-casting technique to interweave the cellulose nanocrystal (CNC) lamellas with PAN electrospun nanofiber networks which shows better performance than the commercial one and are lightweight. Moreover, the materials exhibited obvious enhancement in the low-frequency sound absorption band with all kinds of densities. Therefore, composite nanofibrous aerogels are a potential choice for sound absorption in the fields of vehicles, buildings, and indoor reverberation [57].

10.5 Effect of Electrospinning Parameters on Sound Absorption

Diameter and surface density are important to the flow resistivity and natural resonance frequency of the materials. The diameter usually influences the pore size and porosity of the nanofibers, while the surface density determines the distribution of the fibers. The sound absorption is mainly controlled by the flow resistance as it determines the curvature range of the sound wave transmission channel in the material. For example, materials with low-frequency inherent frequencies exhibit excellent sound absorption properties at low frequencies. Ma et al. [58] prepared PAN nanofibrous membranes with different surface densities and concluded that reducing the pore size of the materials could increase the average sound absorption coefficient of the materials in the middle- and low-frequency bands, and widen the sound absorption frequency bands of the materials. Gao et al. [59] prepared a composite material formed by the PVA electrospun nanofibrous film and the polyester needle-punched nonwoven material that was treated to resist water solubilization, changing the fiber diameter of the composite, and the result showed that the absorption curve of the composite shifted toward the low and medium frequencies. Wang [60] used a mixture of PVP and ethanol as polymer solution to electrospun at different flow rates, different voltages, and different concentrations. The average diameter of the nanofibers increased significantly with increasing solution concentration when the polymer solution concentrations were 0.04, 0.06, and 0.08 $g\,ml^{-1}$. Moreover, the distance between electrodes during the electrostatic spinning process determines the average diameter of the nanofibers, and the outlet velocity of the material determines its area density [9].

As the thickness of the material increases, the pore channels within the material become longer, and the overall dynamic range of sound energy consumption increases while the number of interactions between the sound waves and the fibers increases. There is research indicating that the sound absorption performance of the prepared composites increases significantly with increasing thickness in the low frequency, but the change in thickness has little effect on the sound absorption in the high-frequency band [61].

The porosity of acoustic materials is usually greater than 95%. If the porosity of the material is too large, the interaction between the fiber and the sound wave will be weakened, so that it is difficult to transmit sound waves and ultimately leading to a decline in the material's acoustic performance. If the porosity is too small, the material texture will be tight and the sound wave is not easy to enter the internal, meanwhile, the reflection of the sound wave in the material surface increases, making the acoustic performance decline accordingly [4]. According to Ma [62], the spinning time was set at 6, 8, and 12 h, and PAN nanofibrous membranes with different porosities were obtained, indicating that the absorption coefficient of the material increases with increasing surface density and porosity.

In addition, voltage, flow rate, and polymer concentration play important roles in the sound absorption performance of nanofibrous materials. There has been indicated that the high voltage and polymer concentration were observed to be the most significant parameters at 95% and 99% confidence levels [35]. With the increase in voltage and the decrease in flow rate, ever-increasing numbers and ever-decreasing sizes of the nanoporous microspheres have appeared [63].

10.6 Future Development of Sound-Absorbing Electrospun Materials

Because of their specific properties, electrospun nanofibers will certainly become promising materials for acoustic applications in many fields. To improve application performances, further research needs to be continued. First, the sound absorption properties of electrospun nanofibers and their composites at low frequencies need further improvement. Theoretical studies can reduce the cost and contamination during experiments, and the factors affecting the sound absorption properties of electrospun nanofibers can be investigated. By developing a sound absorption model, the sound absorption properties of the material can be predicted. The accuracy of the prediction can be improved by modifying the model according to the material and structure of the electrospun micro-/nanofibers. Secondly, conventional electrospun nanofibers have many limitations in their application in several different conditions, such as humid and high-temperature environments, due to their poor thermal stability, water resistance, and mechanical properties. Therefore, multifunctional modifications of the fibers are necessary. Electrospun nanofibers can be modified by modifying the material or preparing it for use in innovative processes to obtain properties such as good water resistance and thermal stability, which facilitates the expansion of their application in sound absorption. Finally, most electrospun nanofibers are manufactured by solution electrospinning, which exhibits low efficiency and the solvents used are often corrosive and toxic. These characteristics of solution electrospinning are detrimental to the industrial production of electrospun nanofibers and may lead to environmental problems. Therefore, there is an urgent need to find solutions to the problems inherent in electrospinning. The use of water-soluble polymers in solution electrospinning or the future choice of environmentally friendly melt electrospinning as an alternative to the former are suggestions to solve these problems. In conclusion, further exploration of electrospun nanofibers will facilitate their sound absorption applications in many fields [8].

References

1. Peng, M. and Zhao, X. (2019). Advances in the fiber-based sound-absorbing materials. *Materials Reports* 33: 3669–3677.
2. Wei, H. (2021). Sound absorption mechanism and research progress of textile materials. *Melliand China* 49: 35–38.
3. Na, Y., Lancaster, J., Casali, J., and Cho, G. (2007). Sound absorption coefficients of micro-fiber fabrics by reverberation room method. *Textile Research Journal* 77: 330–335. https://doi.org/10.1177/0040517507078743.
4. Xiao, L., Yuanjun, L., and Xiaoming, Z. (2022). Research progress of electrospinning nanofiber-based sound-absorbing materials. *Advanced Textile Technology* 30: 246–258, https://doi.org/10.19398/j.att.202201033.
5. Tang, X. and Yan, X. (2017). Acoustic energy absorption properties of fibrous materials: a review. *Composites Part a-Applied Science and Manufacturing* 101: 360–380. https://doi.org/10.1016/j.compositesa.2017.07.002.
6. Liu, Y.P. and Hu, H. (2010). Sound absorption behavior of knitted spacer fabrics. *Textile Research Journal* 80: 1949–1957. https://doi.org/10.1177/0040517510373639.
7. Li, H. et al. (2020). Research progress of noise reduction by nanofibers. *Journal of Textile Research* 41: 168–173.
8. Li, X. et al. (2022). Research progress on sound absorption of electrospun fibrous composite materials. *Nanomaterials* 12: 1123. https://doi.org/10.3390/nano12071123.

9 Kalinova, K. (2011). Nanofibrous resonant membrane for acoustic applications. *Journal of Nanomaterials* 1–6. https://doi.org/10.1155/2011/265720 (2011).

10 Zhou, W. et al. (2022). Environmentally friendly waterborne polyurethane nanofibrous membranes by emulsion electrospinning for waterproof and breathable textiles. *Chemical Engineering Journal* 427. https://doi.org/10.1016/j.cej.2021.130925: 130925.

11 Cai, X.M. et al. (2021). Facile preparation of superhydrophobic membrane inspired by chinese traditional hand-stretched noodles. *Coatings* 11. https://doi.org/10.3390/coatings11020228: 228.

12 Liu, F.Q. et al. (2022). Preparation and application of Ultralight nanofiber aerogels. *Progress in Chemistry* 34: 1384–1401. https://doi.org/10.7536/pc210805.

13 Tang, W., Fu, S.R., Luo, N.A. et al. (2022). Para-aramid nanofiber membranes for high-performance and multifunctional materials. *ACS Applied Nano Materials* 5: 747–758. https://doi.org/10.1021/acsanm.1c03488.

14 Peng, M. and Zhao, X.M. (2017). Application of nanomaterial in sound absorption material. *Journal of Chengdu Textile College* 34: 232–237.

15 Cao, L., Fu, Q., Si, Y. et al. (2018). Porous materials for sound absorption. *Composites Communications* 10: 25–35. https://doi.org/10.1016/j.coco.2018.05.001.

16 Liang, L.S., Guo, W.L., and Zhang, Y. (2020). Research progress of noise reduction by nanofibers. *Journal of Textile Research* 51: 5013–5019.

17 Feng, Y. et al. (2021). Gradient structured micro/nanofibrous sponges with superior compressibility and stretchability for broadband sound absorption. *Journal of Colloid and Interface Science* 593: 59–66. https://doi.org/10.1016/j.jcis.2021.03.013.

18 Wu, C.M. and Chou, M.H. (2020). Acoustic-electric conversion and piezoelectric properties of electrospun polyvinylidene fluoride/silver nanofibrous membranes. *Express Polymer Letters* 14: 103–114. https://doi.org/10.3144/expresspolymlett.2020.10.

19 Rahimabady, M. et al. (2017). Hybrid local piezoelectric and conductive functions for high performance airborne sound absorption. *Applied Physics Letters* 111. https://doi.org/10.1063/1.5010743: 241601.

20 Ozturk, M.K., Kalinova, K., Nergis, B., and Candan, C. (2013). Comparison of resonance frequency of a nanofibrous membrane and a homogeneous membrane structure. *Textile Research Journal* 83: 2204–2210. https://doi.org/10.1177/0040517513490064.

21 Kucuk, M. and Korkmaz, Y. (2012). The effect of physical parameters on sound absorption properties of natural fiber mixed nonwoven composites. *Textile Research Journal* 82: 2043–2053. https://doi.org/10.1177/0040517512441987.

22 Ren, Y.H., Sun, X.N., and Song, H. (2011). In *International Conference on Textile Engineering and Materials (ICTEM 2011)*. 959–962.

23 Sun, F., Chen, H., Wu, J., and Feng, K. (2010). Sound absorbing characteristics of fibrous metal materials at high temperatures. *Applied Acoustics* 71: 221–235. https://doi.org/10.1016/j.apacoust.2009.09.001.

24 Ao, Q.B., Wang, J.Z., and Li, A.J. (2017). Sound absorption properties of fibrous porous metals. *Rare Metal Materials And Enginerring* 46: 387–391.

25 Wang, X., Li, Y., Chen, T., and Ying, Z. (2015). Research on the sound absorption characteristics of porous metal materials at high sound pressure levels. *Advances in Mechanical Engineering* 7. https://doi.org/10.1177/1687814015575429.

26 Altay, P. and Ucar, N. (2022). Improvement of insulation properties of glass fiber fabric/epoxy composites modified by polymeric and inorganic fillers. *Polymer Composites* 43: 225–238. https://doi.org/10.1002/pc.26369.

27 Yahya, M.N. and Chin, D.D.V.S. (2017). In *International Research and Innovation Summit (IRIS).*

28 Yang, T. et al. (2020). Sound absorption properties of natural fibers: a review. *Sustainability* 12. https://doi.org/10.3390/su12208477: 1–27.

29 Liu, X.T., Li, L., Yan, X., and Zhang, H.P. (2013). In *3rd International Conference on Textile Engineering and Materials (ICTEM 2013).* 329–332.

30 Putra, A. et al. (2018). Sound absorption of extracted pineapple-leaf fibres. *Applied Acoustics* 136: 9–15. https://doi.org/10.1016/j.apacoust.2018.01.029.

31 Sakthivel, S. et al. (2021). Sound absorbing and insulating properties of natural fiber hybrid composites using sugarcane bagasse and bamboo charcoal. *Journal of Engineered Fibers and Fabrics* 16. https://doi.org/10.1177/15589250211044818: 1–10.

32 Tang, X. (2022). Recent advances on nanofibrous acoustic materials for noise reduction applications. *Journal of the Textile Institute*. https://doi.org/10.1080/00405000.2022.2131339 2022: 1–11.

33 Gao, B., Zuo, L.Y., and Zuo, B.Q. (2016). Sound absorption properties of spiral vane electrospun PVA/nano particle nanofiber membrane and non-woven composite material. *Fibers and Polymers* 17: 1090–1096. https://doi.org/10.1007/s12221-016-6324-z.

34 Selvaraj, S., Jeevan, V., Rao, J.R., and Fathima, N.N. (2019). Conversion of tannery solid waste to sound absorbing nanofibrous materials: a road to sustainability. *Journal of Cleaner Production* 213: 375–383. https://doi.org/10.1016/j.jclepro.2018.12.144.

35 Shen, J., Lee, H.P., and Yan, X. (2022). Sound absorption performance and mechanism of flexible PVA microperforated membrane. *Applied Acoustics* 185. https://doi.org/10.1016/j.apacoust.2021.108420: 108420.

36 Wu, C.M., Chen, J.C., Chou, M.H., Lee, J.Y., and Cheng, K.B. (2015). In *20th International Conference on Composite Materials (ICCM).*

37 Wu, C.M. and Chou, M.H. (2016). Polymorphism, piezoelectricity and sound absorption of electrospun PVDF membranes with and without carbon nanotubes. *Composites Science and Technology* 127: 127–133. https://doi.org/10.1016/j.compscitech.2016.03.001.

38 Wu, C.M. and Chou, M.H. (2016). Sound absorption of electrospun polyvinylidene fluoride/graphene membranes. *European Polymer Journal* 82: 35–45. https://doi.org/10.1016/j.eurpolymj.2016.07.001.

39 Fang, K.Y. et al. (2018). Hybridizing CNT/PMMA/PVDF towards high-performance piezoelectric nanofibers. *Journal of Physics D: Applied Physics* 51: 265305. https://doi.org/10.1088/1361-6463/aac600.

40 Shao, W. et al. (2022). Electrospun PS/PAN nanofiber membranes formed from doped carbon nanotubes with a fluffy and multi-scale construction for air-filtration materials. *Fibers and Polymers* 23: 1197–1205. https://doi.org/10.1007/s12221-022-4652-8.

41 Liu, H. et al. (2014). Application of electrospinning fibres on sound absorption in low and medium frequency range. *Materials Research Innovations* 18: 888–891. https://doi.org/10.1179/1432891714z.000000000797.

42 Khan, W.S., Asmatulu, R., and Yildirim, M.B. (2012). Acoustical properties of electrospun fibers for aircraft interior noise reduction. *Journal of Aerospace Engineering* 25: 376–382. https://doi.org/10.1061/(asce)as.1943-5525.0000118.

43 Yoon, C.K., Park, B.K., and Lee, W.I. (2018). Characteristics of micro-glass bead/PLA porous composite prepared by electrospinning. *Advanced Composite Materials* 27: 183–193. https://doi.org/10.1080/09243046.2017.1405508.

44 Jia, W. and Qin, X. (2014). Acoustic absorption property of electrospun PU nanofibers membrane. *Journal of Donghua University. Natural Science Edition* 40: 509–514.

45 Marques, D.V. et al. (2018). Recycled polyethylene terephthalate-based boards for thermal-acoustic insulation. *Journal of Cleaner Production* 189: 251–262. https://doi.org/10.1016/j.jclepro.2018.04.069.

46 Liu, H. and Zuo, B. (2018). Structure and sound absorption properties of spiral vane electrospun PVA/PEO nanofiber membranes. *Applied Sciences-Basel* 8 (2): 296. https://doi.org/10.3390/app8020296.

47 Ozturk, M.K., Nergis, F.B., and Candan, C. (2018). Design of electrospun polyacrylonitrile nanofiber-coated nonwoven structure for sound absorption. *Polymers for Advanced Technologies* 29: 1255–1260. https://doi.org/10.1002/pat.4236.

48 Xiang, H.F. et al. (2011). Sound absorption behavior of electrospun polyacrylonitrile nanofibrous membranes. *Chinese Journal of Polymer Science* 29: 650–657. https://doi.org/10.1007/s10118-011-1079-x.

49 Hajimohammadi, M., Soltani, P., Semnani, D. et al. (2022). Nonwoven fabric coated with core-shell and hollow nanofiber membranes for efficient sound absorption in buildings. *Building and Environment* 213. https://doi.org/10.1016/j.buildenv.2022.108887: 108887.

50 Ji, G. et al. (2019). Nano-fibrous composite sound absorbers inspired by owl feather surfaces. *Applied Acoustics* 156: 151–157. https://doi.org/10.1016/j.apacoust.2019.06.021.

51 Park, M. et al. (2019). Sound absorption and insulation properties of a polyurethane foam mixed with electrospun nylon-6 and polyurethane nanofibre mats. *Journal of Nanoscience and Nanotechnology* 19: 3558–3563. https://doi.org/10.1166/jnn.2019.16128.

52 Davoudabadi Farahani, M., Jamshidi Avanaki, M., and Jeddi, A.A.A. (2022). Sound absorption of warp knitted spacer fabrics based on knit structure and nanofiber enhancement. *Journal of Industrial Textiles* 51: 1539–1557. https://doi.org/10.1177/1528083720903411.

53 Kucukali-Ozturk, M., Ozden-Yenigun, E., Nergis, B., and Candan, C. (2017). Nanofiber-enhanced lightweight composite textiles for acoustic applications. *Journal of Industrial Textiles* 46: 1498–1510. https://doi.org/10.1177/1528083715622427.

54 Liu, H. and Zuo, B. (2020). Sound absorption property of PVA/PEO/GO nanofiber membrane and non-woven composite material. *Journal of Industrial Textiles* 50: 512–525. https://doi.org/10.1177/1528083719832857.

55 Cao, L.T. et al. (2019). Ultralight, superelastic and bendable lashing-structured nanofibrous aerogels for effective sound absorption. *Nanoscale* 11: 2289–2298. https://doi.org/10.1039/c8nr09288e.

56 Si, Y., Yu, J.Y., Tang, X.M. et al. (2014). Ultralight nanofibre-assembled cellular aerogels with superelasticity and multifunctionality. *Nature Communications* 5. https://doi.org/10.1038/ncomms6802: 5802.

57 Cao, L. et al. (2021). Hierarchically maze-like structured nanofiber aerogels for effective low-frequency sound absorption. *Journal of Colloid and Interface Science* 597: 21–28. https://doi.org/10.1016/j.jcis.2021.03.172.

58 Ma, S., Liu, X., Xie, C., and Su, X. (2020). Sound absorption properties of polyacrylonitrile electrospinning nanofiber membrane and its laminated materials. *Journal of Silk* 57: 13–19.

59 Gao, B. (2017). *Sound Absorption Properties of Electrospun PVA Nano Fiber Membrane*. Soochow University.

60 W., W. C (2016). *Study on Preparation Process of Nanofiber and Its Application Based on Electrospinning*. Soochow University.

61 LiHua, L., Li, Z., and Duoduo, Z. (2022). Preparation and properties of sound absorbing composites based on use of waste straw/polycaprolactone. *Journal of Textile Research* 43: 28–35, https://doi.org/10.13475/j.fzxb.20210901708.

62 Shengnan, M., Xinjin, L., Chunping, X., and Xuzhong, S. (2020). Sound absorption properties of polyacrylonitrule electrospinning nanofiber membrane and its laminated materials. *Journal of Silk* 57: 13–19.

63 Xu, L., Liu, F., and Faraz, N. (2012). Theoretical model for the electrospinning nanoporous materials process. *Computers & Mathematics with Applications* 64: 1017–1021. https://doi.org/10.1016/j.camwa.2012.03.019.

11

Electrospun Nanofiber-Based Triboelectric Nanogenerator

Chentian Zhang and Xueping Zhang

Donghua University, College of Textiles, Key Laboratory of Textile Science & Technology, Ministry of Education, 2999 Renmin North Road, Songjiang District, Shanghai 201620, China

11.1 Introduction

At present, renewable energy is mainly collected from sunlight, wind, and hydroelectric power. However, some low-frequency energy and micro/nano energy collection (such as ocean wave energy and biomechanical energy) is difficult to achieve on a large scale and has not been effectively utilized. Aiming to add the collection methods of renewable energy, extensive research has been conducted in low-frequency and micro/nano energy.

In 2012, the concept of nanogenerators was proposed, and the harvesting of micro energy was enabled by various nanogenerators. To the present time, the development of piezoelectric nanogenerators (PENG) [1–3], triboelectric nanogenerators (TENGs) [4, 5], and pyroelectric nanogenerators (PyNG) [6, 7] has enabled energy harvesting from multiple sources and is expected to replace traditional chemical batteries in many fields as a durable and self-sufficient source of energy [8–11]. Furthermore, TENGs have received greater attention for their lightweight, high power density, and versatile manufacturing methods.

TENG is a type of nanogenerator that collect the triboelectrification and electrostatic induction charge. TENG is composed of a wide range of materials to meet different frictional charging properties, including fabrics and membranes. Among the choices of constituents, nanofibers are outstanding choices for increasing tribological efficiency due to the large specific surface area and multiple nanostructures [12, 13]. One of the most versatile methods for preparing nanofibers is electrospinning, which can process different materials such as polymers, ceramics, and inorganic nanoparticles.

11.2 Triboelectric Nanogenerator

Electrostatic charge is commonly considered a negative phenomenon, one of the most frequent safety hazards in industrial manufacturing, and an important

Electrospinning: Fundamentals, Methods, and Applications, First Edition.
Edited by Liming Wang and Xiaohong Qin.

indicator of a garment's performance. The invention of TENGs eliminates the challenge of harvesting electrostatic energy. TENG is a promising nanogenerator with the advantages of flexibility, wide operating range, diverse material selection, and high power density.

11.2.1 Working Mechanism

Most insulators or less conductive materials have a strong ability to be charged by friction. TENG harvests electrical energy generated by the triboelectrification and electrostatic induction of materials with different contact electrically charged properties [14, 15]. While the continuous operation of opposite and equal electrical potentials forms between the two layers, alternating potentials drive the flow of electrons back and forth in the external load [16]. Briefly, triboelectrification provides a static polarized charge on the material surface, and electrostatic induction drives the conversion of mechanical energy into electrical energy through mechanically triggered changes in electrical potential. This dynamic current is based on Maxwell's displacement current [17].

11.2.2 Four Fundamental Working Modes

As shown in Figure 11.1, four different operation modes of the TENG have been proposed based on the working principles of TENG, including vertical contact–separation (CS) mode, lateral-sliding (LS) mode, single-electrode (SE) mode, and freestanding triboelectric-layer (FT) mode.

11.2.2.1 Vertical Contact–Separation Mode

The vertical contact–separation mode is the most basic type of TENGs. Generally, it is two different dielectric materials facing each other with electrodes attaching to each dielectric (Figure 11.1a). The opposite charges are generated according to the respective polarities when two materials frictional contact each other. When the

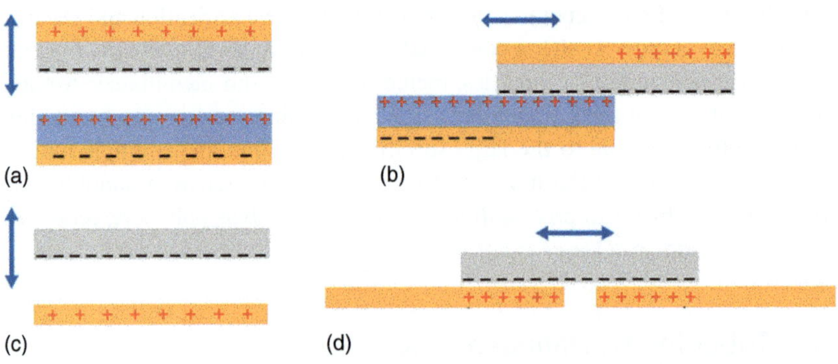

Figure 11.1 Four fundamental working modes of TENG. (a) Vertical CS mode, (b) LS mode, (c) SE mode, (d) FT mode. Reproduced with permission Wu et al. [17]. Copyright 2018, John Wiley and Sons.

two layers separate, free electrons flow from one electrode to the other to balance the potential difference and generate the AC output. During the process of repeated contact and separation, the transferred charges flow through the external load.

11.2.2.2 Lateral-Sliding Mode

The lateral-sliding mode is based on the respective sliding of the top layer to the bottom layer (Figure 11.1b). The top and bottom layers are not completely separated, and they produce different effective contact areas by sliding. The sliding process generates a tribological charge in the two layers, while a potential difference between the two electrodes allows the flow of electrons. Then the reverse motion generates the reverse motion of electrons until the electrostatic equilibrium. Due to the periodic change of the effective contact area, this process generates the AC output.

11.2.2.3 Single-Electrode Mode

In a single-electrode configuration, one output node is connected to a single electrode and the other is virtually grounded [18] (Figure 11.1c). The single-electrode mode can work in both lateral-sliding and contact–separation configurations [19]. Since the other contact layer is free to move, the single-electrode mode has a wider degree of freedom of motion, endows it the advantage of working freely and making full use of the ground/skin/hair as electrodes, but its output performance is lower than other modes.

11.2.2.4 Freestanding Triboelectric-Layer Mode

The freestanding triboelectric-layer mode was developed based on the single-electrode mode, with higher efficiency than single-electrode mode (Figure 11.1d). It can also work in a noncontact mode, but instead of using the ground as the reference electrode, it uses a pair of symmetrical electrodes.

The movement of the rubbing layer along the underlying layer produces an unbalanced charge distribution. This induces electron flow between the two electrodes to balance the local potential distribution.

TENG devices could be designed according to the applications, including four fundamental working modes and the hybrid mode with variable structural design. Here, we will not discuss the application of TENG according to the working mode, but rather as an example of a specific application in Section 11.4.

11.3 Electrospun Nanofiber-Based TENG

TENG has been considered the next generation of energy source to replace traditional rigid chemical batteries, so it needs to be flexible, breathable, and stretchable to meet the requirements of multiple applications. Currently, various materials have been used to fabricate TENG, such as membranes, fabrics, and nonwovens, but they are limited either by poor flexibility or low friction efficiency. Electrospun nanofibers, with excellent flexibility, high friction efficiency, and tunability provide a strong capability to fabricate and enhance the output performance of TENG. Furthermore, tunable fiber aspect ratios and multidimensional structures, as well as special structures such as core–shell, hollow, and oriented, could be easily achieved by electrospun nanofibers.

11.3.1 Enhancement of the Output Performance

Under identical environmental and intermediate conditions, the selection of two materials with different electron affinities is the most straightforward method to enhance the output performance of TENG. The triboelectric series developed by AlphaLab are shown in Table 11.1 [20]. The data were acquired at 22 °C and 35% relative humidity (RH) by using a surface voltmeter (AlphaLab).

Interestingly, commonly used negative materials such as polyvinylidene fluoride (PVDF), polytetrafluoroethylene (PTFE), and cellulose acetate (CA), as well as

Table 11.1 Triboelectric tendency series.

Insulator name	Charge affinity (nC/J)	Insulator name	Charge affinity (nC/J)
Polyurethane foam	+60	Sorbothane	+58
Hair, oily skin	+45	Solid polyurethane	+40
Magnesium fluoride	+35	Nylon, dry skin	+30
Machine oil	+29	Nylatron	+28
Glass (soda)	+25	Paper	+10
Wood (pine)	+7	Cotton	+5
Nitrile rubber	+5	Wool	+0
Polycarbonate	−5	Acrylic	−10
Epoxy	−32	Styrene-butadiene rubber	−35
PET (mylar) solid	−40	EVA rubber	−55
Gum rubber	−60	Polystyrene	−70
Polyimide	−70	Silicones	−72
Vinyl: flexible	−75	LDPE	−90
Polypropylene	−90	HDPE	−90
Cellulose nitrate	−93	UHMWPE	−95
Polychloroprene	−98	PVC (rigid vinyl)	−100
Latex (natural) rubber	−105	Viton, filled	−117
Epichlorohydrin rubber	−118	Santoprene rubber	−120
Hypalon rubber, filled	−130	Butyl rubber, filled	−135
EDPM rubber, filled	−140	PTFE (Teflon)	−190

PET, polyethylene terephthalate; EVA, ethylene vinyl acetate; LDPE, low-density polyethylene; HDPE, High-density polyethylene; UHMWPE, ultra-high molecular weight polyethylene.
Source: Adapted from Chen et al. [20].

commonly used positive materials such as polyacrylonitrile (PAN), polyamide (PA), polyurethane (PU), could all be processed easily by electrospinning. Thus, the structure and frictional electrical properties could be controlled by adjusting the spinning process. A typical example is the formation of highly oriented crystalline β-phase of PVDF by tuning the spinning parameters and adding inorganic substances such as carbon nanotubes [21] and silver nanowires [22]. During the electrospinning process, uniaxial stretching, electric field polarization, and inorganic substances promote the formation of crystalline β-phase [23, 24]. Compared to the α-phase, the crystalline β-phase is characterized by a larger degree of polarization per unit cell and allows a higher surface charge density obtained at the surface of the β-phase [25].

11.3.2 Enhancement of the Charge Generation

11.3.2.1 Physical Modification

The charge density could be easily and effectively increased by physical and chemical modification of the friction layer. Physical modification methods are generally effective in improving friction efficiency by constructing rough surfaces or increasing the friction area [26–29]. It was one of the first methods used and is still an effective tool today. Back in 2012, Fan et al. enhanced the friction effect by designing the polydimethylsiloxane (PDMS) array with different patterns [30]. Electrospun nanofibrous mats with high specific surface area and multilayered structure represent a simple way to build rough surfaces. Wang et al. reported the formation of nanofibrous mats with high surface roughness by controlled electrospinning, collected in Marangoni flow-driven water. The single-electrode mode TENG prepared by hierarchical roughness morphology has a high output power density of $3.37\ \mathrm{W\ m^{-2}}$ [31]. Moreover, the construction of porous structures is also considered an effective way to improve TENG output performance by physical methods [32].

11.3.2.2 Chemical Modification

Modifications of the functional groups have been considered a facile method to improve the electrical polarity of triboelectric materials. Enhancing the polarity of the tribological layer could maximize the polarity gap of the friction layer. First, the frictional charged nature of the polymer was determined by the polar groups on its molecular chain segments. For example, the obvious electrical polarity was brought by fluorine groups [33, 34], oxygen groups [35], amino groups [36], and other polar groups. Zhang et al. [37] modified electrospun poly-ε-caprolactone (PCL) by poly(ethylene glycol) methyl ether (mPEG), which possesses more O atoms, as shown in Figure 11.2. The results show that not only mPEG but also many oxygen-rich polymers have a stronger electron-donating ability, which is quite suitable for enhancing the positive trip materials.

11.3.2.3 Enhancement of the Dielectric Polarization

Dielectric polarization can be divided into electronic, vibrational (or atomic), orientational (or dipolar), ionic, and interfacial polarizations [38, 39]. The

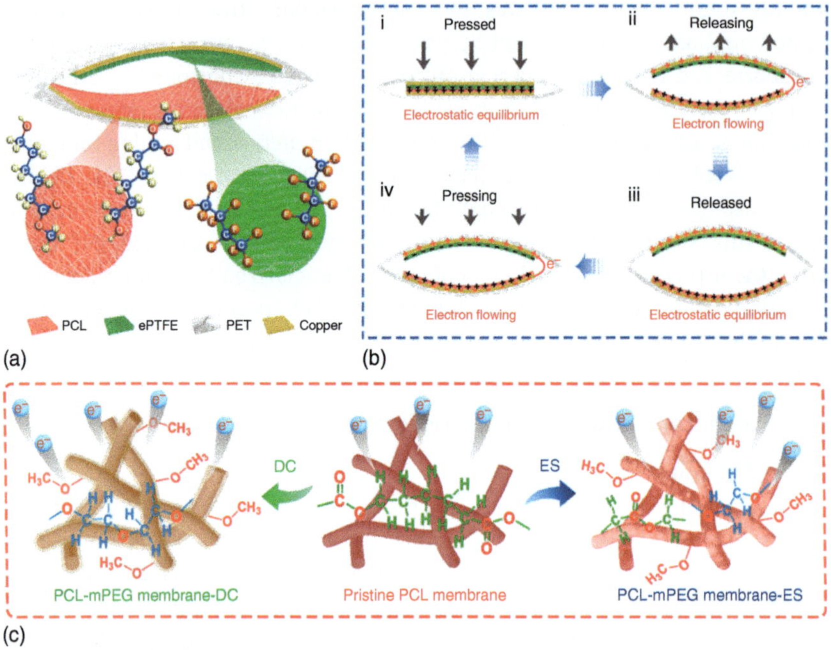

Figure 11.2 Schematics of the structure and working principles of the TENGs. (a) Structure of a PCL/ePTFE TENG. (b) Electricity generation mechanism of the TENG. (c) Schematic illustration of the pristine PCL membrane and PCL-mPEG membranes using dip coating (DC) and electrospinning (ES) methods. Source: Reproduced with permission Zhang et al. [37]. Copyright 2019, American Chemical Society.

dielectric permittivity of polymer nanocomposites would be enhanced by utilizing the polarization of high dielectric permittivity nanoparticles at the polymer–nanoparticle interface [40].

Considering the influence of the dipole formed between the surface charges on the two contact films, the electric field direction of the ferroelectric dipole should match the electric field direction of the frictional electric surface charges to achieve the maximum electric field strength and electrostatic induction. When the direction of the electric field generated by the friction charge and the ferroelectric dipole is matched, the electric field strength summarizes and amplifies the electrostatic induction on the TENG electrode [41]. In the case of PVDF derivatives, the dipole orientation leads to the formation of β-phase, which increases the dielectric constant and thus improves the triboelectric properties [42].

11.3.3 Reduce the Charge Loss

11.3.3.1 Introduce the Charge Trap Layer

In the process of generating charge by friction, some of the charges are dissipated; for example, an interfacial electric field is formed between the bottom electrode and the negative friction material, so that electrons could easily diffuse into the negative friction material and neutralize each other with the induced positive charge [43, 44]. The construction of a charge trap layer for charge capture can effectively reduce charge loss, enhancing the output performance [45].

11.3.3.2 Circuit Finishing

In many cases, the electric output of the TENG could not be directly used for charging and discharging the capacitor. Moreover, a certain adjustment is required to the electric output of the TENG, such as performing a conversion from AC to DC. The TENG could be a stable power source [41, 46, 47], but in such a process, it will inevitably lead to a part of energy loss.

Typically, TENG has high-voltage and low-output current, which makes it uniquely advantageous in areas such as adsorption and filtration but not applicable for low-voltage electronics. Therefore, buck conversion of TENG devices is an attractive direction for power output management. Zhu et al. constructed a power management circuit consisting of a transformer, rectifier, voltage regulator, and capacitor, which can provide a DC output of 5 V constant voltage in less than 0.5 s after the TEG starts to operate [48]. Finally, the integration of the supercapacitor with TENG into a complete charge–discharge recyclable energy system is a promising endeavor [49, 50].

11.4 Electrospun Nanofiber-Based TENG for Energy Harvesting

TENGs are envisioned as an advanced technology for harvesting diverse environmental energy into electrical energy due to the ease of fabrication and versatile material selection, and also the advantages such as simple structure, stability, and environmental friendliness [50, 51]. In addition, TENGs generate high output voltage and high power density, especially efficient collection of low-frequency energy, making it possible to collect some widespread and hardly collected energy, such as human motion energy [52] and wave energy [53].

With high energy density and high adaptability, electrospun nanofiber-based TENG can fully satisfy energy harvesting efforts in multiple scenarios by tuning the preparation process and a variety of structural designs [54, 55].

11.4.1 Human Motion Energy

The internet of things (IoT) era has not only brought about huge changes in the manufacturing industry but is also reflected in the rise of wearable devices. Utilizing human motion energy to power these wearable devices, which can achieve long-lasting wearability and avoid the problem of battery life. Particularly the high flexibility of electrospinning nanofibers, which are suitable for a wide range of limb movements in the human body.

11.4.1.1 Body Movement

The human body is a complex and ordered system with the external limbs moving in different ways. Researchers have focused on the main action limbs of the human body, such as the neck, knee, finger, and feet, which produce larger motion energy [15].

For neck, Lou et al. designed an all-fiber structure pressure sensor based on vertical CS mode TENG [56]. The triboelectric layer consisted of electrospun PVDF/AgNW nanofibrous membranes and ethylcellulose nanofibrous membranes. This pressure sensor demonstrates a pressure sensitivity of 1.67 and 0.20 V kPa^{-1} in the pressure range of 0–3 and 3–32 kPa, respectively, and it showed mechanical stability after continuous operation after 7200 working cycles. The self-powered pulse sensor was capable of being placed on the neck for real-time pulse monitoring.

For fingers, Graham et al. utilized the biocompatible and biodegradable electrospinning polymers polyvinyl alcohol (PVA) and PCL as positive and negative friction materials for fabricated TENG-based sensors [57]. The triboelectric energy harvester and sensor (TEHS) has a high sensitivity and a fast response time of 1.7 ms, and it is also sensitive to simple and small pressure changes generated by finger toughness. In addition, a matrix consisting of self-powered sensors has been proven in various real-time sensing applications and wireless transmission, and it has also been integrated into various health monitoring systems to track and monitor the physical condition of patients.

In addition, the fingers have been commonly used as a friction material for TENGs like electronic skin that could harvest electricity and achieve interactive functions through the touch of the fingers. Zhou et al. designed a flexible self-powered electronic skin (e-skin) based on an ultrastretchable frictional electric nanogenerator (STENG) using electrospun multilayer thermoplastic polyurethane (TPU)/silver nanowires (AgNWs)/reduced graphene oxide (rGO) [58]. The e-skin remains at a full angularity on the surface of the finger even when the finger was bent to a certain degree, exhibiting great flexibility and skin comfort and very sensitive to the intensity of touch. While attached to human skin, it could sense the touch of the finger and obtain electrical energy.

For foot, Kim et al. [59] prepared electrospun ferroelectric poly(vinylidene fluoride trifluoride) (PVDF-TrFE) nanofibrous mats and then partially doped the PDMS layer with the nanofibrous mats as a charge-trapping layer to create an integrated negative friction interface assembly. The nanofiber-doped friction surface significantly improves the frictional charge output performance through the charge trapping effect, minimizing the intrinsic loss of frictional charge. Moreover, with three-dimensional (3D) printing adapted to the plantar interface, the synergy of 3D curvature and soft/hard mismatch of the friction interface improves the friction efficiency and durability. The 3D customized TENG generated a high output voltage and current of 880 V and 3.75 mA, respectively.

At present, TENG is a very promising device for wearable electronics that could achieve a long-lasting energy supply. However, researchers still need to overcome the high cost and short lifetime of the corresponding chemical batteries in the future. On the other hand, how to effectively integrate energy storage systems, such as utilizing supercapacitors and loads to form an energy harvesting–storage–consumption cycle is a challenge that needs to be addressed in the future.

11.4.1.2 Human Breath

Based on frictional electricity and electrostatic induction, TENG could generate high open circuit voltages and not ionize the surrounding air, so no ozone is produced [60, 61]. Which makes TENG a significant application prospect in air purification. Meanwhile, the in situ high-voltage polarization during the electrospinning process enables the nanofibrous filter to exhibit a stronger electrostatic effect, thus increasing the efficiency of capturing airborne particles [62].

Hao et al. [60] introduced a self-powered triboelectric air filter (STAF) consisting of a PU conductive sponge electroplated with copper, nickel, and electrospun PAN/PTFE fibers, which constitutes a respiratory triboelectric nanogenerator (R-TENG) by coupling the electrical difference between the two materials and the breathing airflow. This R-TENG works in vertical contact–separation modes driven by breathing airflow. STAF could maintain a filtration efficiency of 98% for $PM_{2.5}$ and 91.5% for $PM_{0.5}$, which exceeds the level of commercial masks.

11.4.2 Renewable Energy

Since the twenty-first century, with a strong awareness of environmental protection, efficient collection of renewable energy is considered as the most critical technology. Currently, the harvesting of renewable energy is dominated by electromagnetic generators, but TENG has its unique advantage in a part of low-frequency energy. First, the output voltage of an electromagnet generator (EMG) is extremely low at low frequencies, while the output voltage of TENGs is independent of frequency, which is much more convenient for disordered energy collection than EMG [8]. Second, the study demonstrates the existence of a threshold frequency for the preexisting one, below which the output power of TENG is higher than EMG [63]. In 2014, theoretical comparison and experimental verification established that TENG could serve as the basis for new energy technology, possibly alongside or possibly equally important to EMG [64], with the prospect of serving as an effective device for low-frequency energy harvesting in the future.

Here, we have to mention the characteristics of electrospun nanofiber-based TENG: easy preparation, flexible structure, and high efficiency in the work scenario of high sensitivity needs, but also the existence of the disadvantages of mechanical properties, and poor abrasion resistance for long-term operation. Therefore, the basic structure of large-scale energy harvesting still needs certain improvements.

11.4.2.1 Airflow Energy

Wind power is one of the most abundant sources of renewable energy on earth. Traditional wind generators produce significant electrical power with huge turbine fan blades driving electromagnetic generators, which is an important way of collecting renewable energy at present. However, the frequency of wind power in nature varies greatly, and there are many areas where low wind speed limits wind power harvesting. TENGs enable a flexible collection of wind energy with variable frequency and small energy compared to large-scale EMG power collection, which we called wind energy as airflow energy, which is more appropriate for the collection of TENG.

Yang et al. designed a sandwich-like TENG for wind energy harvesting and utilized the high voltage, low current characteristics of the TENG for emulsion separation applications of water and oil [65]. Electrospun PVDF nanofibers are used as a flexible vibrating layer and nylon nanofibers as a bonding layer. The PI film is inserted as a supporting and transition layer between the conductive layer and the basic triboelectric layer. The output voltage was 1400 V and the trigger frequency was 28 Hz, driven by the wind with a wind speed of $8\ m\ s^{-1}$. The emulsion consisted of 0.25 g of sodium dodecyl benzene sulfonate (SDBS) dissolved in 50 ml of deionized water and then blended into 450 ml of lubricating oil. After 30 minutes of separation under the TENG drive, a large number of droplets were separated from the water–oil (W/O) emulsion and deposited at the bottom of the vessel with a free water volume of about 49.1 mL, and the color of the W/O emulsion changed from milky white to light yellow, close to the color of pure lubricating oil.

Phan et al. utilized aerodynamic and aeroelastic flutter-driven TENGs successfully to harvest broadband airflow energy [66]. The single-unit component of the chattering diaphragm consists of a separate thin aluminum foil electrode covered with electrospun polyvinyl chloride (PVC) nanofibrous mats on both sides of the electrode to provide a beneficial triboelectric surface and enhance the friction area. The output performance of a flutter-membrane-based triboelectric nanogenerator (FM-TENG) could be enhanced by stacking multiple layers of flexible tremolo with contact and separation to drive the electrons back and forth.

Furthermore, wind energy has also been considered as a convenient source of energy for outdoor areas, such as the self-powered monitoring of plants in agriculture [67].

11.4.2.2 Rain Droplet Energy

Like wind power, large-scale hydrokinetic energy harvesting through EMG is commonly used in rivers, reservoirs, and so on. Raindrops, which are often dispersed and have relatively low kinetic energy, have been considered difficult to harvest energy. TENG makes this energy available for harvesting and has been explored [68, 69].

Zeng et al. developed a liquid–solid-TENG (LS-TENG) for efficient water/rain droplet energy harvesting and analyzed the response signals generated by droplets, water flow, and pouring water in contact with the LS-TENG [70]. The impact of droplets on the LS-TENG and the electrostatic induction of droplets flowing down the FEP film generate two response signals. In addition, the LS-TENG can be used as a velocity sensor to monitor the droplet's falling speed. Finally, a self-powered window-closing system was designed with the potential for application in smart homes.

11.4.2.3 Sound Energy

Sound energy is a kind of renewable energy that could be found everywhere, such as in factory machinery and vehicle engines. It is probably one of the most underestimated energies. It is feasible to design a reasonable sound harvesting device by taking advantage of the low-frequency energy collected by TENG, and unique applications in specific areas, such as cochlear implants [71] and waveform recognized [31].

The high specific surface area and high flexibility of electrospun nanofibers allow them to respond to slight environmental changes, making them more applicable than other generators for harvesting weak and variable energy such as sound. Shao et al. composed two gold-plated PET films sandwiched between electrospun PAN films to form an acoustoelectric [72]. Under 117 dB noise, a 3–4 cm^2 nanofibrous membrane device could generate up to 58 V with 210.3 μW (surface power density of 17.53 $\mu W\ cm^{-2}$), the generated power was sufficient to power commercial electronics.

11.4.3 Mechanical Vibration Energy

The development of IoT technology brings a lot of innovation, such as intelligent workshops and intelligent driving. Although the power required for operating each sensor is small, typically in the microwatt-to-watt range, the number is so large. One possible solution is to make each device self-powered by harvesting energy from vibrations [73] such as machine vibrations and vehicle movements. Varghese et al. reported a TENG based on electrospun cellulose acetate nanofibers and fabricated surface-modified polydimethylsiloxane that was capable of harvesting vibrational energy to power commercial sensors [29]. This self-powered vibration sensor could clearly distinguish the operation of the sewing machine at different frequencies, and the complete mapping of the vibration of equipment could be easily achieved. In addition, vibration analysis using a self-powered vibration sensor is also feasible for the detection of faults in different machinery and equipment; for example, error detection for hard disk drives mainly comes from the fan or the drive itself.

Yang et al. proposed a self-powered triboelectric sensor (CN-STS) fabricated by electrospun composite nanofibers for intelligent traffic monitoring and management [74]. The sensors are sensitive enough to sense differences in output signals and distinguish between pedestrian and motor vehicle signal outputs and detect unbalanced moving vehicles and overlapping vehicles. The average speed of a vehicle could be detected by two CN-STS device systems, and the recorded speed is more reasonable than the instantaneous speed from a radar speedometer or GPS speed detection of complex roads or communities. Moreover, it is synchronized with the database and constitutes a very promising IoT component.

11.5 Conclusion and Prospect

TENGs constitute an important breakthrough in the field of energy, which could effectively complement the absence of micro-/nanoscale energy in the existing components of energy harvesting. Electrospun nanofiber-based TENG has the advantages of wide material selection, high flexibility, and large specific surface area to prepare highly sensitive self-powered devices to collect some renewable but often wasted energy such as human activities, wind energy, and mechanical vibration energy. Meanwhile, it can also be scaled up to macroscopic energy harvesting by combining reasonable material selection and structural design.

As a promising energy technology, more research is still needed to support the large-scale application. On the one hand, the output performance of TENG needs to be further improved. On the other hand, the durability of electrospun TENG needs to be optimized for practical applications and needs enhancement. It is the common expectation of all researchers that TENG will come into people's lives soon.

References

1 Mahapatra, B., Kumar Patel, K., Vidya, and Patel, P.K. (2021). *Materials Today: Proceedings* 46: 5523.
2 Zhang, C., Fan, W., Wang, S. et al. (2021). *ACS Applied Electronic Materials* 3: 2449.
3 Scheffler, S. and Poulin, P. (2022). *ACS Applied Materials & Interfaces* 14: 16961.
4 Khandelwal, G. and Dahiya, R. (2022). *Advanced Materials* 2200724.
5 Zhang, R. and Olin, H. (2020). *EcoMat* 2: https://doi.org/10.1002/eom2.12062.
6 Korkmaz, S. and Kariper, İ.A. (2021). *Nano Energy* 84: 105888.
7 Askari, H., Xu, N., Groenner Barbosa, B.H. et al. (2022). *Materials Today* 52: 188.
8 Wang, Z.L., Jiang, T., and Xu, L. (2017). *Nano Energy* 39: 9.
9 Win Zaw, N.Y., Yun, J., Goh, T.S. et al. (2022). *Energy* 247: 123422.
10 Chen, P., An, J., Shu, S. et al. (2021). *Advanced Energy Materials* 11: 2003066.
11 Lei, R., Shi, Y., Ding, Y. et al. (2020). *Energy & Environmental Science* 13: 2178.
12 Arica, T.A., Isık, T., Guner, T. et al. (2021). *Macromolecular Materials and Engineering* 306: 2100143.
13 Fang, J. and Lin, T. (ed.) (2019). *Energy Harvesting Properties of Electrospun Nanofibers*. IOP Publishing.
14 Wang, Z.L. and Wang, A.C. (2019). *Materials Today* 30: 34.
15 Zou, Y., Raveendran, V., and Chen, J. (2020). *Nano Energy* 77: 105303.
16 Fan, F.R., Tang, W., and Wang, Z.L. (2016). *Advanced Materials* 28: 4283.
17 Wu, C., Wang, A.C., Ding, W. et al. (2019). *Advanced Energy Materials* 9: 1802906.
18 Kim, W.-G., Kim, D.-W., Tcho, I.-W. et al. (2021). *ACS Nano* 15: 258.
19 Khandelwal, G., Raj, N.P.M.J., and Kim, S.J. (2020). *Nano Today* 33: 100882.
20 Chen, J. and Wang, Z.L. (2017). *Joule* 1: 480.
21 Shaikh, M.O., Huang, Y.-B., Wang, C.-C., and Chuang, C.-H. (2019). *Micromachines* 10: 438.
22 Kim, S.-R., Yoo, J.-H., and Park, J.-W. (2019). *ACS Applied Materials & Interfaces* 11: 15088.
23 Zhang, J., Wang, H., Blanloeuil, P. et al. (2020). *Composites Communications* 22: 100535.
24 Sha, Z., Boyer, C., Li, G. et al. (2022). *Nano Energy* 92: 106713.

25 Cheon, S., Kang, H., Kim, H. et al. (2018). *Advanced Functional Materials* 28: 1703778.

26 Zhao, Z., Wei, B., Wang, Y. et al. (2022). *Nanomaterials* 12: 721.

27 Bai, L., Li, Q., Yang, Y. et al. (2021). *Research* 2021: 1.

28 Fahad, M.M., Reza, M.S., Prasad, G. et al. (2022). *Fibers and Polymers*. https://doi.org/10.1007/s12221-022-4969-3.

29 Varghese, H., Hakkeem, H.M.A., Chauhan, K. et al. (2022). *Nano Energy* 98: 107339.

30 Fan, F.-R., Lin, L., Zhu, G. et al. (2012). *Nano Letters* 12: 3109.

31 Wang, S., Tian, M., Hu, S. et al. (2022). *Nano Energy* 97: 107149.

32 Mi, H.-Y., Jing, X., Meador, M.A.B. et al. (2018). *ACS Applied Materials & Interfaces* 10: 30596.

33 Rana, S.M.S., Rahman, M.T., Salauddin, M. et al. (2021). *ACS Applied Materials & Interfaces* 13: 4955.

34 He, Y., Wang, H., Sha, Z. et al. (2022). *Nano Energy* 98: 107343.

35 Saqib, Q.M., Chougale, M.Y., Khan, M.U. et al. (2021). *Nano Energy* 89: 106458.

36 Zhang, J., Hu, Y., Lin, X. et al. (2022). *Carbohydrate Polymers* 291: 119586.

37 Zhang, Z., Gong, W., Bai, Z. et al. (2019). *ACS Nano* 13: 12787.

38 Baer, E. and Zhu, L. (2017). *Macromolecules* 50: 2239.

39 Zhu, L. (2014). *Journal of Physical Chemistry Letters* 5: 3677.

40 Kim, M.P., Um, D.-S., Shin, Y.-E., and Ko, H. (2021). *Nanoscale Research Letters* 16: 35.

41 Wang, Y., Jin, X., Wang, W. et al. (2021). *ACS Applied Electronic Materials* 3: 532.

42 Choi, G.-J., Baek, S.-H., Lee, S.-S. et al. (2019). *Journal of Alloys and Compounds* 797: 945.

43 Park, H.-W., Huynh, N.D., Kim, W. et al. (2018). *Nano Energy* 50: 9.

44 Kim, D.W., Lee, J.H., You, I. et al. (2018). *Nano Energy* 50: 192.

45 Hwang, H.J., Hong, H., Cho, B.G. et al. (2021). *Nano Energy* 90: 106647.

46 Qian, Z., Li, R., Guo, J. et al. (2019). *Nano Energy* 64: 103900.

47 Qiao, G., Wang, J., Yu, X. et al. (2021). *Nano Energy* 79: 105408.

48 Zhu, G., Chen, J., Zhang, T. et al. (2014). *Nature Communications* 5: 3426.

49 Mao, Y., Li, Y., Xie, J. et al. (2021). *Nano Energy* 84: 105918.

50 He, W., Fu, X., Zhang, D. et al. (2021). *Nano Energy* 84: 105880.

51 Cheng, K., Wallaert, S., Ardebili, H., and Karim, A. (2022). *Carbon* 194: 81.

52 Goswami, S., dos Santos, A., Nandy, S. et al. (2019). *Nano Energy* 60: 794.

53 Rodrigues, C., Ramos, M., Esteves, R. et al. (2021). *Nano Energy* 84: 105890.

54 Abbasipour, M. and Khajavi, R. (2013). *Advances in Polymer Technology* 32.

55 Reddy, V.S., Tian, Y., Zhang, C. et al. (2021). *Polymers* 13: 3746.

56 Lou, M., Abdalla, I., Zhu, M. et al. (2020). *ACS Applied Materials & Interfaces* 12: 1597.

57 Graham, S.A., Patnam, H., Manchi, P. et al. (2022). *Nano Energy* 107455.

58 Zhou, K., Zhao, Y., Sun, X. et al. (2020). *Nano Energy* 70: 104546.

59 Kim, Y.W., Lee, H.B., Yoon, J., and Park, S.-H. (2022). *Nano Energy* 95: 107051.

60 Hao, R., Yang, S., Yang, K. et al. (2021). *ACS Applied Energy Materials* 4: 14700.

61 Gu, G.Q., Han, C.B., Lu, C.X. et al. (2017). *ACS Nano* 11: 6211.

62 Wang, L., Bian, Y., Lim, C.K. et al. (2021). *Nano Energy* 85: 106015.
63 Zi, Y., Guo, H., Wen, Z. et al. (2016). *ACS Nano* 10: 4797.
64 Zhang, C., Tang, W., Han, C. et al. (2014). *Advanced Materials* 26: 3580.
65 Yang, D., Feng, Y., Wang, B. et al. (2021). *Nano Energy* 90: 106641.
66 Phan, H., Shin, D.-M., Heon Jeon, S. et al. (2017). *Nano Energy* 33: 476.
67 Li, X., Jiang, C., Zhao, F. et al. (2019). *Nano Energy* 61: 78.
68 Yoo, D., Park, S.-C., Lee, S. et al. (2019). *Nano Energy* 57: 424.
69 Cheng, B., Niu, S., Xu, Q. et al. (2021). *ACS Applied Materials & Interfaces* 13: 59975.
70 Zeng, Y., Luo, Y., Lu, Y., and Cao, X. (2022). *Nano Energy* 98: 107316.
71 Liu, Y., Zhu, Y., Liu, J. et al. (2018). *Nanoscale Research Letters* 13: 191.
72 Shao, H., Wang, H., Cao, Y. et al. (2020). *Nano Energy* 75: 104956.
73 Wang, Z.L. (2019). *Nano Energy* 58: 669.
74 Yang, X., Liu, G., Guo, Q. et al. (2022). *Nano Energy* 92: 106757.

12

Preparation and Application of Thermoelectric Materials and Devices Based on Electrospun Fibers

Xinyang He and Liming Wang

Donghua University, Key Laboratory of Textile Science & Technology, Ministry of Education, College of Textiles, 2999 Renmin North Road, Songjiang District, Shanghai 201620, China

12.1 Introduction

With the rapid development of wearable electronic devices, the demand for flexible and wearable energy sources was increasing [1–3]. Wearable electronics mostly focus on portability, high flexibility, and certain ductility and are expected to operate in a self-powered mode by converting human energy into electrical energy [4–7]. Therefore, emerging energy conversion technologies such as piezoelectric [8], triboelectric [9], thermoelectric [10], and photovoltaic [11] are expected to become important ways of energy supply for wearable electronic devices. Among them, thermoelectric materials, which can directly convert low-grade heat into electricity without involving moving parts or dangerous working fluids, have attracted extensive attention in the field of energy harvesting [12–15]. In particular, thermoelectric materials can directly convert the temperature difference between the human body and the environment into electrical energy and drive the operation of wearable electronic devices. The Figure of merit (zT) is used to evaluate the thermoelectric properties of a material and is defined as $zT = S^2\sigma T/\kappa$, where S is the Seebeck coefficient, σ is the electrical conductivity, κ is the thermal conductivity, and T is the absolute temperature. In the past few decades, conventional inorganic thermoelectric materials (composed of inorganic thermoelectric alloys such as Bi_2Te_3, PbTe, and $CoSb_3$), conducting polymers and derivatives, including poly(3,4-ethylenedioxythiophene):poly(styrene sulfonate) (PEDOT:PSS), polyaniline (PANI), polypyrrole (PPy), and poly(3-hexylthiophene) (P3HT) have been intensively studied [16–20].

At present, due to the rapid development of Internet of Things (IoT) technology and comfortable electronic devices, the urgent need for environmentally friendly, stable, and portable power sources has driven the exploration of flexible wearable thermoelectric devices [21, 22]. Most of the traditional thermoelectric materials and devices have been difficult to meet the needs of portable devices for lightness, flexibility, and comfort. Fiber-based thermoelectric materials with outstanding

Electrospinning: Fundamentals, Methods, and Applications, First Edition.
Edited by Liming Wang and Xiaohong Qin.

flexibility and mechanical properties have received increasing attention [23–25]. In addition, fiber-based thermoelectric devices can be tightly attached to the human skin and can be directly converted by the temperature difference between the human body and the environment to generate maintenance-free electrical energy. Wu et al. [26] used polyester/cotton yarn as the substrate and coated it with a composite water-soluble thermoelectric coating of polyurethane (PU), multiwall carbon nanotubes (MWCNT) and PEDOT:PSS to obtain a thermoelectric fiber with good flexibility. At room temperature, the thermoelectric fiber exhibits a maximum conductivity of $138.26\,S\,cm^{-1}$ and a maximum Seebeck coefficient of $10\,\mu V\,K^{-1}$. Thermoelectric fibers prepared by solution coating method can obtain excellent physical and mechanical properties with the choice of substrate. Wen et al. [27] reported a high-performance PEDOT:PSS thermoelectric fiber that was prepared by a modified wet spinning method. The optimum Seebeck coefficient of PEDOT:PSS fiber reaches $19.2\,\mu V\,K^{-1}$ at room temperature, and has a maximum elongation at break of 30.5%. The thermoelectric generator, composed of five pairs of PEDOT:PSS fibers and nickel wires, can achieve an output power density of $0.323\,\mu W\,cm^{-2}$ at a temperature difference (ΔT) of 10 K.

Electrospinning technology has been shown to be an efficient method for the continuous production of nanofibers with diameters in the range of 100–1000 nm. Nanofibers prepared by electrospinning have many excellent properties such as high aspect ratio, large specific surface area, highly tunable surface morphology, and excellent mechanical properties. At present, electrospun nanofibers have been widely studied and applied in the field of wearable electronic devices based on energy harvesting and conversion. In particular, in recent years, researchers have begun to focus on the intersection of electrospinning and thermoelectric fields, because the unique flexibility of electrospinning fibers can help thermoelectric materials fit to complex human surfaces, and good air permeability and comfort give great experience for the wearer.

12.2 Design and Fabrication of Thermoelectric Materials Based on Electrospinning

Recently, electrospinning technology is widely combined with physical vapor deposition, vacuum pumping, spraying, and high-temperature calcination to prepare thermoelectric fiber. Due to the high aspect ratio, large specific surface area, highly adjustable surface shape, and excellent mechanical properties of the thermoelectric fibers prepared by electrospinning technology, it exhibits great application prospects.

12.2.1 Vacuum Filtration

He et al. [28] reported flexible PEDOT:PSS/CNT thermoelectric composite films with extremely high stretchability. The PEDOT:PSS/CNT composite films were prepared onto electrospun PU/PCL nanofiber films by vacuum filtration method followed by hot pressing (Figure 12.1a). The PU nanofiber film was used as a stretchable flexible framework, while the PCL nanofiber was used as a binder to improve the interfacial stability between PEDOT:PSS/CNT and PU nanofiber films.

The thermoelectric properties of the composites were influenced by the weight ratios of PEDOT:PSS/CNT. The highest electrical conductivity reaches 1581 S m^{-1} when the weight ratio of PEDOT:PSS/CNT was 7 : 3 (Figure 12.1b). The change law of power factor for composite films was basically the same as the conductivity, and the maximum value is 1.9 μW m^{-1} K^{-2} (Figure 12.1c). All the fabricated PEDOT:PSS/CNT composite films showed an ultrahigh fracture strain of more than 400% which was similar to the pristine PU nanofiber film (Figure 12.1d). The resistance of the sample showed a positive response with increasing strain, indicating its potential as a strain sensor (Figure 12.1e).

12.2.2 Alternate Spraying

He et al. [29] reported an advanced preparation strategy combining electrospinning and spraying technology to prepare CNT/polyvinyl pyrrolidone (PVP)/PU composite thermoelectric fabrics. Composite fabrics were obtained via manipulating electrospinning and spraying alternately (Figure 12.2a). Among them, PU nanofiber film was used as a stretchable skeleton, and CNTs were sprayed on its surface as thermoelectric materials. Due to the dispersion effect of PVP, it can be seen that CNTs are uniformly dispersed on the fiber surface (Figure 12.2b). And the fabric exhibits excellent breathability and flexibility (Figure 12.2c,d).

The maximum electrical conductivity of the composite fabric was 20 S cm^{-1}, which was obtained when the CNT/PVP ratio of the composite fabric arrived at 7 : 3 (Figure 12.2e). The composite fabric with CNT/PVP ratio of 7 : 3 was chosen for subsequent testing because of its best thermoelectric performance. A device composed of five composite fabrics (5 × 15 mm) in series was designed to further demonstrate its power generation capability. When the temperature difference of 4, 8, and 12 K was applied, the maximum output voltage of the device can reach 1, 2, and 3 mV, respectively (Figure 12.2f). As shown in Figure 12.2g, when the temperature difference of 4, 8, and 12 K was applied, the maximum output power can reach 64, 261, and 586 pW, respectively, and the corresponding load resistance (R_{load}) was matched well with the internal resistance of the device.

12.2.3 Coagulation-Bath Electrospinning

Coagulation-bath electrospinning is an emerging method combining electrospinning and wet spinning [30]. Usually, the polymer is dissolved in an appropriate solvent to obtain a solution with a certain viscosity and good spinnability, and then a voltage is applied to the spinneret to deposit the nanofibers into a coagulation bath to obtain continuous fibers.

He et al. [31] reported that an advanced fabrication approach combining coagulation-bath electrospinning and self-assembly strategies is proposed to efficiently and continuously fabricate CNT/PEDOT:PSS thermoelectric nanofiber

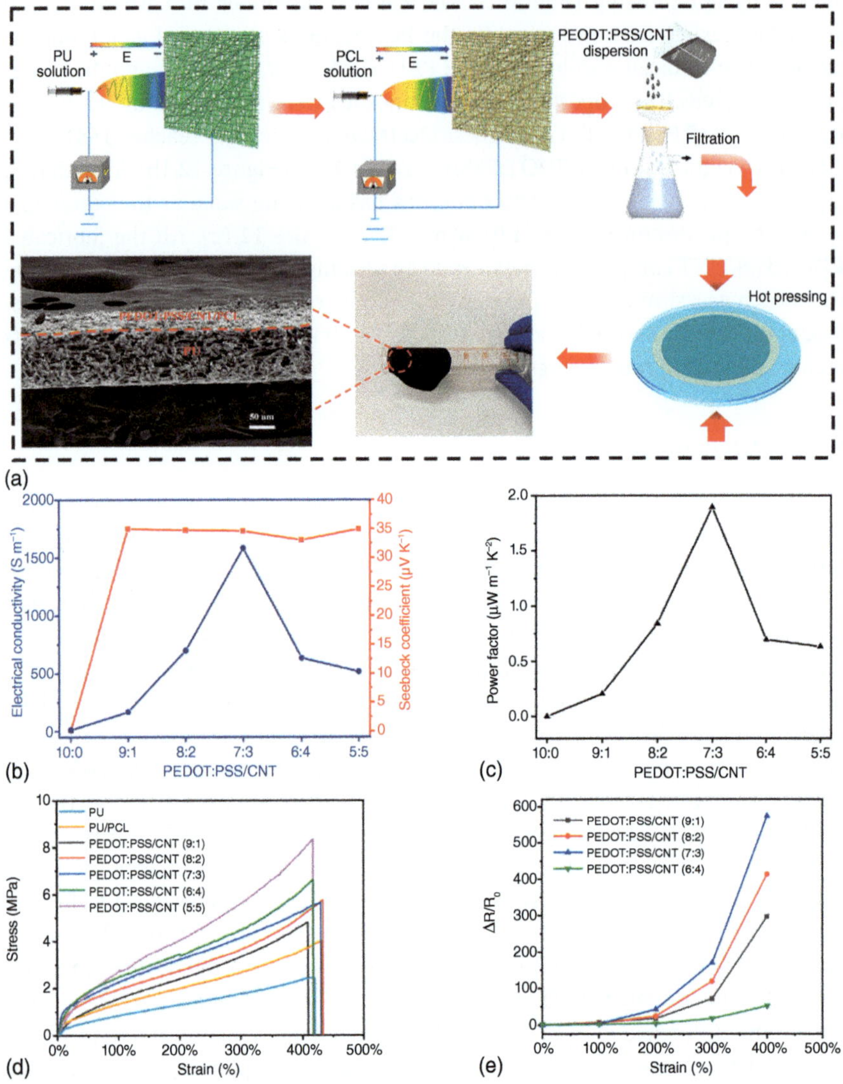

Figure 12.1 (a) Preparation procedure and cross-sectional scanning electron microscopy (SEM) image of the stretchable PEDOT:PSS/CNT composite film. (b) The electrical conductivity and Seebeck coefficient of prepared PEDOT:PSS/CNT composite films with different weight ratios of PEDOT:PSS/CNT. (c) The power factor of prepared PEDOT:PSS/CNT composite films with different weight ratios of PEDOT:PSS/CNT. (d) Stress–strain curves of PU nanofiber films, PU/PCL nanofiber films, and PEDOT:PSS/CNT composite films. (e) Resistance changes of PEDOT:PSS/CNT composite films under different strains, where $\Delta R = R - R_0$, R is the resistance under strain and R_0 is the initial resistance. Source: Reproduced with permission from He et al. [28] / Elsevier.

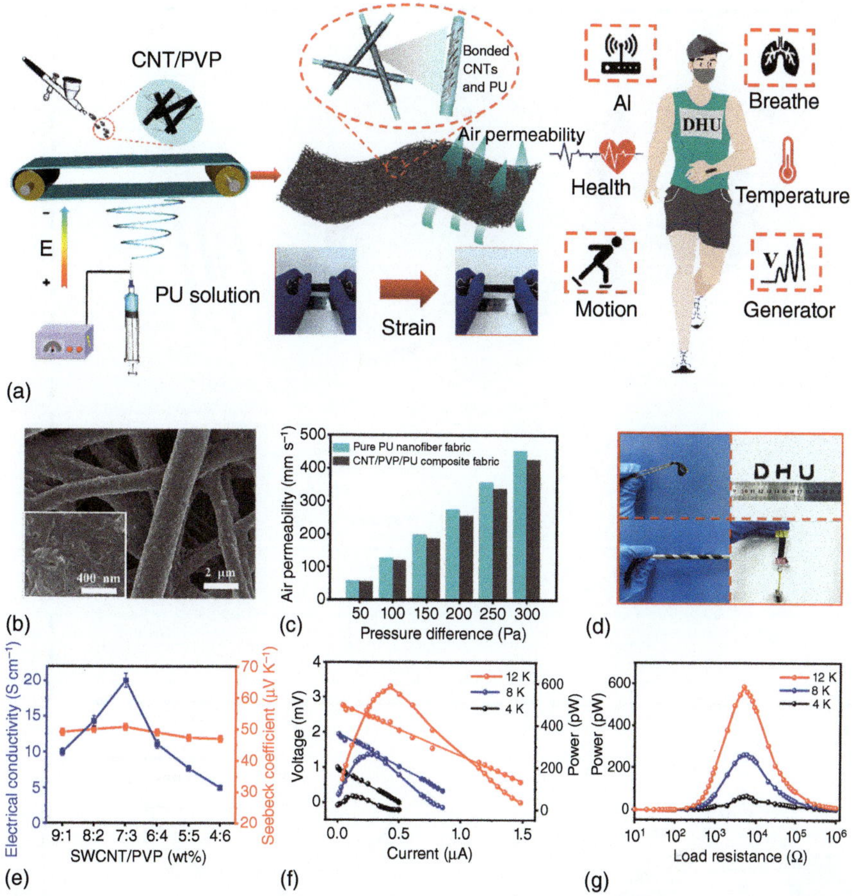

Figure 12.2 (a) The stretchable CNT/PVP/PU composite fabric is prepared by a typical electrospinning process combined with spraying technology and can be applied to wearable thermoelectric generators, human–computer interaction, and health monitoring. (b) SEM of CNT/PVP/PU composite fabric and CNTs distribution on a single fiber. (c) Air permeability of pure PU nanofiber fabric and CNT/PVP/PU composite fabric under different pressures. (d) CNT/PVP/PU composite fabrics can be bent, wound, stretched, and cut into any shape. (e) Electrical conductivity and Seebeck coefficient of CNT/PVP/PU composite fabrics prepared with different CNT/PVP weight ratios. (f) The relationship between output voltage and output power versus current of the device at different temperature differences. (g) The output power of the device as a function of load resistance. Source: Reproduced with permission from He et al. [29] / Elsevier.

yarns. The nanofibers were collected into a coagulation bath, and CNTs were deposited on each nanofiber through a non-solvent-induced phase separation between the residual DMF (*N*,*N*-Dimethylformamide) and the water in the bath (Figure 12.3a). In addition, the cationic polymer PEI (Polyethylenimine) and anionic polymer PSS have unique self-assembly effects in the bath, while PEDOT assembles into thermoelectric nanofiber yarns via adsorption on PSS long chains.

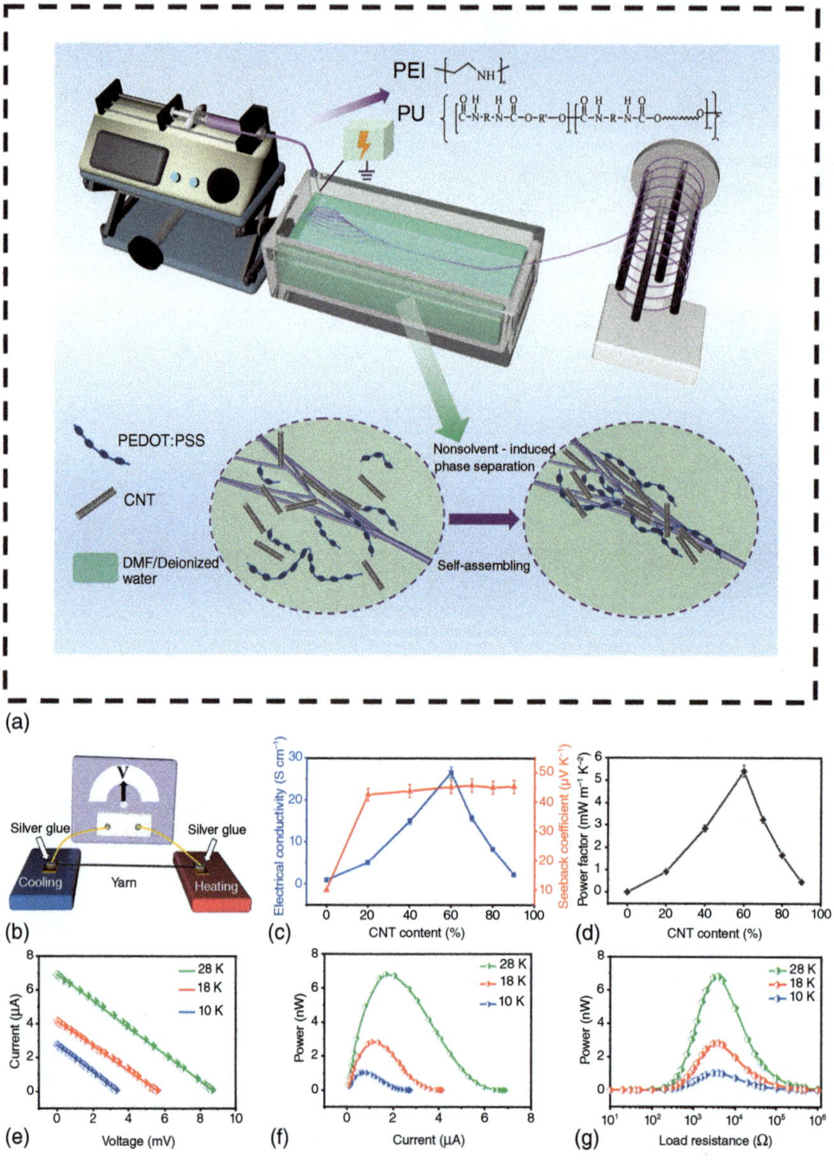

Figure 12.3 (a) Physical images of the preparation procedure of the stretchable thermoelectric nanofiber yarns. (b) Schematic diagram of yarn thermoelectric performance test. (c) Seebeck coefficient and electrical conductivity of thermoelectric nanofiber yarn with different CNT contents. (d) Power factor of thermoelectric nanofiber yarn with different CNT contents. (e) The current–voltage curve of the device at different temperature differences. (f) The power–current curve of the device at different temperature differences. (g) The output power of the device as a function of load resistance at different temperature differences. Source: He et al. [29] / with permission of Elsevier.

A thermoelectric performance testing platform was used to evaluate the thermoelectric performance of the stretchable thermoelectric nanofiber yarns (Figure 12.3b). The thermoelectric properties of nanofiber yarns strongly depended on the ratio of CNT/PEDOT:PSS. The electrical conductivity and the power factor reached the highest when the weight ratio of CNT/PEDOT:PSS was 4 : 6 (Figure 12.3c,d). A thermoelectric device composed of eight thermoelectric nanofiber yarns (15 mm) in series was designed to demonstrated its power generation capability. The current was inversely proportional to the voltage, and the output voltage increased with the increase in the temperature difference (Figure 12.3e). When the device was given a temperature difference of 10, 18, and 28 K, the maximum output powers were 1.02, 2.80, and 6.78 nW, respectively, and the corresponding load resistance (R_{load}) was matched well with the internal resistance of the device (Figure 12.3f,g).

12.2.4 High-Temperature Calcination

High-temperature calcination of the prepared electrospun nanofibers is one of the important methods to prepare high-performance inorganic thermoelectric fibers.

Yin et al. [32] reported a novel technique to process nanocrystalline $Ca_3Co_4O_9$ ceramics with much enhanced thermoelectric properties. Fiber templates ready for calcination were first prepared by sol–gel electrospinning. Inorganic thermoelectric fibers are then obtained at 850 °C. The calcined fibers are in good condition and can be assembled with electrodes for thermoelectric testing. The Seebeck coefficient and conductivity of the samples prepared by the electrospinning method are better than those prepared by the sol–gel method.

Morata et al. [33] proposed an industrially scalable fabrication method for the fabrication of large-area, cost-effective electrospun fiber-based thermoelectric materials. A silicon layer was deposited on the substrate of electrospun carbon nanofibers, and P-type silicon nanotubes were prepared by high-temperature calcination. In a typical process, the electrospun polyacrylonitrile (PAN) nanofibers were annealed to obtain carbon nanofibers (Figure 12.4a). Then, they were used as a sacrificial template, and a polysilicon layer was deposited (Figure 12.4b). After calcination, the carbon core and the silicon oxide shell were removed to generate silicon oxide nanotubes (Figure 12.4c). Finally, a layer of active polysilicon is deposited on the silicon oxide nanotubes (Figure 12.4d). The zT value of this thermoelectric material is 0.34 at 550 °C, which is the highest zT value among silicon-based materials.

12.2.5 Physical Vapor Deposition

Deposition of semiconductor particles on electrospun fiber substrates is a common method for preparing high-performance flexible thermoelectric fibers. Physical vapor deposition, developed in the early twentieth century, uses physical methods (such as evaporation and sputtering) to vaporize the coating material and deposit a film on the surface of the substrate.

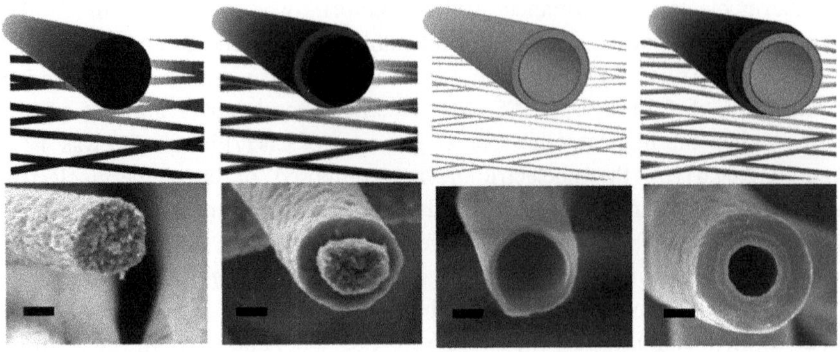

(a) (b) (c) (d)

Figure 12.4 (a) Electrospun carbon fiber. (b) Polysilicon layer coating the carbon fiber. (c) Silicon oxide substrate remaining after the annealing of the carbon core. (d) Polysilicon layer surrounding the silicon oxide substrate. Scale bars are 200 nm long. Source: Reproduced from Morata et al. [33] / Springer Nature / CC BY 4.0.

Lee et al. [15] reported the preparation of flexible thermoelectric yarns based on electrospinning technology and magnetron sputtering. Electrospinning was performed on two parallel wire collectors to prepare PAN nanofibrous membranes with highly aligned structures (Figure 12.5a). Alternating strips of Sb_2Te_3 and Bi_2Te_3 were then sputtered on both sides of the sheet substrate, followed by thermal annealing at 200 °C to crystallize the semiconductors (Figure 12.5b,c). These thermoelectric strips are then interconnected using gold as the metal electrode. Finally, the fiber film is twisted to obtain a flexible thermoelectric yarn with stable structure (Figure 12.5d). The optimal Seebeck coefficients of p-type Sb_2Te_3 fibers and n-type Bi_2Te_3 yarns at room temperature are 178 and $-176\,\mu V\,K^{-1}$, respectively.

Wang et al. [34] reported a stretchable and highly aligned PEDOT:PSS/Bi_2Te_3 hybrid thermoelectric nanofiber film constructed by combining electrospinning and surface coating methods. In a typical process, PU nanofiber films with highly aligned structures were collected on a high-speed rotating roller by electrospinning, followed by thermal evaporation of Bi_2Te_3 layers and dip-coating of PEDOT:PSS layers to prepare aligned PEDOT: PSS/Bi_2Te_3@PU nanofiber film (Figure 12.5e). The electrical conductivity of the nanofiber films increased with increasing dip-coating times, while the Seebeck coefficient gradually decreased (Figure 12.5f,g). The power factor of the aligned hybrid nanofiber films peaked at $50\,\mu W\,m^{-1}\,K^{-2}$ when dipped twice in PEDOT:PSS solution (Figure 12.5h). Compared with the power factor of the random PEDOT:PSS/Bi2Te3@PU nanofiber film, this value is improved by a factor of about 7.

12.2.6 In Situ Synthesis

Jin et al. [35] proposed a new method to optimize the thermoelectric and mechanical properties of the PEDOT:PSS/polyvinyl alcohol (PVA) composite nanofiber

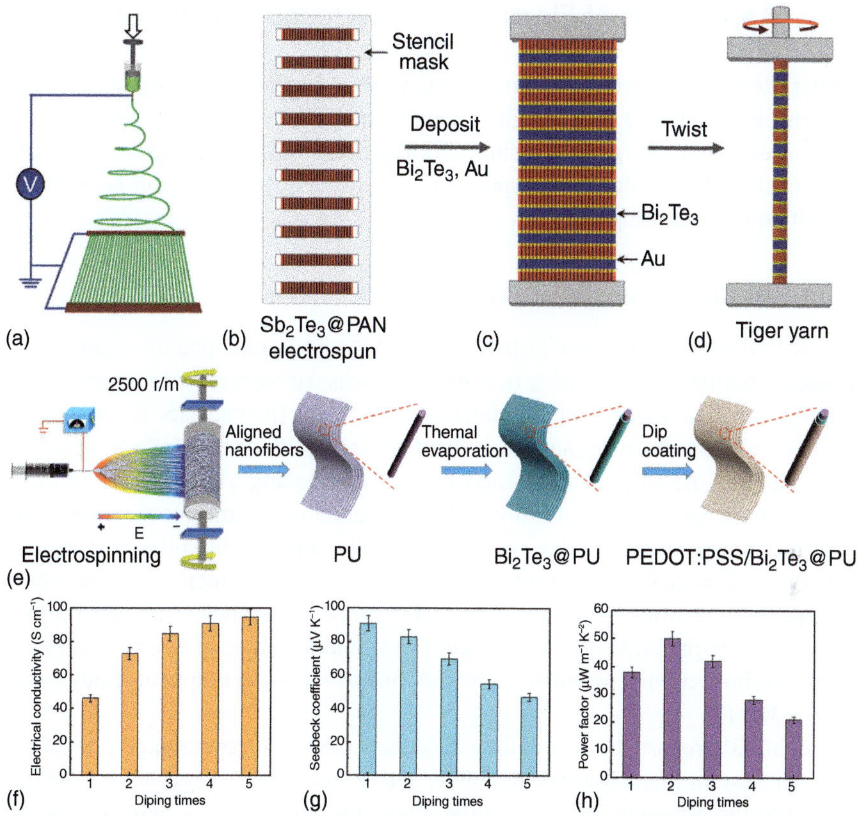

Figure 12.5 (a) Illustration of the electrospinning process. (b and c) Illustration of the stencil mask-based method used to provide alternating p-Au-n segments for tiger yarns by sequentially sputtering Sb_2Te_3-gold-Bi_2Te_3 on both sides of a PAN nanofiber sheet. Source: Lee et al. [15] / with permission of John Wiley & Sons. (d) Illustration of the twist spinning process used to convert the nanofiber sheet into a tiger yarn. (e) The fabrication process of stretchable and highly aligned PEDOT:PSS/Bi_2Te_3@PU nanofiber films. (f) Electrical conductivity, (g) Seebeck coefficient, and (h) power factor of aligned hybrid nanofiber films with various dipping times in PEDOT:PSS solution. Source: Wang et al. [34] / with permission of Springer Nature.

film-based wearable thermoelectric devices using electrospinning combined with in situ synthesis of silver nanoparticles (Ag NPs) as a post-treatment. In situ synthesis of Ag NPs on the surface of PAN nanofibers immersed in PEDOT:PSS solution to prepare flexible PEDOT:PSS/PVA@Ag NPs nanofiber films. After Ag NP coating, the electrical conductivity of the nanofiber film was significantly improved, and the maximum electrical conductivity could reach ~41.5 $S\,cm^{-1}$, which is 60 times higher than that of the impregnated film. With the increase of Ag NP coating, the Seebeck coefficient of the composite film did not decrease significantly, but only decreased from 25.1 to 17 $\mu V\,K^{-1}$.

12.3 Application of Electrospun Thermoelectric System

The flexible thermoelectric fibers and their devices prepared by electrospinning have excellent characteristics such as light, excellent mechanical properties, comfortable wearing, excellent energy conversion performance, and high integration, which show considerable application prospects in the fields of flexible wearable generators and self-powered sensing system.

12.3.1 Flexible Thermoelectric Generator

Jin et al. [35] connected several PEDOT:PSS/PVA@Ag NPs nanofiber films prepared by electrospinning to fabricate a flexible thermoelectric generator to improve the output power. The thermoelectric material layer on the fiber surface has excellent flexibility, is self-supporting, and can be bent arbitrarily. Based on the Seebeck effect of the material, the output voltage increases with increasing ΔT. Likewise, the output voltage increases monotonically with the number of thermoelectric units, and when the thermoelectric device consists of 5 strips, the output voltage can reach 3.79 mV at $\Delta T = 31$ K. These results show that it is feasible to increase the number of thermoelectric units to optimize the performance of thermoelectric generators. In addition, the prepared flexible generator can generate an open-circuit voltage of 3.43 mV, a short-circuit current of 0.0087 mA, and a real power output at steady state at $\Delta T = 30.1$ K.

He et al. [31] reported a wearable flexible thermoelectric generator composed of eight electrospun thermoelectric nanofiber yarns. One side of the device is directly exposed to the air environment, and the other side is close to the skin through a fabric, creating a temperature difference (Figure 12.6a). The designed thermoelectric device is directly fixed on the wrist to demonstrate its practical energy harvesting and self-powering properties on the human body surface. Infrared images show that when the room temperature is 24.6 °C, the temperature difference between the two sides of the device due to the thickness of the fabric is about 4 K (Figure 12.6b). When worn on the wrist, it can generate a sustained output voltage of about 1.1 mV (Figure 12.6c). This demonstrates the ability of thermoelectric devices to harvest energy at room temperature when worn on the surface of the human body.

Wang et al [34] designed a solar-driven thermoelectric device with radiative-cooling photothermal-heating electrodes based on the unique highly aligned nanofiber structure of the PEDOT:PSS/Bi_2Te_3@PU films prepared by electrospinning to realize stable in-plane temperature difference between the two sides of the aligned nanofiber films and high-voltage output under sunlight (Figure 12.6d). The radiative-cooling electrode consisted of electrospun PEO nanofiber film and silver paste, while the PEDOT:PSS/CNT composite was coated on the other side as the photothermal-heating electrode. Under 1 sun, the heating rate of the PEDOT:PSS/CNT electrode is significantly faster than that of the PEO (Polyethylene oxide)/Ag electrode, and both reach a stable value after 100 s and generate a constant temperature difference $T = \sim 24.9$ K, which corresponds to their solar absorption ability (Figure 12.6e). The solar-driven thermoelectric device was placed

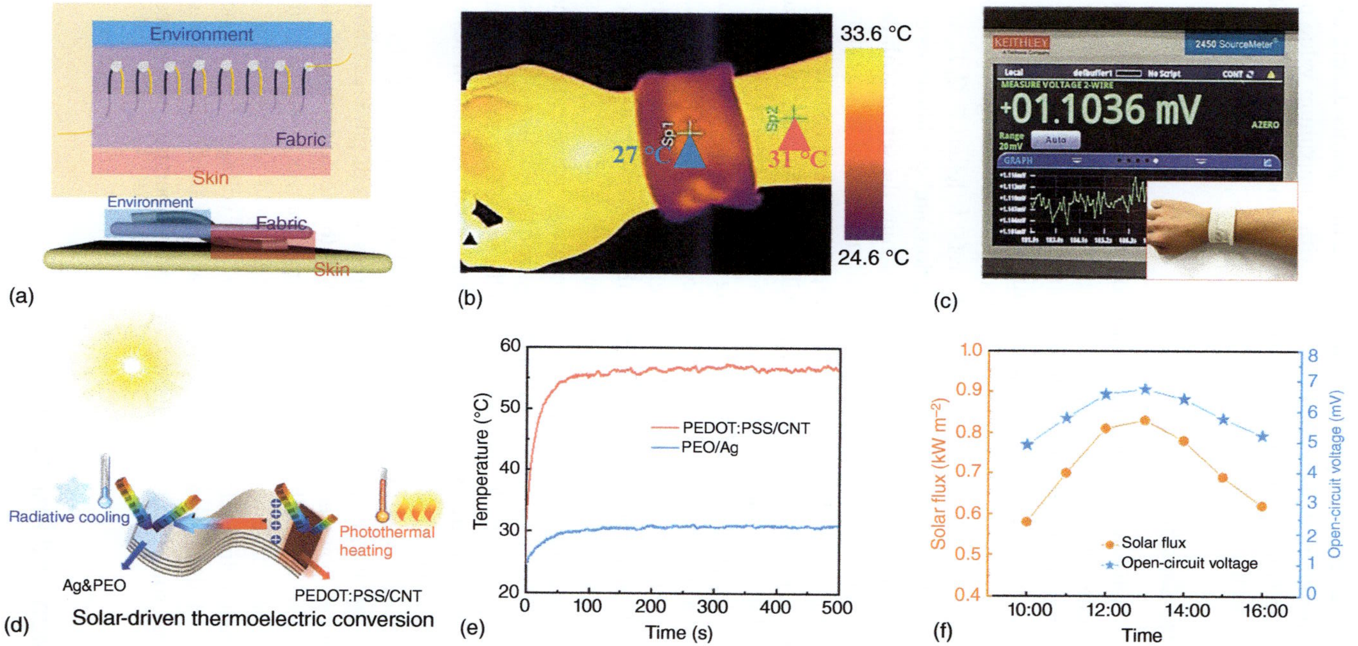

Figure 12.6 (a–c) Power generation performance of PEDOT:PSS/PVA@Ag NPs nanofiber-based flexible thermoelectric generator. (a) Schematic of top and side views of thermoelectric devices mounted on human skin. (b) Schematic diagram of the thermoelectric device mounted on the wrist. One end of the device is close to the heat source (skin), and the other end is close to the cold source (air). (c) The thermoelectric device can collect about 1.1 mV when fixed on the human body at room temperature. Source: Adapted from He et al. [31]. (d–f) Solar-driven thermoelectric device with radiative-cooling and photothermal-heating electrodes to realize stable in-plane temperature difference between the two sides of the aligned nanofiber films and high-voltage output under sunlight. (d) Schematical illustration showing the application of highly aligned thermoelectric nanofiber films in solar-driven thermoelectric conversion. (e) The temperature response curve of solar-driven thermoelectric device under 1 sun illumination. (f) Real-time solar flux and corresponding open circuit voltage of the device in an outdoor environment during the day. Source: Adapted from Wang et al. [34].

in an outdoor environment and exposed to direct sunlight to further evaluate the practical significance of its energy harvesting performance. Under outdoor conditions, the real-time solar flux generated at different time points during the day can generate corresponding open-circuit voltages, with a maximum value of ~6.8 mV at around 1 pm (Figure 12.6f). This work provided a new design concept for wearable solar-driven thermoelectrics.

12.3.2 Self-Powered Sensing System

He et al. [28] reported a self-powered sensor consisting of five 5 × 15 mm electrospun thermoelectric composite fabrics. By touching different numbers of fingers, corresponding output thermal voltage signals can be generated to realize the conversion of thermal voltage and words. For example, when a finger touches, the voltage of the sensor increases significantly, we define this output thermal voltage signal as the letter A (Figure 12.7a,b). Accordingly, 2, 3, 4, and 5 refer to letters B, C, D, and E, respectively. Based on this, some interesting permutations and combinations can be made with this self-powered sensor. For example, touching the sensor with two fingers, then one finger, and then four fingers, the resulting thermal voltage signal corresponds to the word "bad" (Figure 12.7c). This self-powered temperature sensor hopefully can help some people who lack language skills in the future or to realize some applications in human–computer interaction. In addition, He et al. installed this thermoelectric device in a mask to detect breathing frequency in a self-powered mode to realize human health monitoring. The device is embedded in the mask with one end directly exposed to the air and the other end close to the upper lip (Figure 12.7d). When the suction action is performed, the temperature of the detection end shows a downward trend. At this time, the temperature of both ends of the device is close, and the output voltage is small. On the contrary, when exhaling, the temperature of the detection terminal increased rapidly, and the output voltage also increased (Figure 12.7e). The mask can operate in a self-powered mode to differentiate breathing rates during exercise and walking, showing great potential in health monitoring and physical danger warning (Figure 12.7f).

He et al. [30] reported a smart glove stitched with electrospun thermoelectric yarn for distinguishing between hot and cold sources in self-powered mode (Figure 12.7g). The self-powered sensor was attached to the fingertips of the glove for temperature sensing. When it touched the air, there was no temperature difference across the sensor. But when it was exposed to cold and hot water, it would output different voltage signals (Figure 12.7h). The voltage of the sensor changes very fast, and the heating response time was only 0.5 s, which was of great significance for some dangerous alarms in high-temperature workplaces (Figure 12.7i).

12.4 Conclusion and Prospects

In recent years, with the rapid development of wearable thermoelectric devices, various flexible thermoelectric materials have emerged one after another, and the overall

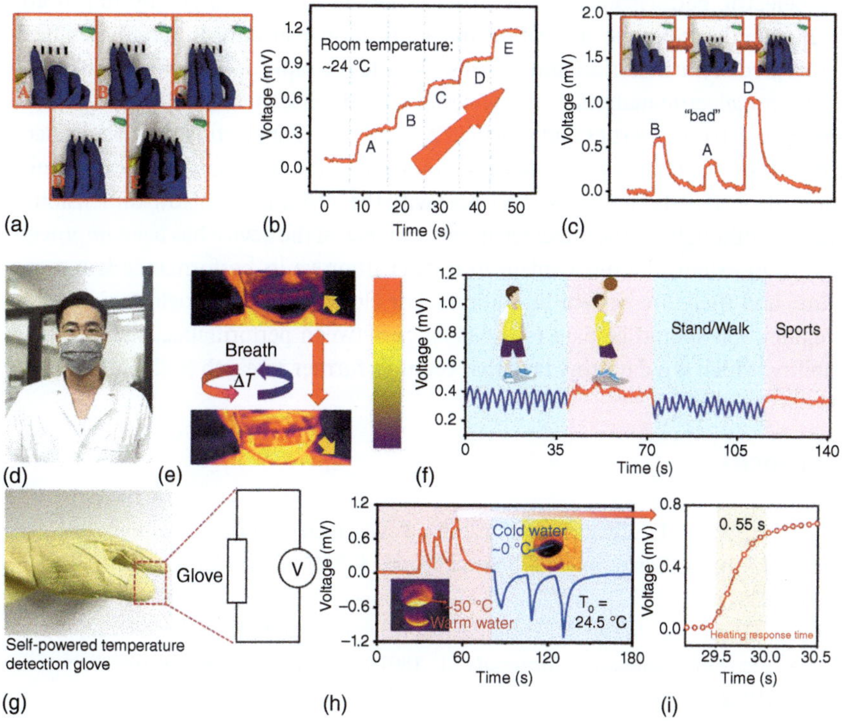

Figure 12.7 (a–c) Temperature sensing characteristics of CNT/PVP/PU composite fabric and temperature sensing application which can convert thermal voltage signal into language in self-powered mode. (a) Physical image of a self-powered temperature sensor composed of five serial composite fabrics. (b) The output thermal voltage generated by the touch of 1–5 fingers corresponding to the self-powered temperature sensor. (c) The self-powered temperature sensor converts the thermal voltage generated by the temperature difference into a word, take 'bad' as an example. Source: Adapted from He et al. [28]. (d–f) The application of flexible wearable thermoelectric device in human health monitoring. (d) Physical image of a mask for respiratory rate detection. The device is installed in the mask with one end directly exposed to the air and the other end close to the upper lip. (e) Infrared images of exhalation and inhalation while wearing a mask. When inhaling, the temperature of the end of the device near the upper lip decreases, and the temperature rises when exhaling. (f) At room temperature ($T_0 = 24\,°C$), the thermal voltage output when standing/walking shows regular fluctuations. The output of thermal voltage tends to be stable during running/strength exercise. (g–i) The applications of stretchable thermoelectric yarns for temperature recognition in self-powered mode. (g) Smart gloves with temperature recognition ability based on thermoelectric yarns, where the yarns are integrated into the gloves in the form of stitching. (h) The thermal voltage output by the smart glove when exposed to hot and cold water, respectively, and T_0 is the room temperature. (i) Heating response time of glove thermal voltage. Source: Adapted from He et al. [30].

thermoelectric performance and wearing experience have also greatly improved. This chapter describes the fabrication and application of electrospinning-based flexible thermoelectric materials and devices. Despite the rapid development of electrospun thermoelectric materials and devices, they still face great challenges in terms of thermoelectric properties and practical applications. For electrospun thermoelectric materials, the bonding force between thermoelectric materials and flexible substrates still needs to be actively solved. And with the introduction of low-dimensional materials, although the thermoelectric performance of the device has been improved to a certain extent, the mechanism to further improve its performance is not yet mature, and there are certain limitations. For electrospun thermoelectric devices, more attention should be paid to the balance between performance, stability, and flexibility, which is a daunting task that requires further research in the future.

References

1 Gao, M., Wang, P., Jiang, L. et al. (2021). Power generation for wearable systems. *Energy & Environmental Science* 14: 2114–2157. https://doi.org/10.1039/D0EE03911J.

2 Wang, C., Xia, K., Wang, H. et al. (2019). Advanced carbon for flexible and wearable electronics. *Advanced Materials* 31: 1801072. https://doi.org/10.1002/adma.201801072.

3 Gao, W., Ota, H., Kiriya, D. et al. (2019). Flexible electronics toward wearable sensing. *Accounts of Chemical Research* 52: 523–533. https://doi.org/10.1021/acs.accounts.8b00500.

4 Xu, S., Qin, Y., Xu, C. et al. (2010). Self-powered nanowire devices. *Nature Nanotechnology* 5: 366–373. https://doi.org/10.1038/nnano.2010.46.

5 Wang, S., Lin, L., and Wang, Z.L. (2015). Triboelectric nanogenerators as self-powered active sensors. *Nano Energy* 11: 436–462. https://doi.org/10.1016/j.nanoen.2014.10.034.

6 Zhou, Y., Xiao, X., Chen, G. et al. (2022). Self-powered sensing technologies for human Metaverse interfacing. *Joule* 6: 1381–1389. https://doi.org/10.1016/j.joule.2022.06.011.

7 Son, D., Lee, J., Qiao, S. et al. (2014). Multifunctional wearable devices for diagnosis and therapy of movement disorders. *Nature Nanotechnology* 9: 397–404. https://doi.org/10.1038/nnano.2014.38.

8 Bowen, C.R., Kim, H.A., Weaver, P.M., and Dunn, S. (2014). Piezoelectric and ferroelectric materials and structures for energy harvesting applications. *Energy & Environmental Science* 7: 25–44. https://doi.org/10.1039/c3ee42454e.

9 Wang, Z.L. (2013). Triboelectric nanogenerators as new energy technology for self-powered systems and as active mechanical and chemical sensors. *ACS Nano* 7: 9533–9557. https://doi.org/10.1021/nn404614z.

10 Wang, L.M., Zhang, Z.M., Geng, L.X. et al. (2018). Solution-printable fullerene/TiS_2 organic/inorganic hybrids for high-performance flexible n-type

thermoelectrics. *Energy & Environmental Science* 11: 1307–1317. https://doi.org/10.1039/c7ee03617e.

11 Park, S., Heo, S.W., Lee, W. et al. (2018). Self-powered ultra-flexible electronics via nano-grating-patterned organic photovoltaics. *Nature* 561: 516–521. https://doi.org/10.1038/s41586-018-0536-x.

12 He, J. and Tritt Terry, M. (2017). Advances in thermoelectric materials research: Looking back and moving forward. *Science* 357: eaak 9997. https://doi.org/10.1126/science.aak9997.

13 Wang, L., Zhang, Z., Liu, Y. et al. (2018). Exceptional thermoelectric properties of flexible organic-inorganic hybrids with monodispersed and periodic nanophase. *Nature Communications* 9: 3817. https://doi.org/10.1038/s41467-018-06251-9.

14 Minnich, A.J., Dresselhaus, M.S., Ren, Z.F., and Chen, G. (2009). Bulk nanostructured thermoelectric materials: current research and future prospects. *Energy & Environmental Science* 2: 466–479. https://doi.org/10.1039/B822664B.

15 Lee, J.A., Aliev, A.E., Bykova, J.S. et al. (2016). Woven-Yarn thermoelectric textiles. *Advanced Materials* 28: 5038–5044. https://doi.org/10.1002/adma.201600709.

16 Zhang, C., Zhang, C., Ng, H., and Xiong, Q. (2019). Solution-processed n-type Bi_2Te_3−xSex nanocomposites with enhanced thermoelectric performance via liquid-phase sintering. *Science China Materials* 62: 389–398. https://doi.org/10.1007/s40843-018-9312-5.

17 Heremans, J.P., Jovovic, V., Toberer, E.S. et al. (2008). Enhancement of thermoelectric efficiency in PbTe by distortion of the electronic density of states. *Science* 321: 554–557. https://doi.org/10.1126/science.1159725.

18 Tan, G., Zhao, L.-D., and Kanatzidis, M.G. (2016). Rationally designing high-performance bulk thermoelectric materials. *Chemical Reviews* 116: 12123–12149. https://doi.org/10.1021/acs.chemrev.6b00255.

19 Yao, Q., Chen, L., Zhang, W. et al. (2010). Enhanced thermoelectric performance of single-walled carbon nanotubes/polyaniline hybrid nanocomposites. *ACS Nano* 4: 2445–2451. https://doi.org/10.1021/nn1002562.

20 Deng, W.J., Deng, L., Li, Z.P. et al. (2021). Synergistically boosting thermoelectric performance of PEDOT:PSS/SWCNT composites via the Ion-exchange effect and promoting SWCNT dispersion by the Ionic liquid. *ACS Applied Materials & Interfaces* 13: 12131–12140. https://doi.org/10.1021/acsami.1c01059.

21 Jia, F., Wu, R., Liu, C. et al. (2019). High thermoelectric and flexible PEDOT/SWCNT/BC nanoporous films derived from aerogels. *ACS Sustainable Chemistry & Engineering* 7: 12591–12600. https://doi.org/10.1021/acssuschemeng.9b02518.

22 Xu, S.H., Fan, Z., Yang, S.T. et al. (2021). Highly flexible, stretchable, and self-powered strain-temperature dual sensor based on free-standing PEDOT:PSS/carbon nanocoils-poly(vinyl) alcohol films. *ACS Sensors* 6: 1120–1128. https://doi.org/10.1021/acssensors.0c02390.

23 Zhang, C., Zhang, Q., Zhang, D. et al. (2021). Highly stretchable carbon nanotubes/polymer thermoelectric fibers. *Nano Letters* 21: 1047–1055. https://doi.org/10.1021/acs.nanolett.0c04252.

24 Kim, J.-Y., Lee, W., Kang, Y.H. et al. (2018). Wet-spinning and post-treatment of CNT/PEDOT:PSS composites for use in organic fiber-based thermoelectric generators. *Carbon* 133: 293–299. https://doi.org/10.1016/j.carbon.2018.03.041.

25 Zeng, W., Shu, L., Li, Q. et al. (2014). Fiber-based wearable electronics: a review of materials, fabrication, devices, and applications. *Advanced Materials* 26: 5310–5336. https://doi.org/10.1002/adma.201400633.

26 Wu, Q. and Hu, J. (2016). Waterborne polyurethane based thermoelectric composites and their application potential in wearable thermoelectric textiles. *Composites Part B: Engineering* 107: 59–66. https://doi.org/10.1016/j.compositesb.2016.09.068.

27 Wen, N., Fan, Z., Yang, S. et al. (2020). Highly conductive, ultra-flexible and continuously processable PEDOT:PSS fibers with high thermoelectric properties for wearable energy harvesting. *Nano Energy* 78: 105361. https://doi.org/10.1016/j.nanoen.2020.105361.

28 He, X., Shi, J., Hao, Y. et al. (2021). PEDOT:PSS/CNT composites based ultra-stretchable thermoelectrics and their application as strain sensors. *Composites Communication* 27: 100822. https://doi.org/10.1016/j.coco.2021.100822.

29 He, X.Y., Shi, J., Hao, Y.N. et al. (2022). Highly stretchable, durable, and breathable thermoelectric fabrics for human body energy harvesting and sensing. *Carbon Energy* 4: 621–632. https://doi.org/10.1002/cey2.186.

30 Levitt, A., Seyedin, S., Zhang, J. et al. (2020). Bath electrospinning of continuous and scalable multifunctional MXene-infiltrated nanoyarns. *Small* 16: 2002158. https://doi.org/10.1002/smll.202002158.

31 He, X., Gu, J., Hao, Y. et al. (2022). Continuous manufacture of stretchable and integratable thermoelectric nanofiber yarn for human body energy harvesting and self-powered motion detection. *Chemical Engineering Journal* 450: 137937. https://doi.org/10.1016/j.cej.2022.137937.

32 Yin, T., Liu, D., Ou, Y. et al. (2010). *Journal of Physical Chemistry C* 114: 10061–10065. https://doi.org/10.1021/jp1024872.

33 Morata, A., Pacios, M., Gadea, G. et al. (2018). Large-area and adaptable electrospun silicon-based thermoelectric nanomaterials with high energy conversion efficiencies. *Nature Communications* 9: 4759. https://doi.org/10.1038/s41467-018-07208-8.

34 Wang, L., He, X., Hao, Y. et al. (2023). Rational design of stretchable and highly aligned organic/inorganic hybrid nanofiber films for multidirectional strain sensors and solar-driven thermoelectrics. *Science China Materials* 66: 707–715. https://doi.org/10.1007/s40843-022-2160-8.

35 Jin, S., Sun, T., Fan, Y. et al. (2019). Synthesis of freestanding PEDOT:PSS/PVA@Ag NPs nanofiber film for high-performance flexible thermoelectric generator. *Polymer* 167: 102–108. https://doi.org/10.1016/j.polymer.2019.01.065.

13

Electrospun Nanofiber-Based Water-Induced Electric Generation

Zhaoyang Sun and Liming Wang

Donghua University, College of Textiles, Key Laboratory of Textile Science & Technology, Ministry of Education, 2999 Renmin North Road, Songjiang District, Shanghai 201620, China

13.1 Introduction

With the rapid depletion of fossil resources and increasing environmental pollution, it is urgent to develop renewable energy technologies to replace traditional fossil fuels and satisfy environmental needs [1]. Nanofibrous materials have been widely used in energy convention systems in recent years and are considered promising candidates to address these critical issues because of their excellent properties, such as extremely large surface area, high length/diameter ratio, good flexibility, high porosity, and multiple functionalities [2]. Electrospinning is a fabrication method to produce continuous ultrafine fibers with diameters in the range of a few tens of nanometers to a few micrometers in the form of nonwoven mats and yarns. Therefore, electrospun nanofibers have been widely used in many energy systems, including triboelectric [3], piezoelectric [4], thermoelectric [5], and water-induced power generation [6–13]. In this chapter, we focus on the application of electrospun nanofibers in the field of water-induced power generation.

Water-induced electric generators that can directly harvest electricity from ubiquitous moisture or water evaporation are one of the most fascinating and promising candidates to supply renewable and clean power. Existing water-induced power generation technology can be divided into two forms according to the existence of water. One is to use the liquid water system. It needs to construct an electric double layer in micro-nano-structured materials, ions will move with the driving force that comes from water evaporation. The materials that can be used for liquid water-induced electric generation include nanostructured carbon materials and metal oxides. The other one is used in the gaseous water system, that is, moisture in the air. We also call it moisture-electric generation. The power generation usually employs the asymmetric structure to construct the moisture gradient, and at the same time, it can also form the electric double layer by constructing the micro-/nanostructures and generating electric current driven by the moisture flow. Materials currently used for moisture-electric generation must be containing oxygen-containing functional

Electrospinning: Fundamentals, Methods, and Applications, First Edition.
Edited by Liming Wang and Xiaohong Qin.

groups, which can generate freely mobile protons under the induction of moisture. Electrospinning nanofibrous membranes have abundant micro- and nano-channel structures and are easy to prepare, making them one of the ideal methods for constructing water-induced generators. This chapter introduces water-induced electric generation from liquid and gaseous water systems and also discusses device setup, materials selection principle, suggested mechanisms, and applications.

13.2 Liquid Water System

13.2.1 Device Setup and Materials Selection Principle

The classic water-inducing electric device usually consists of calcinated electrospun nanofibrous membrane, a water tank, and two electrodes (Figure 13.1). Typically, one side of the electrode is submerged in the water tank, creating an asymmetrical water gradient on both sides of the electrode. The key component of the water-induced electric generator is the applied nanofibrous membrane, which naturally possesses a porous micro-/nanostructure that could form numerous micro-/nanochannels across the two electrodes. But not all nanofibrous membranes can be used for power generation. In general, the required materials must have electronegativity on the surface to adsorb hydrogen ions or hydroxide ions in water so that negative ions can be enriched in the pores. At present, only inorganic nanofibers obtained by high-temperature calcination (Figure 13.1) have such characteristics; typical examples like developed silica [6] and carbon nanofibers [7] will meet this demand. When water flows from these electronegative nanochannels, a double layer will be formed in the channels, and counterions will move with driving force (normally evaporation force), thus producing electricity.

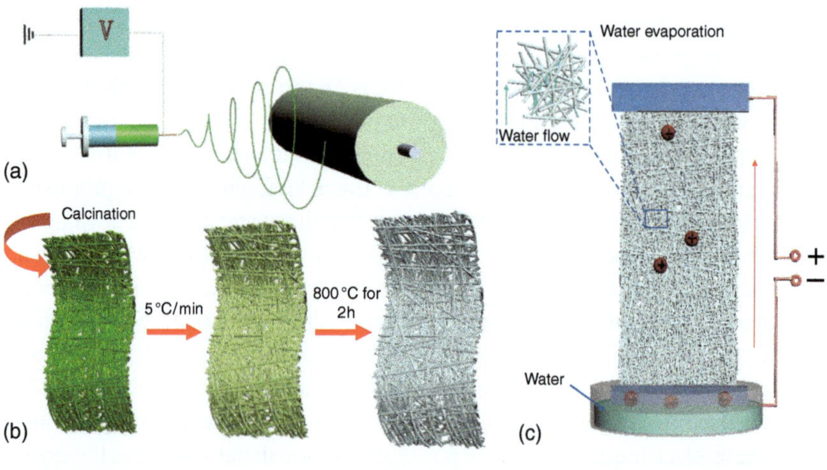

Figure 13.1 A typical water-induced electric generation device and the preparation method. (a) Sol-gel electrospinning followed by (b) calcination to fabricate a flexible inorganic nanofiber membrane. (c) Schematic of the device setup of the water-induced nanofiber-based generator. Source: Sun et al. [6] / with permission of American Chemical Society.

13.2.2 Effect of Changing Various Structural Parameters

For calcined electrospun membranes, the membrane thickness, length, and width of the macroscope fiber membrane, as well as the fiber diameter, all affect the final electrical output. In addition, ambient temperature and humidity are also key factors that affect hydropower generation.

The voltage and current output will increase with the increased thickness of the membrane, while further increasing the thickness will cause the current and voltage output to decrease. The initial increase is likely attributed to the improved water flow in the membrane as more pores become available for water transport. However, the nanofibrous membrane's thickness surpasses a certain value as thickness will also affect the evaporation rate, which may decrease in a thick membrane. As for the width, the current output will be increased continuously with increased width due to the growing number of channels for water rising and the increased surface area becoming available for water evaporation.

Similarly, with thickness, both voltage and current output initially increase and then decrease as height increases. Voltage and current increase initially with the height because of the increased surface area for water evaporation, which should increase water flow in the nanofibrous membrane. The power output eventually decreases because water can rise only to a certain height governed by the capillary force in the nanofibrous membrane.

According to the previous literature [14–18], decreasing the fiber diameter also decreases the size of the pores surrounded by the fibers. As the pore size decreases, the capillary force or the channel pressure that drives water to rise increases. Additionally, the ion selectivity of the channels improves due to increases in the ratio between the Debye length and the pore radius. The improved ion selectivity of the channels facilitates the transport of one type of ions over the other type and boosts net current flow. Besides, decreasing the fiber diameter also increases the specific surface area of the nanofibrous membrane and increases the rate of water evaporation. In summary, decreasing the fiber diameters facilitates power generation from the water-induced electric generators.

Environmental factors also affect power generation efficiency differently [6, 19, 20]. As the temperature increases, the evaporation rate increases, the ions move faster, and the corresponding voltage increases. When humidity increases, it inhibits the evaporation of water, which reduces the speed of ion movement and affects the ability to generate electricity from water.

13.2.3 Suggested Mechanism for Nanofiber-Based Water-Induced Electric Generator

The origin of the electrokinetic effect in a cylindrical capillary channel that is filled with water. In the absence of water flow, the dissociation of surface functional groups, such as carboxyl and phenolic groups, leads to the radial separation of charges, resulting in the formation of an electric double layer and its associated zeta potential ζ. However, on average, free ions do not deviate left or right along the axial

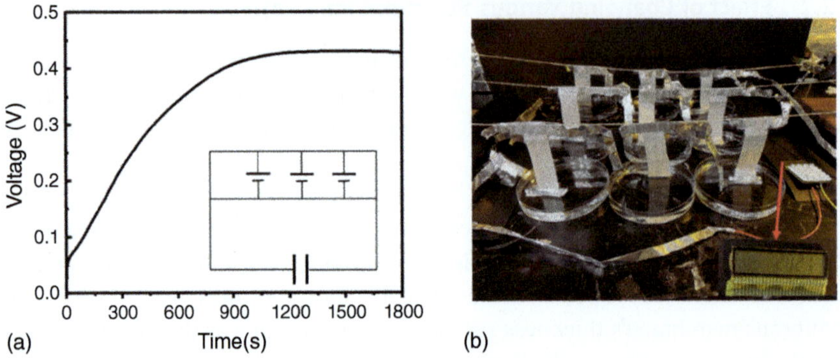

Figure 13.2 Liquid water system connected to power capacitor (a), and calculator (b). Source: Sun et al. [6] / with permission of American Chemical Society.

direction from their grafted counterions. Therefore, no net dipoles or potentials are created in the axial direction. This changes when water flows through this capillary. The flow of water pushes along counterions in the moving region away from the surface, creating a streaming current. The net migration of the free ions in the presence of water flow leads to charge polarization and thus generates an electric potential V The potential drives the back convection of ions, producing a conduction current opposing the flow.

Application

Currently, the application of nanofiber-based water-induced electric generators is mainly applied to the field of direct power supply. Also, a single unit can only generate low current and voltage; thus, multiple devices in series and parallel are generally required to supply power generation devices. Developed devices show that multiple units of series–parallel nanofiber-based water-induced electric-generating devices can be used to power capacitor or small electronic devices (Figure 13.2a,b) [6], as well as to light LEDs. [7].

13.3 Gaseous Water System

13.3.1 Device Setup and Materials Selection Principle

Different from the aforementioned water-induced electric generator, the moist electric generator (MEG) consists of a pair of electrodes and hygroscopic materials, and there is no need to have an extra water tank. The top electrode is usually dug with holes to guarantee that moisture comes in. Conductive wires are connected to the top and bottom electrodes (Figure 13.3a). When MEG is exposed to moisture flow, spontaneous absorption and subsequent hydration of the hygroscopic materials result in plenty of free-ion (H^+) carriers by an ionization effect. Then, charge ions will diffuse under the concentration difference and, with the help of moist flow, give rise to the electric potential between the two electrodes, thus inducing a voltage output.

For a MEG, the most critical component is its middle layer that must be hygroscopic materials with functional groups. So, materials with functional groups or

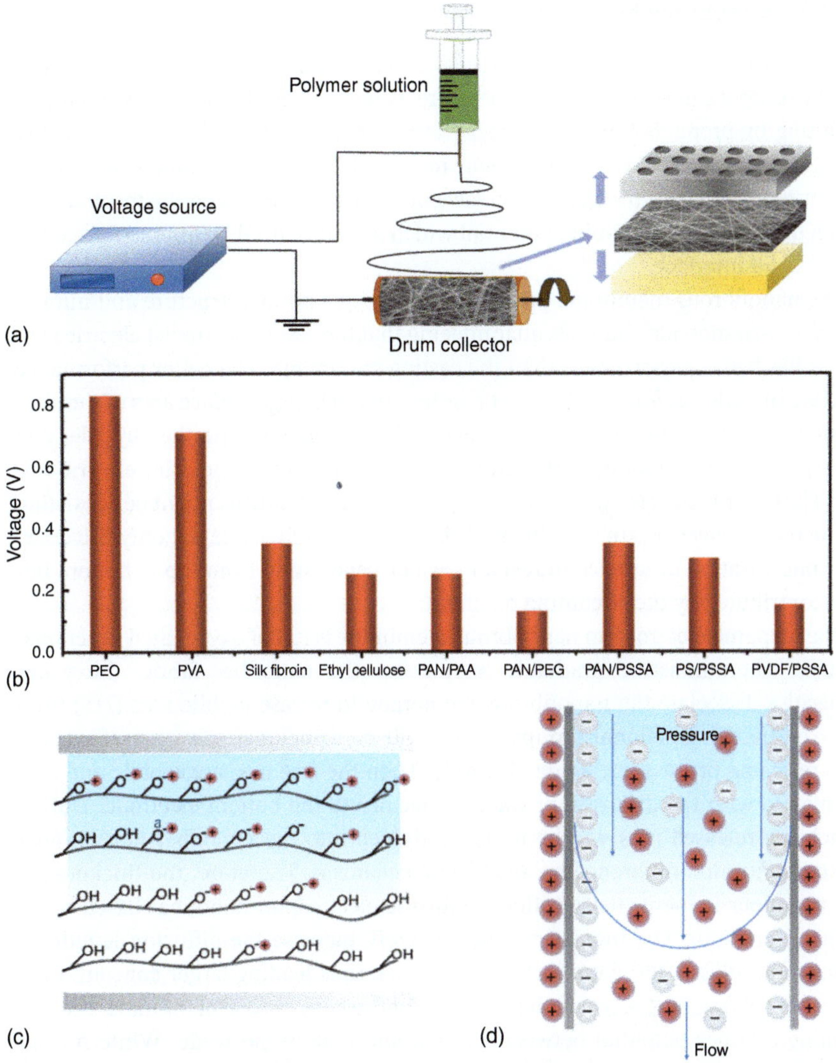

Figure 13.3 (a) A typical moisture electric generating device and its preparation method. (b) Materials can be fabricated to assemble MEG and their voltage output. Source: Sun et al. [9] / with permission of Elsevier. (c, d) Scheme diagram of suggested mechanisms for MEG (c) ions diffusion and (d) streaming potential.

can be treated to be possessed with functional groups (normally –OH, –COOH) can be directly fabricated to nanofibers with electrospinning technology was first chosen to construct MEG [8, 9]. We can see that polyvinyl alcohol (PVA), polyethylene oxide (PEO), silk fibroin, ethyl cellulose, sodium alginate (SA), and acid cellulose were used for MEG. Except the abovementioned materials, we also blended functional materials with electrospun polymers to realize moist electric generation (Figure 13.3b).

13.3.2 Established Mechanism for Moist Electric Generation

The moisture energy conversion process usually occurs spontaneously (e.g. diffusion, evaporation), without temperature variation, mechanical movement, or polluting by-products (chemical energy loss). Remarkably, atmospheric moisture can generate electricity without being restricted by the environment or region. Moisture-enabled electric generation developed to date is based on the mechanisms of ion diffusion and streaming potential which are influenced by interactions at the water–solid interface.

The nanofibrous membrane possesses with unique porous structure and numerous micro-/nanoscale channels guaranteeing that the assembled moist electric generator has better performance than the casting membrane. The better performance can be concluded as follows. The first reason is that the large surface area of the electrospun nanofibers provides more oxygenated functional groups, thereby releasing more ions when interacting with water molecules and largely promoting electric output. The other reason is that two power generation mechanisms might be coexisting in the porous electrospun nanofiber fabrics. The ion diffusion is not only driven by the concentration difference induced by water penetrating from top to bottom but also contributed by the streaming potential.

The prepared electrospun nanofibrous membrane is full of oxygenated functional groups, when moisture gradually approaches the nanofiber fabric, the water molecules dissociate the nanofibrous membrane to release mobile ions (H^+) from the oxygenated functional groups. There will be a regional solvation effect and ions releasing process in the top electrode. Then the free ions are mobile and able to migrate with the diffusion of water molecules to the bottom electrode. During which the released ions will be transferred deep into the nanofiber fabric under the concentration difference by a vehicle mechanism. Therefore, the thickness of the nanofiber is the primary influence for the electrospun nanofiber-based moist electric generator. The increasing thickness will increase the diffusion length for water molecules in the nanofibrous membrane, thus leading larger concentration difference of free ions between the two sides of the nanofibrous membrane, thereby forming a larger potential between the top and bottom electrode. While further increasing the membrane thickness, the infiltration of moisture and movement of H^+ ions become weak, thereby the potential reaches equilibrium at the thickness in a certain value.

Besides ions diffusion, owing to the unique porous structure with ideal micro-/nanoscale channels, similar to water-induced electric generation, streaming potential, a classic electrokinetic phenomenon induced by the driving of ionic solutions through narrow channels under a pressure gradient, also plays a vital role in electric generation for nanofiber fabric (Figure 13.3c). Generally, the surface of micro-/nanochannels in electrospun nanofiber fabric will be electrically charged with H^+ because of the amphoteric dissociation of surface oxygenated functional groups. The released ions will attract the anions in water, so a charge distribution is formed on the surface of the vertical channels (Figure 13.3d). Finally, counterions will be mobile with the help of moist flow, thereby producing a streaming potential.

13.3.3 Different Types of Nanofiber-Based MEG

Current moist electric generators have developed various types of forms to realize self-powered sensing or power generation. For example, a neuron-like Nb2CTx/sodium alginate composite membrane prepared by an electrospinning approach was developed to achieve mutisensing application. The doped Nb2CTx significantly enhanced water absorption property allowing excellent response to moisture. Besides, changing electrodes to active metal can significantly increase the power density of nanofiber-based MEGs [10]. Compared with typical MEG, assembled active electrode applied MEG can produce a sustained voltage output of 1.1 V for 40000 s without any weak signs, reaching the highest level among all reported nanofiber-based MEGs (Figure 13.4a) [11]. This remarkable performance mainly arises from the higher concentration difference induced by the introduced active electrode, which enhances ion diffusion through the porous nanofibrous membrane. In addition, some researchers realized moist electric generation via optimized porous structures of nanofibrous membranes. They found the pore size and the porosity of cellulose acid (CA) membranes can be readily tuned via a facile compression and annealing process. The smaller pore can build more nanochannels around the electrode, thus leading to higher output voltages [12]. Except for the abovementioned strategy, a combined sol–gel electrospinning and calcination process was developed to prepare flexible and freestanding TiO_2/ZrO_2

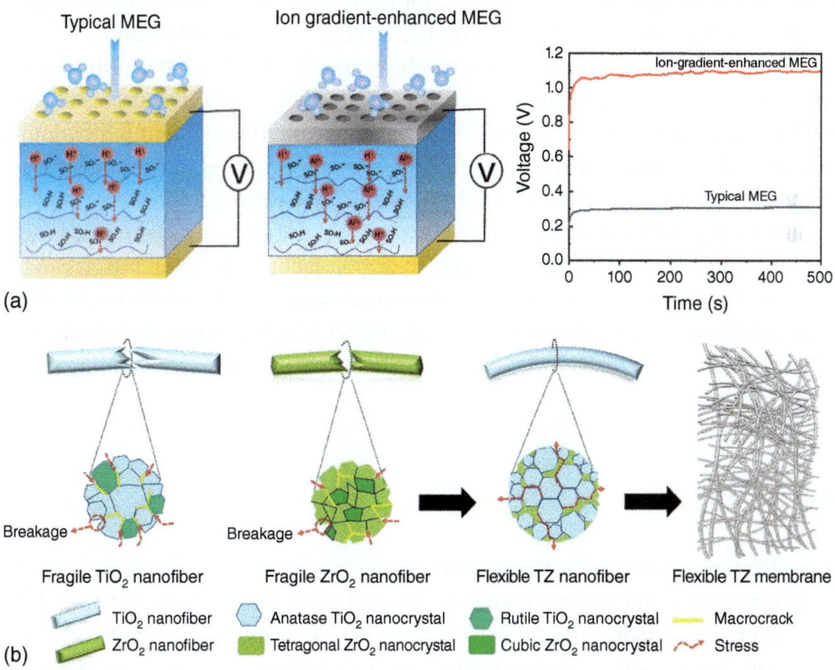

Figure 13.4 (a) 4(a) Active electrode-applied MEG. Source: Sun et al. [11] / with permission of Royal Society of Chemistry. (b) Inorganic nanofiber-based MEG with extremely flexible properties. Source: Wang et al. [13] / with permission of Springer Nature.

(TZ) composite nanofiber-based MEGs (Figure 13.4b). The excellent flexibility of TZ nanofibrous membranes can be attributed to the suppression of crystal structure transitions, dispersion of stress concentrations, and reduction of crack propagation through interfacial engineering. The porous structure of electrospun nanofibrous membrane supplies abundant charged narrow channels for water molecules diffusion and generates a streaming potential [13].

13.3.4 Applications Based on Electrospun Nanofiber-Based MEG

Benefiting from its breathability and natural softness, the moist electric generator not only can be used to power small electronics but can also be used in multifunctional applications such as respiration assessment, noncontact monitoring, and intelligent alarm. When integrating the MEG with a mask, it can be used to test human respiration. Developed devices display real-time breath signals under eupnea and tachypnea conditions. Of note, the semi-enclosed mask is not conducive to the humidity diffusion of exhaled gas, resulting in the MEG being in a high-humidity environment. In addition, it also has potential to use in noncontact devices. The voltage will increased with closing finger and then decreases when the finger is moved away, showing a potential application of contactless switch to enhance the safety of its electrical appliances. The simulation test also verified the application prospect of the field of moisture power generation in the field of diapers. When water is dropped on the diaper, the moisture generator attached to the diaper will spontaneously generate voltage with the increase in ambient humidity.

Application in disaster warning also can be realized with MEG, when the device contact with ammonia gas, the voltage dropped sharply, which may indicate leakage of ammonia gas (Figure 13.5a). When a fire occurs, the ambient temperature increases and the humidity decreases, resulting in a sharp drop in the voltage output, indicating that a fire has occurred (Figure 13.5b). Although there are applications in power generation and self-powered sensors, some inorganic nanofiber-based MEGs also show good self-cleaning ability in degrading organic pollutants. It can be seen, when TiO_2 particles containing MEG were exposed to UV in a photochemical reactor for photocatalysis, the RhB solution gradually faded to colorless under the catalysis of the TiO_2 nanofibrous membrane, clearly showing the self-cleaning effect (Figure 13.5c).

13.4 Outlook

Water-induced electric generation as a novel energy-conversion technology that can harvest energy from the environment has attracted tremendous attention [21–23]. In this chapter, we have discussed this topic from two systems, water-induced electric system and moist-induced electric system. We also discussed every aspect of device setup, materials selection principle, mechanisms, and application.

As an emerging new research field, water-induced power generation system still needs more exploration. Compared with other material systems, which are often

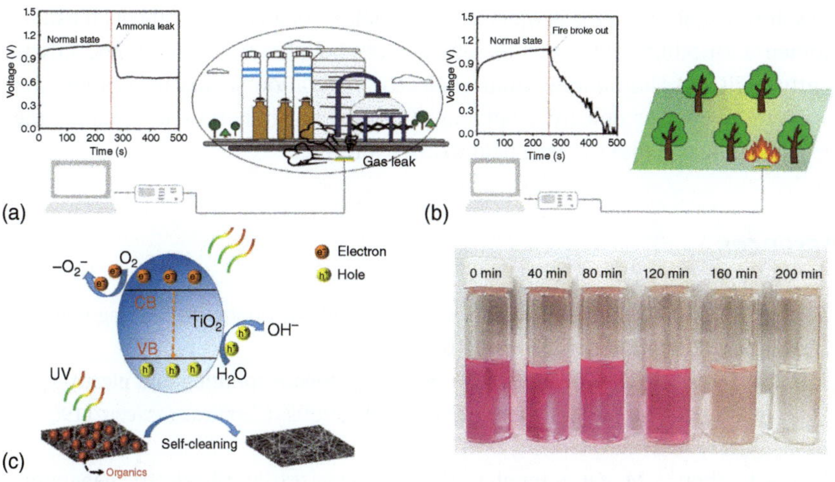

Figure 13.5 Applications of MEG in the disaster warning. (a) For ammonia gas leak and (b) fire detection. Source: Sun et al. [11] / with permission of Royal Society of Chemistry. (c) TiO_2-blended nanofiber is used for self-cleaning. Source: Wang et al. [13] / with permission of Springer Nature.

composed of metal-organic frameworks (MOF) materials deposited on substrate materials, inorganic nanofibers can be independently self-supporting as core devices, broadening their application scenarios. On the other hand, the synthesis of MOFs on the surface of nanofibers by hydrothermal growth and other methods can also be used in the future development direction of nanofiber-based hydroelectric generators. Besides, current generators all suffer from low energy output, and increasing power density is a future direction. With continuous development, this environmentally friendly generator, with its low cost, provides excellent potential for future green energy utilization and opens up new possibilities for portable electronics.

Despite intensive efforts, nanofiber-based moist-induced electric generator development is still in the early stages, with plenty of opportunities and challenges for future advancement. Some of these are as follows: first, a primary guidance mechanism for nanofiber-based MEG is still urgently needed. The interactions between functional materials and water should be further investigated for ion diffusion from experimental and theoretical perspectives. The abovementioned interactions are strongly determined by adsorption dynamics, transport kinetics, and the primary processes during water evaporation and electricity generation. Previous research has mostly studied liquid water for streaming potential, so the guidance mechanism for gaseous water needs further investigation. Second, enhancing the power density of existing MEGs to meet the practical use requirements of commercial wearable devices or electronics is in demand. The voltage and current output of the water-induced electric generators are still low. Increasing power density by interacting with MOF materials or modifying their functional groups might develop. Finally, as a new energy conversion technology, it provides novel solutions for problems in other research fields. As mentioned above, it can be combined with other fields, such as self-cleaning benefits from the photocatalytic property of TiO_2. As liquid water induces electric generation, increasing evaporation force increases power density. Applying photothermal materials to enhance the device's evaporation rate can be a field worth exploring.

References

1 Tentzeris, M.M., Georgiadis, A., and Roselli, L. (2014). Energy harvesting and scavenging. *Proceedings of the IEEE* 102: 1644–1648.

2 Xue, J.J., Wu, T., Dai, Y.Q., and Xia, Y.N. (2019). Electrospinning and electrospun nanofibers: methods, materials, and applications. *Chemical Reviews* 119: 5298–5415.

3 Tao, X.J., Zhou, Y.M., Qi, K. et al. (2022). Wearable textile triboelectric generator based on nanofiber core-spun yarn coupled with electret effect. *Journal of Colloid and Interface Science* 608: 2339–2346.

4 Selleri, G., Gino, M.E., Brugo, T.M. et al. (2022). Self-sensing composite material based on piezoelectric nanofibers. *Materials & Design* 219: 110787.

5 Lee, J.A., Aliev, A.E., Bykova, J. et al. (2016). Woven-yarn thermoelectric textiles. *Advanced Materials* 28: 5038–5044.

6 Sun, Z.Y., Feng, L.L., Wen, X. et al. (2021). *ACS Applied Materials & Interfaces* 47: 56226–56232.

7 Tabrizizadeh, T., Wang, J., Kumar, R. et al. (2021). Water-evaporation-induced electric generator built from carbonized electrospun polyacrylonitrile nanofiber mats. *ACS Applied Materials & Interfaces* 43: 50900–50910.

8 Sun, Z.Y., Feng, L.L., Xiong, C.D. et al. (2021). Electrospun nanofiber fabric: an efficient, breathable and wearable moist-electric generator. *Journal of Materials Chemistry A* 9: 7085–7093.

9 Sun, Z.Y., Wen, X., Wang, L.M. et al. (2022). Emerging design principles, materials, and applications for moisture-enabled electric generation. *eScience* 2: 32–46.

10 Zhao, Q.N., Jiang, Y.D., Duan, Z.H. et al. (2022). A Nb_2CT_x/sodium alginate-based composite film with neuron-like network for self-powered humidity sensing. *Chemical Engineering Journal* 438: 135588.

11 Sun, Z.Y., Feng, L.L., Wen, X. et al. (2021). Nanofiber fabric based ion-gradient-enhanced moist-electric generator with a sustained voltage output of 1.1 volts. *Materials Horizons* 8: 2303–2309.

12 Lyu, Q.Q., Peng, B.L., Xie, Z.J. et al. (2020). Moist-induced electricity generation by electrospun cellulose acetate membranes with optimized porous structures. *ACS Applied Materials & Interfaces* 12: 57373–57381.

13 Wang, L.M., Feng, L.L., Sun, Z.Y. et al. (2022). Flexible, self-cleaning, and high-performance ceramic nanofiber-based moist-electric generator enabled by interfacial engineering. *Science China Technological Sciences* 65: 450–457.

14 Olthuis, W., Schippers, B., Eijkel, J., and van den Berg, A. (2005). Energy from streaming current and potential. *Sensors and Actuators B: Chemical* 111–112: 385–389.

15 van der Heyden, F.H.J., Bonthuis, D.J., Stein, D. et al. (2006). Electrokinetic energy conversion efficiency in nanofluidic channels. *Nano Letters* 6: 2232–2237.

16 van der Heyden, F.H.J., Bonthuis, D.J., Stein, D. et al. (2007). Power generation by pressure-driven transport of ions in nanofluidic channels. *Nano Letters* 7: 1022–1025.

17 Jiang, S.H., Chen, Y.M., Duan, G.G. et al. (2018). Electrospun nanofiber reinforced composites: a review. *Polymer Chemistry* 9: 2685–2720.

18 Hussain, D., Loyal, F., Greiner, A., and Wendorff, J.H. (2010). Structure property correlations for electrospun nanofiber nonwovens. *Polymer* 51: 3989–3997.

19 Zhang, Z.H., Li, X.M., Yin, J. et al. (2018). Emerging hydrovoltaic technology. *Nature Nanotechnology* 13: 1109–1119.

20 Xue, G., Xu, Y., Ding, T. et al. (2017). Water-evaporation-induced electricity with nanostructured carbon materials. *Nature Nanotechnology* 12: 317–321.

21 Zhao, F., Cheng, H.H., Zhang, Z.P. et al. (2015). Direct power generation from a graphene oxide film under moisture. *Advanced Materials* 27: 4351–4357.

22 Wang, H.Y., Sun, Y.L., He, T.C. et al. (2021). Bilayer of polyelectrolyte films for spontaneous power generation in air up to an integrated 1,000 V output. *Nature Nanotechnology* 16: 811–819.

23 Liu, X.M., Gao, H.Y., Ward, J. et al. (2020). Power generation from ambient humidity using protein nanowires. *Nature* 578: 550–554.

14

Electrospun Nanofibers for Flexible Sensors

Fei Wang and Xueping Zhang

Donghua University, College of Textiles, Key Laboratory of Textile Science & Technology, Ministry of Education, 2999 Renmin North Road, Songjiang District, Shanghai 201620, China

14.1 Introduction

Sensors, as a kind of sensing electronic device, can convert the detected stimulations into other forms of signals, for example, electrical signals, and then realize detection, conversion, and information transmission [1]. However, traditional sensing materials for electronic devices, such as silicon [2], germanium, gallium arsenide, and other inorganic semiconductor materials, are rigid [3] and have a large volume [4]. This greatly limits their potential applications in terms of portability and deformation requirements. Thus, the most significant challenge is to develop effective ways of designing and manufacturing flexible sensors that offer good wearability, including dynamic surface conformability (twistability, bendability, and stretchability), breathability, washability, thermo-comfortability, and high sensing performance. Textile-based sensors are a promising solution for addressing this challenge toward creating truly breathable daily-wear sensor devices because they can be produced using traditional textile fibers and sensing fibers [5]. These sensors simulate skin perception and response to environmental stimuli [6], which has attracted significant attention for promising applications in real-time monitoring of vital personal signs, such as blood sugar value, heart rate, breathing rate, temperature, and respiratory moisture. Textile-based sensors have the potential to become highly accurate and reliable wearable devices, offering real-time monitoring of physiological signals.

With the rapid development of nanoscience and nanotechnology, there has been a growing interest in the use of nanofiber-based sensors. Electrospun nanofibers are particularly attractive due to their softness, light weight, extensibility, good moisture absorption, and air permeability, in contrast to rigid materials [7]. These sensors can be integrated into clothing or attached to the skin surface to monitor various physical and biochemical signals and movements in real time. While common nanofibers

Electrospinning: Fundamentals, Methods, and Applications, First Edition.
Edited by Liming Wang and Xiaohong Qin.

are insulating, they can be made conductive by adding conductive materials to form a conductive path [8], which can detect external stimuli and convert them into electrical signals. Electrodes for nanofiber-based sensors have been developed using various materials, such as carbon-based nanomaterials (e.g. carbon black [9], carbon nanotubes [CNTs] [10], graphene [11]), metal nanoparticles (e.g. silver nanowires [AgNWs] [12], Au), and conductive polymers (e.g. polypyrrole [13], polyaniline [PANI] [14], poly(3,4-ethylenedioxythiophene): poly(styrene sulfonate) (PEDOT:PSS) [15]). Electrospun nanofiber materials have been widely used as an ideal carrier in smart wearable fields, such as flexible sensors, electronic skin, and energy collection and storage, due to their excellent characteristics, such as extremely fine size, large specific surface area [16], high density of sensitive sites, and ease of functional modification by doping [17]. This chapter will focus on the recent progress of some electrospun nanofiber-based flexible sensors, including their categories, fabrication, structures, properties, and applications.

14.2 Mechanical Sensor

Mechanical sensors can convert mechanical signals or deformation, such as pressure, tension, vibration, bending, and shear, into readable electrical signals [18]. Electrospun nanofiber-based mechanical sensors, with their softness, flexibility, and high sensitivity, have shown great potential in a variety of applications including health monitoring, human motion detection, and intelligent robotics. Two common types of mechanical sensors are strain sensors and pressure sensors, both of which can be fabricated using electrospun nanofibers.

14.2.1 Strain Sensor

Strain sensors are a critical category of flexible sensors with significant potential in various fields. The working principle of typical strain sensors is to detect electrical signals under varying strains. As illustrated in Figure 14.1, strain sensors can be classified into different categories, namely resistive, capacitive, and piezoelectric sensors [20–22]. The resistive and capacitive strain sensors are more frequently employed and studied due to their facile manufacturing process and ease of use.

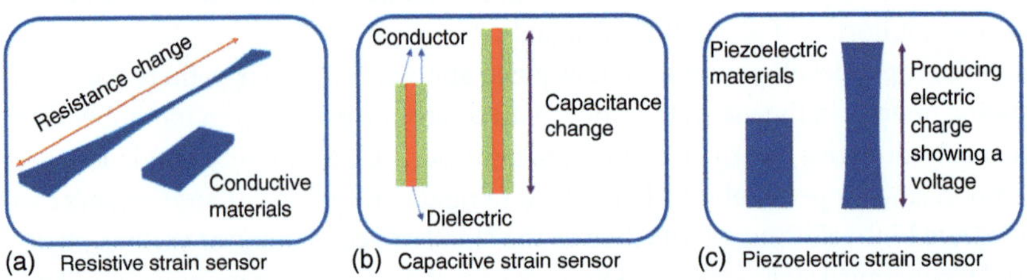

Figure 14.1 Classification of strain sensors: (a) resistive strain sensor; (b) capacitive strain sensor; and (c) piezoelectric strain sensor. Source: Wang et al. [19]. Reproduced with permission of Wiley.

14.2.1.1 Resistive Strain Sensor

Resistive strain sensor is the most widely studied sensor, which can detect strain deformation by measuring the electrical resistance change of conductive materials in the sensors [23]. When the strain is released, the electrical resistance of the sensor returns to its original state as conductive materials restore the original structures [19]. This type of strain sensor is mainly composed of active materials and a flexible substrate. Wang et al. [24] developed a novel flexible electrically resistive strain sensor with a special three-dimensional (3D) conductive network. They fabricated thermoplastic polyurethane (TPU) fibrous mats by electrospinning and used reduced graphene oxide (rGO) as the conductive filler. The nanosheets overlapped with each other and formed rGO conductive paths on the TPU fibers' surface. These conductive fibers connected and constructed an excellent 3D conductive network by using the flexible TPU mat as a skeleton. The special hierarchical conductive network endowed the rGO/TPU strain sensor with desirable integration of good stretchability and high sensitivity (gage factor of 11 in a strain of 10% and 79 in a strain of 100% in reversible strain regime), good durability and stability (stretch/release test of 6000 cycles), and fast response speed.

14.2.1.2 Capacitive Strain Sensor

Capacitive strain sensors are electronic components composed of two opposite electrodes made of active materials, which are separated by a dielectric layer of insulating material. When a direct current voltage is applied to the capacitive strain sensor, a capacitance is generated with opposite charges accumulated on the two textile electrodes due to the dielectric layer preventing current flow between the electrodes. As the shape of the sensor changes under mechanical deformation, the distance between the two electrodes varies, causing the capacitance to change. The capacitance of capacitive strain sensors increases linearly with the amount of induced strain.

14.2.1.3 Piezoelectric Strain Sensor

Mechanical deformation can also be converted into electrical energy through the use of piezoelectric strain sensors, which utilize piezoelectric materials to generate a voltage difference when external stimuli, such as pressure, tensile forces, compressive forces, and torsion, are applied [25]. Among polymers, poly(vinylidene fluoride) (PVDF) offers the highest piezoelectric coefficient [26], with electrospun PVDF nanofibers exhibiting a very high piezoelectric coefficient that can be used directly in sensor applications. PVDF fibers have excellent mechanical strength, low acoustic impedance, a flat frequency response, and a broad dynamic response. To monitor human motion, Asadnia et al. [27] proposed a fabrication strategy using electrospinning to develop a stretchable and sensitive PVDF nanofibrous strain sensor (Figure 14.2a). The strain sensor has a good response to strain measurement (Figure 14.2b). A smart glove assembled with stretchable strain sensors in each finger was fabricated and used for real-time motion detection of fingers (Figure 14.2c). As an application, the proposed strain sensors have been used in posture detection and the control of a robotic hand using the smart glove device.

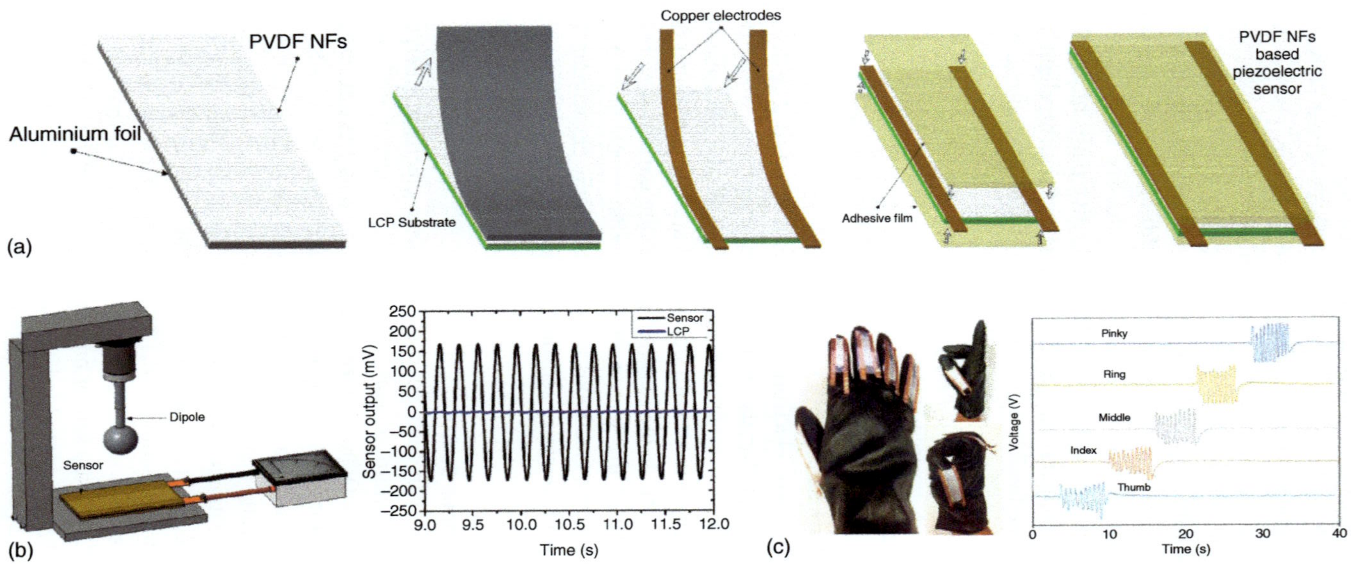

Figure 14.2 (a) Fabrication process of the strain sensor based on PVDF electrospun nanofiber on a flexible liquid crystal polymer (LCP) substrate. (b) Testing the sensor under low-frequency dynamic loads. (c) Flexible electrospun strain sensors for human-motion detection. Source: Khan et al. [27]/MDPI/CC BY 4.0.

14.2.2 Pressure Sensor

Strain and pressure sensors are widely used for human body detection, even though they differ in mechanism, composition, performance characteristics, and applications. Pressure sensors can directly convert external stimuli into electrical signals, and can be categorized into three types: piezoresistive, capacitive, and piezoelectric sensors (see Figure 14.3). Flexible pressure sensors have wide applications on account of their excellent electrical and mechanical properties, such as high flexibility, sensitivity, resolution ratio, and rapid response [28].

14.2.2.1 Piezoresistive Pressure Sensor

Pressure sensors are widely used for detecting external stimuli and can be divided into different types based on their sensing mechanism. One such type is the piezoresistive sensor, which works by changing the distribution and contact status of conductive fillers inside a composite material, resulting in the resistance of the composite changing. Elastic substrates such as polyurethane (PU), polyamide 6 (PA6), polyvinyl alcohol (PVA), and polyacrylonitrile (PAN) are commonly used for the fabrication of flexible pressure sensors. Conductive fillers such as carbon-based fillers, metal nanoparticles, and conductive polymers are assembled with the elastic substrate to form conductive materials. Various microstructures and nanoscale geometries, such as microstructures, porous structures, and nanonetwork structures, have been constructed to enhance the sensor's sensitivity and response time. Nanofiber structures, with their high flexibility, thinness, and abundant contact points, are particularly well-suited for the fabrication of high-sensitivity and rapid-response wearable pressure sensors.

Zhou et al. [29] developed a piezoresistive pressure sensor using a multilayer nanofiber network structure that is prepared through a simple electrospinning technique. This sensor comprises upper and lower flexible electrodes made of

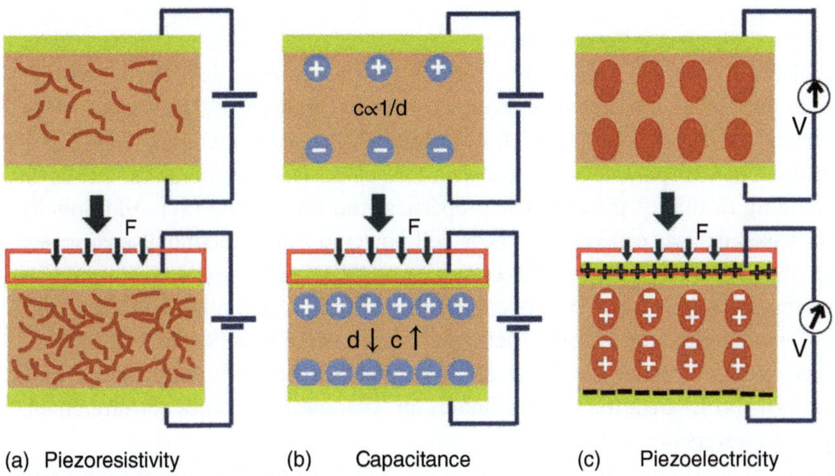

Figure 14.3 Classification of strain sensors: (a) piezoresistive pressure sensor; (b) piezocapacitive pressure sensor; and (c) piezoelectric pressure sensor. Source: Xu et al. [28]/MDPI/CC BY 4.0.

Au-deposited PA6 nanofiber networks, as well as a sensing layer composed of a PEDOT:PSS/PA6 composite nanofiber network. The conductive pathway is formed by interweaving PEDOT:PSS fibers as conductive fillers with PA6 fibers in the composite nanofiber network. The sensor exhibits a significant variation in electrical signal in response to weak external force, demonstrating an ultrahigh sensitivity (6554.6 kPa^{-1} at 0–1.4 kPa) and fast response (53 ms).

14.2.2.2 Piezocapacitive Pressure Sensor

Piezocapacitive sensors are a popular type of high-sensitivity sensor due to their excellent sensitivity, low power consumption, outstanding temperature insensitivity, and rapid dynamic response. Piezocapacitive pressure sensors can be considered as parallel plate capacitors, typically consisting of two flexible electrodes at the top and bottom and a flexible dielectric layer in between. The capacitance of the parallel plate capacitor is proportional to the dielectric constant and the corresponding overlapping surface of the two electrodes and inversely proportional to the distance between the electrodes. When an external force is applied to the sensor, the thickness of the piezocapacitive sensor decreases accordingly, resulting in a change in capacitance. By monitoring these piezocapacitive signals, the amount of external pressure applied can be determined.

Wang et al. [30] have developed a flexible piezocapacitive sensor based on a bionic komochi konbu structure elastomer (BKKE). The sensor has a Cu/ Ni nanofiber network prepared on flexible polyester (PET) films as the top and bottom electrodes, coupled with BKKE as the dielectric layer. The elastomer structure has significant protruding structures and internal micropores that increase the effective dielectric constant under compressive force. This increases the sensor's sensitivity and provides an ultrafast response and recovery time (162 ms). The sensor's response time is 250 times faster than a conventional flat polydimethylsiloxane (PDMS) capacitive sensor, which has a response time of 40.6 s.

14.2.2.3 Piezoelectric Pressure Sensor

Piezoelectric polymers are highly desirable for their mechanical flexibility, light weight, and stable operation without an external battery. PVDF, in particular, has been extensively researched for various applications, including nonvolatile memory, touch display devices, and wearable piezoelectric generators, due to its outstanding flexibility, piezoelectric properties, and nontoxicity [31]. Additionally, PVDF-based devices can easily achieve self-powered multifunctional sensing as a result of their piezoelectricity. The copolymer of PVDF, poly(vinylidene fluoride-co-trifluoroethylene) (P(VDF-TrFE)), can readily form a pure b-phase with the addition of a small amount of trifluoroethylene, resulting in superior crystallinity and piezoelectricity compared to PVDF [32, 33]. Consequently, piezoelectric polymers like P(VDF-TrFE) are attractive candidates for wearable electronics and stretchable energy harvesters in our daily lives.

Zhu et al. [34] constructed a wafer-scale, self-powered pressure sensor based on P(VDF-TrFE) nanofibers by electrospinning technique. The device can convert external pressure directly into electrical signals, allowing for the instantaneous detection of stress stimuli. As shown in Figure 14.4a, the device exhibits high

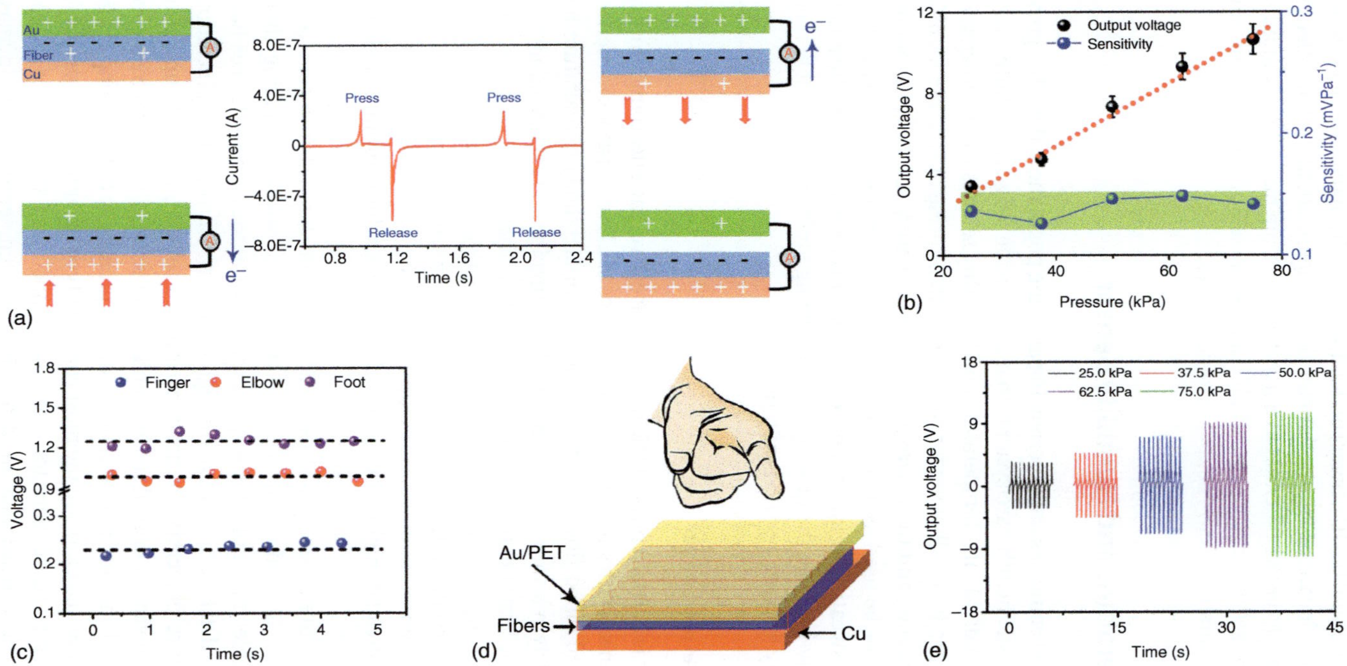

Figure 14.4 (a) Schematic of the charge formation and transport in the device during the pressing and releasing process. (b) Relationship of applied pressure and the output voltage, as well as the corresponding sensitivity. (c) Stability of output voltage induced by human activity movement. (d) Schematic illustration of human movement with the device. (e) Output voltage with different loading forces. Source: Zhu et al. [34]. Reproduced with permission of Royal Society of Chemistry.

sensitivity, fast response speed, excellent operational stability, and self-powered capability (Figure 14.4b,c). The piezoelectric output signals were stabilized at 1.3V (foot), 0.98V (elbow), and 0.23V (finger) under differential human movements when attached to various locations on the human body (Figure 14.4d,e). Piezoelectric polymer nanofibers have the potential to harvest energy from human body motions under various stresses, making them suitable for use in self-powered e-skins and multifunctional wearable micro-/nanoelectronic devices.

14.3 Temperature and Humidity Sensor

As many developed countries are experiencing an aging population, there is a growing need for domestic health-care systems that can provide long-term health monitoring for elderly individuals. Temperature and humidity sensors can be integrated into smart textile products, allowing for continuous monitoring of vital functions in patients, convalescents, and the elderly [35, 36]. Temperature sensors can monitor body temperature or be used as a presence sensor in beds, while humidity sensors can be utilized as sweating or incontinence sensors. This technology has the potential to significantly improve the quality of life for aging individuals and help them live independently for longer periods of time.

14.3.1 Temperature Sensor

Body temperature is a crucial physiological signal that can reveal important information about a person's health status, such as the presence of a fever, changes in thermal regulation, reduced blood flow, and muscle fatigue. In fact, body temperature has become particularly relevant in the context of the ongoing COVID-19 pandemic, where it is often the first sign of infection [37]. Therefore, there is a growing need for flexible temperature sensors that can provide high sensitivity and accuracy for continuous, long-term monitoring of body temperature. Such sensors could enable more personalized care and more efficient postoperative recovery, and they could even help prevent the spread of highly contagious diseases by enabling prompt detection of febrile individuals.

Flexible temperature sensors have garnered significant interest in recent years due to their ability to conform to irregular and curved surfaces, making them suitable for wearable and implantable applications. Numerous sensor devices have been developed to detect temperature through various physical changes, including resistive temperature detectors, thermistors, infrared sensors, thermocouple sensors, field-effect transistors, optical sensors, and silicon sensors, some of which have attracted substantial interest due to their favorable properties [38]. Among them, resistive temperature detectors are the most common type of temperature sensor.

The resistive temperature detector uses the temperature dependence of the material on electrical resistance to determine the temperature. An increase in temperature causes an increase in resistance due to electron vibrations at higher temperatures which prevent the free flow of electrons in the conductive material [39]. Resistive temperature detector sensors are widely used in many applications for their merits in high accuracy, fast response, physiological stability, simplicity of fabrication, and ease of mass production. To date, flexible temperature sensors have been developed mainly based on electrically conductive composites containing various fillers including graphene quantum dot, rGO [40], AgNWs [41], PEDOT:PSS [42], CNTs [43], carbon nanofiber (CNF) [44] and exfoliated graphite [11]. These conductive composites constitute the resistive temperature detector [45]. The sensitivity of the resistive-type temperature sensors is commonly evaluated using the temperature coefficient of resistance (TCR) [46]:

$$\mathrm{TCR} = \frac{\Delta R/R_0}{\Delta T} \times 100\%$$

where ΔR is the total resistance change upon temperature variation, ΔT, and R_0 is the initial resistance of the temperature sensor.

Lee et al. [44] have developed a flexible temperature sensor using electrospinning aligned PAN-based CNF films, which is highly responsive to temperature variation amidst many external stimuli. The temperature sensor made from an aligned CNF film exhibits excellent sensitivity of $1.52\%\,°\mathrm{C}^{-1}$, high accuracy, good linearity, a quick response time of 1.2 s, and excellent durability. Additionally, the aligned CNF temperature sensor has been proven to be insensitive to pressure, bending, and humidity. The highly selective temperature sensing performance is well-maintained even after repeated stretch/release cycles due to the highly aligned structure's ability to recover the sensing network.

In addition to the use in healthcare, flexible temperature sensors have also found applications in sports monitoring, food storage, and aerospace industries. With the rapid advancements in materials science and nanotechnology, it is expected that flexible temperature sensors will continue to play a critical role in real-time, noninvasive health monitoring, disease diagnosis, and personalized medicine.

14.3.2 Humidity Sensor

The subtle human body temperature fluctuation may lead to the change of microenvironment humidity, making humidity detection an important means of monitoring human security. With the development of wearable electronic systems, humidity sensors have garnered significant attention for human body humidity detection in areas such as respiratory behavior, speech recognition, skin moisture, noncontact switch [47], and diaper monitoring [48]. Human body-related humidity sensors can be mainly divided into resistance [49], capacitance [50], impedance, voltage, quartz crystal microbalance, and surface acoustic wave sensors [51], based on different humidity-sensing signal-conversion mechanisms.

As the core component, the choice of humidity-sensing materials plays a critical role in determining the sensing performance of humidity sensors. Various humidity-sensing materials have been reported for fabricating humidity sensors, including cellulose paper, carbon materials (e.g. graphene [52], graphene oxide [GO] [53], CNT, graphitic carbon nitride, rGO, and graphdiyne [54]), polymers, 2D materials (e.g. WS_2, $TiSi_2$, MoS_2, and $Ti_3C_2T_x$), metal oxides [48], and their

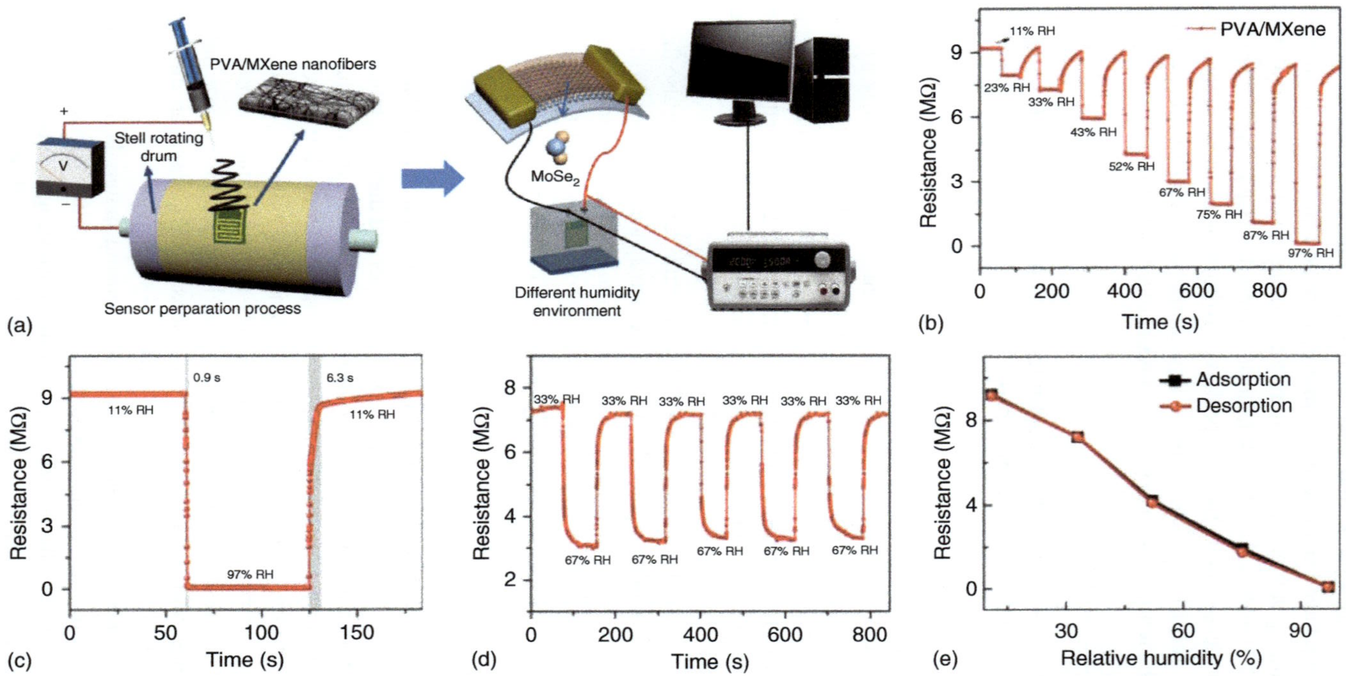

Figure 14.5 (a) Schematic diagram for the fabrication of the humidity sensor and the experimental platform for humidity sensing measurement. (b) Dynamic resistance changes of PVA/MXene film sensor under different RHs. (c) Response and recovery curves of the PVA/MXene sensor between 11 and 97% RHs. (d) Humidity hysteresis curves of the humidity sensor. (e) Repeatability of the humidity sensor. Source: Wang et al. [56]/Springer Nature/CC BY 4.0.

composites. Binary or multicomponent composites have been found to have better humidity sensing performance, as demonstrated by a large number of studies.

Zhu et al. [55] developed a high-sensitivity and durable bilayer humidity sensor using nanoporous cellulose nanofiber and CNT as the humidity-sensitive materials for water molecule exchange. The cellulose paper substrate plays a crucial role as (i) a support for the cellulose nanofiber/CNT layer, (ii) a water molecule reservoir for rapid moisture exchange between the paper fibers and the surrounding air, and (iii) a microporous holder for the adhesion and interlocking of cellulose nanofiber/CNT networks and paper fibers. This unique structural design results in a humidity sensor with a high response (up to 65.0%), excellent linearity ($R^2 = 0.995$), good stability (over 15 days), and outstanding bending (with a maximum curvature of $22.2\,cm^{-1}$), and folding durability (up to 50 times).

Wang et al. [56] have developed a self-powered flexible humidity-sensing device that utilizes a PVA/transition metal carbide or nitride (MXene) nanofiber film as the humidity-sensitive material and a monolayer $MoSe_2$ piezoelectric nanogenerator to convert mechanical energy into electric energy (Figure 14.5a). The PVA/MXene nanofiber film was electrospun onto interdigital electrodes to form the humidity sensor, which exhibited high sensitivity and a fast response/recovery time of 0.9/6.3 seconds, low hysteresis of 1.8%, and excellent repeatability (Figure 14.5b–e). The self-powered flexible humidity sensor was capable of detecting human skin moisture and ambient humidity, making it useful for wearable electronic applications. The device offers the advantage of being self-powered, reducing the need for external power sources, and allowing for greater mobility and flexibility.

14.4 Gas Sensor

Gas sensors play a vital role in detecting the presence of harmful gases in various environments, such as workplaces or homes [7]. The technology for gas sensors has been widely researched and developed for monitoring air quality, breath analysis for disease diagnosis, gas leakage detection, and hazard monitoring. The sensing principle of gas sensors is based on the changes in the electrical resistance of the sensing material when exposed to different gases [57]. A direct interaction occurs between the analytes and the sensing material, causing changes in the sensing material properties. The performance features of gas sensors, including sensitivity, selectivity, time response, stability, durability, reproducibility, and reversibility, are largely influenced by the properties of the sensing materials [58]. A higher surface area of the gas sensor corresponds to better gas adsorption, which leads to higher sensitivity [59]. Therefore, several studies have focused on developing morphologies with a high surface area to enhance the sensing performance of gas sensors. Electrospun fibers with controllable membrane thickness, fine structures, diversity of materials, and large specific surface are expected to be an ideal candidate as the structure of sensing materials. Various materials, such as polymers [60], semiconductors [61], metal nanoparticles, carbon graphite, and organic/inorganic composites [62], have been used as sensing materials to detect the targeted gases.

The majority of conductometric gas sensors are based on semiconducting metal oxides due to their tunable transport properties and the high sensitivity of their surface electronic properties to changes in the composition of the surrounding atmosphere. This is attributed to their large stoichiometry variability, relatively high catalytic activity, and the presence of different cationic oxidation states [63]. Up to now, many promising metal oxide semiconductor sensing materials, such as SnO_2, In_2O_3, ZnO, Fe_2O_3, WO_3, and NiO, have been continuously reported for detecting poisonous and flammable gases. The gas sensor that adopts metal oxide semiconductor as sensitive material has become the focus due to its low cost, high stability, simple operating principle, and easy fabrication [64].

Kou et al. [65] used electrospinning followed by a calcination process to synthesize 1–3 mol% Ru-doped SnO_2 nanofibers and studied the effect of Ru^{4+} doping on the gas sensing properties of SnO_2 nanofibers. They found that the sensors based on Ru-doped SnO_2 nanofibers exhibited improved gas sensing properties for all tested gases. In particular, the 2 mol% Ru-doped SnO_2 nanofibers showed a response of 118.8 to 100 ppm acetone at 200 °C, which was 12 times higher than that of pure SnO_2 nanofibers.

Nanocarbon materials, such as CNTs, graphene, GO, and CNFs, have recently emerged as a new category of gas sensors. Avossa et al. [66] prepared a thin nanofibrous layer comprising polymers polystyrene (PS) and polyhydroxibutyrate

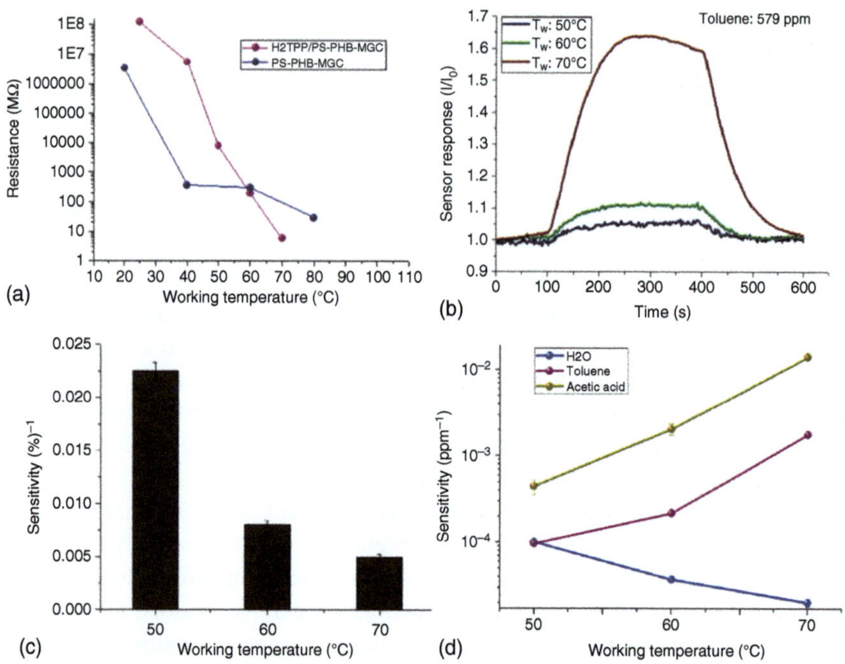

Figure 14.6 (a) The relationship of resistance and temperature for fibers. (b) Transient responses upon injection of 579 ppm of toluene. (c) Response curves of the electrode between different working temperatures. (d) Sensitivity values to water vapors and toluene between the different working temperatures. Source: Avossa et al. [66]/MDPI/CC BY 4.0.

(PHB), nanofillers of mesoporous graphitized carbon (MGC), and a free-based tetraphenylporphyrin (H_2TPP), which was deposited onto an interdigitated electrode by electrospinning technology. The sensor worked reliably and reproducibly between 50 and 70 °C without any significant degradation and revealed nonlinear relationships between the conductivity and the temperature. The electrical conductivity increased as the temperature increased, making it more resistive at lower temperature (Figure 14.6a). At higher temperature, H_2TPP promoted the conductivity of the fibrous layer. Furthermore, porphyrin increased the sensor sensitivity to toluene vapor (Figure 14.6b), and increased with increasing temperature, whereas the sensitivity to water vapor decreased with heating (Figure 14.6c,d).

14.5 Electrochemical Biosensor

Electrochemical biosensors, which combine electrochemical sensors and specific biomolecule identification, have been widely used in clinical diagnosis, analytical chemistry, food detection, and medical analysis. They are favored for their real-time monitoring, high sensitivity, high selectivity, low cost, and easy preparation. These biosensors utilize specific biochemical reactions, mediated by isolated enzymes, immune systems, tissues, organelles, or whole cells, to detect chemical compounds via electrical, thermal, or optical signals. These electrochemical biosensors consist of two sections: a transducer (electrode) and a biological recognition element (bioreceptor) to capture the biomarker. The bioreceptor is a biomolecule that recognizes the target analyte, and the transducer converts the recognition event into a measurable signal. The types of electrochemical biosensors include electrochemical enzyme sensors, electrochemical immunosensors, microbial electrochemical sensors, electrochemical DNA biosensors, and electrochemical tissue and cell sensors.

14.5.1 Electrochemical Enzyme Sensor

Electrochemical enzyme sensors, which immobilize enzymes on the electrode surfaces, can detect target analytics through their reaction or interaction with analytics. Due to their high sensitivity, good selectivity, water solubility, and low toxicity, electrochemical enzyme sensors are a promising tool for the analysis of clinically important molecules. To enhance the electrocatalytic activity and selectivity, many studies have employed enzyme-coated electrochemical sensors to detect a wide variety of molecules from complex samples.

Electrochemical biosensors based on glucose enzyme reactions have played a major role in the continuous monitoring of glucose levels in personal glucometers. Baek et al. [67] developed an electrochemical biosensor of Cu-nanoflower@AuNPs-GO nanofibers for detecting glucose. The GO-PVA solution was coated onto an Au chip by electrospinning, then decorated with gold nanoparticles (AuNPs) and an organic–inorganic hybrid nanoflower (Cu nanoflower-glucose oxidase and horseradish peroxides), fabricating the Cu-nanoflower@AuNPs-GO nanofibers. The nanofibers showed excellent electrochemical catalytic activity and selectivity for the conversion of glucose to gluconic acid. The resulting Au chip coated with Cu-nanoflower@AuNPs-GO nanofibers exhibited a good linear range of 0.001–0.1 mM with a detection limit of 0.018 μM. The use of immobilized enzymes on solid supports provides several advantages, including high enzyme-to-substrate ratio, great enzyme stability, reusability, and good diagnosis, making this electrochemical biosensor a promising tool for the analysis of clinically important molecules.

Electrochemical biosensors based on nonenzymatic reactions have gained increasing attention due to the limited lifetime of enzymes caused by environmental factors such as temperature and pH. Liu et al. [68] developed a novel electrochemical sensor by preparing hollow CuO/PANI nanohybrid fibers for accurate detection of hydrogen peroxide and glucose. The nonenzymatic electrochemical biosensor exhibits a wide detection range, low detection limit, high selectivity, and long-term stability, indicating its potential application in various fields such as food security, biomedicine, environmental detection, and pharmaceutical analysis.

14.5.2 Electrochemical Immunosensor

Electrochemical immunosensors are highly sensitive and specific detection tools that can be used to identify trace levels of various substances, including bacteria, viruses, drugs, hormones, pesticides, and numerous other chemicals. These sensors have numerous applications, such as monitoring food safety in relation to severe allergies, detecting environmental pollutants, identifying biomedical substances, and monitoring biowarfare agents such as toxins, bacteria, viruses, and spores. By using antibodies, antigens, or aptamers as the biological recognition elements, electrochemical immunosensors can detect the target analyte with high selectivity and sensitivity, providing a quick and reliable analysis of the samples.

Adabi et al. [69] developed an electrochemical immunosensor for the detection of human epidermal growth factor receptor 2 (Her-2) in human blood serum samples. The sensor utilized an electrospun CNF mat electrode that was modified with AuNPs, a cysteamine self-assembled monolayer, CNTs, and antibody molecules. The electrospun CNF mat served as both the transducer and matrix for the placement of the antibody. The combination of electrospun CNF mat, CNTs, and AuNPs resulted in a highly sensitive and selective sensor with good repeatability and reproducibility. The sensor showed great potential for the determination of Her-2 in human blood serum samples.

14.5.3 Microbial Electrochemical Sensor

Microbial electrochemical sensors utilize microbial electrochemical technology to sense electroactive microorganisms and/or biofilms in either the anodic or cathodic reactions. In this technology, an oxidation reaction occurs at the anode and a reduction reaction at the cathode, both of which can be catalyzed by microorganisms.

These microorganisms are known as electroactive microorganisms, and through their metabolism, they are capable of releasing electrons to the electrode (anode) or accepting electrons from the electrode (cathode). Both anodic and cathodic biofilms are sensitive to changes in operational conditions, and variations in these conditions can lead to an increase or decrease in current output. A positive or negative response can be observed through the variation of oxidant or reductant concentration provided at the cathode and anode, respectively.

14.5.3.1 Anodic Microbial Electrochemical Sensors

Electroactive microorganisms and biofilms that colonize the anode are able to pair their oxidative metabolisms with electron transfer to the anode, forming the basis of microbial electrochemical technology capable of sensing a metabolic substrate of interest by measuring the electroactive microorganism's metabolic activity via a current response. Anodic microbial electrochemical sensors have various applications, such as measuring biological or chemical oxygen demand, assessing toxicity, and detecting pathogens.

Khan et al. [70] coated a stainless-steel electrode with a mixture of $MgFe_2O_4$, PVDF, and N-methyl-2-pyrrolidone via electrospinning method to create a microbial fuel cell (MFC) biosensor capable of detecting toxicants. The study found a direct correlation between the concentration of the added toxicant and the MFC biosensor's performance. The concentration versus voltage output graph showed an almost perfect linear fit, demonstrating the biosensor's potential as a toxicity detector. The cyclic voltammetry analysis confirmed the bioactivity of the MFC reactor compared to the unacclimated system. Moreover, the cyclic voltammetry analysis of MFCs at different dye concentrations suggested their promising toxicity-sensing potential. Electrochemical impedance spectroscopy data confirmed that biofouling affected the internal resistances of MFCs. The study also compared the voltage outputs of this work with a previous study to establish the effect of extended run time on MFC performance. Compared to the previous study, the voltage output for MFCs with plain and magnesium ferrite-coated stainless-steel mesh declined by nearly 53% and 41%, respectively, indicating fouling of the system. The MFC with a modified electrode showed a lesser drop in performance, suggesting the potential of a modified electrode for more efficient electron transfer in the long run.

14.5.3.2 Cathodic Microbial Electrochemical Sensors

Electroactive microorganisms and biofilms can also populate the cathode and are capable of pairing a biological reduction reaction with electron transfer from the cathode. This reaction often involves the catalyzed oxygen reduction reaction (ORR) either biotically or abiotically. The applications of microbial electrochemical sensors utilizing the cathode include corrosion sensing, dissolved oxygen determination, and toxicity measurements.

Jiang et al. [71] developed a composite cathode by loading a metal–organic framework (MOF) onto the surface of PAN nanofibers, which improved the ORR performance of MFC cathode. The resulting material, called MOF/CNFs, was a self-supporting electrode with a large specific surface area and layered porous structure. The MOF/CNF electrode showed excellent ORR catalytic performance, which led to improved power generation capacity and higher pollutant removal efficiency in MFCs equipped with this cathode. Furthermore, the MOF/CNF electrode accelerated the cathode ORR reaction rate and improved the activity of electricity-producing bacteria in MFC anode, resulting in better power output.

14.5.4 Electrochemical DNA Biosensor

Electrochemical DNA biosensor is a biosensor system that uses DNA as a sensitive element and converts the signals generated during the specific recognition of nucleic acid into electrical signals through transducers. This approach can achieve qualitative or quantitative detection of nucleic acid. Electrochemical DNA biosensor combines the characteristics of both nucleic acid molecular hybridization technology and electrochemical methods, offering advantages in terms of cost, efficiency, and accuracy. As technology continues to advance, electrochemical DNA biosensors are expected to have great potential for development in various fields, such as genetic disease diagnosis, drug mechanism analysis, and environmental pollution monitoring.

Hui et al. [72] prepared novel composite nanofibers by grafting polyethylene glycol (PEG) polymer onto PANI nanofibers to create antifouling electrochemical biosensors. The PANI/PEG nanofibers possess a large surface area and remain conductive, while exhibiting excellent antifouling properties in both single-protein solutions and complex human serum samples. The attachment of DNA probes to the PANI/PEG nanofibers allows for the fabrication of sensitive and low-fouling electrochemical biosensors for the breast cancer susceptibility gene (BRCA1). The biosensor shows a high sensitivity to target BRCA1, with a linear range of 0.01 pM to 1 nM, and can detect DNA mismatches with satisfactory selectivity. Moreover, the DNA biosensor based on the PANI/PEG nanofibers supported the quantification of BRCA1 in complex human serum, indicating the great potential of this novel biomaterial for application in biosensors and bioelectronics.

14.5.5 Electrochemical Tissue and Cell Sensor

Liu et al. [73] designed a stretchable electrochemical sensor for real-time monitoring of cells and tissues by synthesizing Au-nanotube networks through mild galvanic displacement of sacrificial AgNWs on PDMS film. The interlacing network of Au nanotubes provides desirable stability against mechanical deformation, while the Au nanostructure offers excellent electrochemical performance and biocompatibility. By culturing cells on the sensor, the team was able to achieve real-time monitoring of nitric oxide release from mechanically sensitive human umbilical vein endothelial cells in both their releasing and stretching states. Additionally, the sensor was successfully rolled up to interface with the circular lining of elastic human vein and monitored nitric oxide release therein.

14.6 Conclusion and Perspective

The development of flexible wearable and implantable sensors that can accurately detect physiological signals in real time has become a hot topic in both academia and industry. However, traditional planar sensors with 3D architectures and large volumes are difficult to integrate into irregular and soft dynamic tissues, leading to unstable device–tissue interfaces. To address this issue, micro-/nanofiber sensors created through electrospinning have gained popularity due to their large specific surface area, high density of sensitive sites, lightweight nature, and mechanical properties that match those of human tissues. Various types of flexible sensors based on electrospun micro-/nanofibers have been developed, including mechanical, temperature and humidity, gas, and electrochemical biosensors, which could provide valuable information for researchers planning and executing future research.

References

1 Lou, Z., Wang, L., and Shen, G. (2018). Recent advances in smart wearable sensing systems. *Advanced Materials Technologies* 3 (12): 1–17. https://doi.org/10.1002/admt.201800444.

2 Kumar, S.S. and Pant, B.D. (2014). Design principles and considerations for the 'ideal' silicon piezoresistive pressure sensor: a focused review. *Microsystem Technologies* 20 (7): 1213–1247. https://doi.org/10.1007/s00542-014-2215-7.

3 Unno, Y. et al. (2011). Development of n-on-p silicon sensors for very high radiation environments. *Nuclear Instruments and Methods in Physics Research Section A: Accelerators, Spectrometers, Detectors and Associated Equipment* 636 (1 SUPPL): 24–30. https://doi.org/10.1016/j.nima.2010.04.080.

4 Qiu, Q. et al. (2019). Highly flexible, breathable, tailorable and washable power generation fabrics for wearable electronics. *Nano Energy* 58 (February): 750–758. https://doi.org/10.1016/j.nanoen.2019.02.010.

5 Zheng, Y. et al. (2022). Durable, stretchable and washable inorganic-based woven thermoelectric textiles for power generation and solid-state cooling. *Energy & Environmental Science* https://doi.org/10.1039/d1ee03633e.

6 Xiong, J., Chen, J., and Lee, P.S. (2021). Functional fibers and fabrics for soft robotics, wearables, and human–robot interface. *Advanced Materials* 33 (19): 1–43. https://doi.org/10.1002/adma.202002640.

7 Ding, B., Wang, M., Yu, J., and Sun, G. (2009). Gas sensors based on electrospun nanofibers. *Sensors* 9 (3): 1609–1624. https://doi.org/10.3390/s90301609.

8 Zhou, K., Dai, K., Liu, C., and Shen, C. (2020). Flexible conductive polymer composites for smart wearable strain sensors. *SmartMat* 1 (1): 1–5. https://doi.org/10.1002/smm2.1010.

9 Liu, H. et al. (2019). A flexible multimodal sensor that detects strain, humidity, temperature, and pressure with carbon black and reduced graphene oxide hierarchical composite on paper. *ACS Applied Materials & Interfaces* 11 (43): 40613–40619. https://doi.org/10.1021/acsami.9b13349.

10 Nag, A., Mukhopadhyay, S.C., and Kosel, J. (2016). Flexible carbon nanotube nanocomposite sensor for multiple physiological parameter monitoring. *Sensors and Actuators A: Physical* 251: 148–155. https://doi.org/10.1016/j.sna.2016.10.023.

11 Wang, F. et al. (2020). Flexible wearable graphene/alginate composite non-woven fabric temperature sensor with high sensitivity and anti-interference. *Cellulose* 27 (4): 2369–2380. https://doi.org/10.1007/s10570-019-02951-7.

12 Zhu, F. et al. (2020). Integrated multifunctional flexible electronics based on tough supramolecular hydrogels with patterned silver nanowires. *Journal of Materials Chemistry C* 8 (23): 7688–7697. https://doi.org/10.1039/d0tc01011a.

13 Wang, D. et al. (2021). Freestanding silver/polypyrrole composite film for multifunctional sensor with biomimetic micropattern for physiological signals monitoring. *Chemical Engineering Journal* 404 (June 2020): 126940. https://doi.org/10.1016/j.cej.2020.126940.

14 Wang, Y., Mao, H., Wang, Y. et al. (2020). 3D geometrically structured PANI/CNT-decorated polydimethylsiloxane active pressure and temperature dual-parameter sensors for man-machine interaction applications. *Journal of Materials Chemistry A* 8 (30): 15167–15176. https://doi.org/10.1039/d0ta05651k.

15 Sensors, T. and Pss, P. (2019). Development of flexible and wearable. 66 (7): 3129–3133.

16 Zhang, C.L., Lu, B.R., Cao, F.H. et al. (2018). Hierarchically structured Co_3O_4@carbon porous fibers derived from electrospun ZIF-67/PAN nanofibers as anodes for lithium ion batteries. *Journal of Materials Chemistry A* 6 (27): 12962–12968. https://doi.org/10.1039/c8ta03397h.

17 Wang, Y., Yokota, T., and Someya, T. (2021). Electrospun nanofiber-based soft electronics. *NPG Asia Materials* 13 (1): https://doi.org/10.1038/s41427-020-00267-8.

18 Trung, T.Q. and Lee, N.E. (2016). Flexible and stretchable physical sensor integrated platforms for wearable human-activity monitoringand personal healthcare. *Advanced Materials* 28 (22): 4338–4372. https://doi.org/10.1002/adma.201504244.

19 Wang, J., Lu, C., and Zhang, K. (2020). Textile-based strain sensor for human motion detection. *Energy & Environmental Materials* 3 (1): 80–100. https://doi.org/10.1002/eem2.12041.

20 Liao, X., Zhang, Z., Kang, Z. et al. (2017). Ultrasensitive and stretchable resistive strain sensors designed for wearable electronics. *Materials Horizons* 4 (3): 502–510. https://doi.org/10.1039/c7mh00071e.

21 Kim, S.R., Kim, J.H., and Park, J.W. (2017). Wearable and transparent capacitive strain sensor with high sensitivity based on patterned Ag nanowire networks. *ACS Applied Materials & Interfaces* 9 (31): 26407–26416. https://doi.org/10.1021/acsami.7b06474.

22 Ji, Z. and Zhang, M. (2022). Highly sensitive and stretchable piezoelectric strain sensor enabled wearable devices for real-time monitoring of respiratory and heartbeat simultaneously. *Nami Jishu yu Jingmi Gongcheng/Nanotechnology and Precision Engineering* 5 (1): https://doi.org/10.1063/10.0009365.

23 Yamada, T. et al. (2011). A stretchable carbon nanotube strain sensor for human-motion detection. *Nature Nanotechnology* 6 (5): 296–301. https://doi.org/10.1038/nnano.2011.36.

24 Wang, Y. et al. (2018). Flexible electrically resistive-type strain sensors based on reduced graphene oxide-decorated electrospun polymer fibrous mats for human motion monitoring. *Carbon N. Y.* 126: 360–371. https://doi.org/10.1016/j.carbon.2017.10.034.

25 Zhou, J. et al. (2008). Flexible piezotronic strain sensor. *Nano Letters* 8 (9): 3035–3040. https://doi.org/10.1021/nl802367t.

26 Yang, Z. et al. (2021). Ultra-sensitive flexible sandwich structural strain sensors based on a silver nanowire supported PDMS/PVDF electrospun membrane substrate. *Journal of Materials Chemistry C* 9 (8): 2752–2762. https://doi.org/10.1039/d0tc04659k.

27 Khan, H., Razmjou, A., Warkiani, M.E. et al. (2018). Sensitive and flexible polymeric strain sensor for accurate human motion monitoring. *Sensors (Switzerland)* 18 (2): 1–10. https://doi.org/10.3390/s18020418.

28 Xu, F. et al. (2018). Recent developments for flexible pressure sensors: a review. *Micromachines* 9 (11): 1–17. https://doi.org/10.3390/mi9110580.

29 Zhou, Y. et al. (2021). All-nanofiber network structure for ultrasensitive piezoresistive pressure sensors. *ACS Applied Materials & Interfaces* https://doi.org/10.1021/acsami.1c24257.

30 Wang, J., Suzuki, R., Shao, M. et al. (2019). Capacitive pressure sensor with wide-range, bendable, and high sensitivity based on the Bionic Komochi Konbu structure and Cu/Ni Nanofiber network. *ACS Applied Materials & Interfaces* 11 (12): 11928–11935. https://doi.org/10.1021/acsami.9b00941.

31 Yang, Y. et al. (2020). Flexible piezoelectric pressure sensor based on polydopamine-modified $BaTiO_3$/PVDF composite film for human motion monitoring. *Sensors and Actuators A: Physical* 301: 111789. https://doi.org/10.1016/j.sna.2019.111789.

32 Sharma, T., Je, S.S., Gill, B., and Zhang, J.X.J. (2012). Patterning piezoelectric thin film PVDF-TrFE based pressure sensor for catheter application. *Sensors and Actuators A: Physical* 177: 87–92. https://doi.org/10.1016/j.sna.2011.08.019.

33 Jiang, J. et al. (2020). Flexible piezoelectric pressure tactile sensor based on electrospun $BaTiO_3$/poly(vinylidene fluoride) nanocomposite membrane. *ACS Applied Materials & Interfaces* 12 (30): 33989–33998. https://doi.org/10.1021/acsami.0c08560.

34 Zhu, M., Chng, S.S., Cai, W. et al. (2020). Piezoelectric polymer nanofibers for pressure sensors and their applications in human activity monitoring. *RSC Advances* 10 (37): 21887–21894. https://doi.org/10.1039/d0ra03293j.

35 Soukup, R., Hamacek, A., Mracek, L., and Reboun, J. (2014). Textile based temperature and humidity sensor elements for healthcare applications. *Proceedings of the 2014 37th International Spring Seminar on Electronics Technology* 407–411. https://doi.org/10.1109/ISSE.2014.6887634.

36 Liang, T. and Yuan, Y.J. (2016). Wearable medical monitoring systems based on wireless networks: a review. *IEEE Sensors Journal* 16 (23): 8186–8199. https://doi.org/10.1109/JSEN.2016.2597312.

37 Tansey, E.A. and Johnson, C.D. (2015). Recent advances in thermoregulation. *Advances in Physiology Education* 39 (1): 139–148. https://doi.org/10.1152/advan.00126.2014.

38 Arman Kuzubasoglu, B. and Kursun Bahadir, S. (2020). Flexible temperature sensors: a review. *Sensors and Actuators A: Physical* 315: 112282. https://doi.org/10.1016/j.sna.2020.112282.

39 Prajapati, H. and Deshmukh, N.N. (2019). Design and development of thin wire sensor for transient temperature measurement. *Measurement: Journal of the International Measurement Confederation* 140: 582–589. https://doi.org/10.1016/j.measurement.2019.04.020.

40 Liu, G. et al. (2018). A flexible temperature sensor based on reduced graphene oxide for robot skin used in internet of things. *Sensors (Switzerland)* 18 (5): https://doi.org/10.3390/s18051400.

41 Li, S. et al. (2019). High-performance temperature sensor based on silver nanowires. *Materials Today Communications* 20 (June): 100546. https://doi.org/10.1016/j.mtcomm.2019.100546.

42 Bali, C., Brandlmaier, A., Ganster, A. et al. (2016). Fully inkjet-printed flexible temperature sensors based on carbon and PEDOT: PSS. *Materials Today Proceedings* 3 (3): 739–745. https://doi.org/10.1016/j.matpr.2016.02.005.

43 Karimov, K.H.S., Khalid, F.A., Chani, M.T.S. et al. (2012). Carbon nanotubes based flexible temperature sensors. *Optoelectronics and Advanced Materials, Rapid Communications* 6 (1–2): 194–196.

44 Lee, J.H. et al. (2021). Flexible temperature sensors made of aligned electrospun carbon nanofiber films with outstanding sensitivity and selectivity towards temperature. *Materials Horizons* 8 (5): 1488–1498. https://doi.org/10.1039/d1mh00018g.

45 Yang, Z. et al. (2022). Carbonized silk nanofibers in biodegradable, flexible temperature sensors for extracellular environments. *ACS Applied Materials & Interfaces* 14 (16): 18110–18119. https://doi.org/10.1021/acsami.2c00384.

46 Wu, L. et al. (2019). Screen-printed flexible temperature sensor based on FG/CNT/PDMS composite with constant TCR. *Journal of Materials Science: Materials in Electronics* 30 (10): 9593–9601. https://doi.org/10.1007/s10854-019-01293-1.

47 Lu, L., Jiang, C., Hu, G. et al. (2021). Flexible noncontact sensing for human–machine interaction. *Advanced Materials* 33 (16): 1–10. https://doi.org/10.1002/adma.202100218.

48 Duan, Z., Jiang, Y., and Tai, H. (2021). Recent advances in humidity sensors for human body related humidity detection. *Journal of Materials Chemistry C* 9 (42): 14963–14980. https://doi.org/10.1039/d1tc04180k.

49 Su, P.G. and Wang, C.S. (2007). Novel flexible resistive-type humidity sensor. *Sensors and Actuators B: Chemical* 123 (2): 1071–1076. https://doi.org/10.1016/j.snb.2006.11.015.

50 Reddy, A.S.G., Narakathu, B.B., Atashbar, M.Z. et al. (2011). Fully printed flexible humidity sensor. *Procedia Engineering* 25: 120–123. https://doi.org/10.1016/j.proeng.2011.12.030.

51 Farahani, H., Wagiran, R., and Hamidon, M.N. (2014). Humidity sensors principle, mechanism, and fabrication technologies: a comprehensive review. *Sensors* 14 (5): https://doi.org/10.3390/s140507881.

52 Liang, R., Luo, A., Zhang, Z. et al. (2020). Research progress of graphene-based flexible humidity sensor. *Sensors (Switzerland)* 20 (19): 1–17. https://doi.org/10.3390/s20195601.

53 Wang, Y., Zhang, L., Zhang, Z. et al. (2020). High-sensitivity wearable and flexible humidity sensor based on graphene oxide/non-woven fabric for respiration monitoring. *Langmuir* 36 (32): 9443–9448. https://doi.org/10.1021/acs.langmuir.0c01315.

54 Yan, H. et al. (2018). Carbon atom hybridization matters: ultrafast humidity response of graphdiyne oxides. *Angewandte Chemie, International Edition* 57 (15): 3922–3926. https://doi.org/10.1002/anie.201709417.

55 Zhu, P., Ou, H., Kuang, Y. et al. (2020). Cellulose nanofiber/carbon nanotube dual network-enabled humidity sensor with high sensitivity and durability. *ACS Applied Materials & Interfaces* 12 (29): 33229–33238. https://doi.org/10.1021/acsami.0c07995.

56 Wang, D., Zhang, D., Li, P. et al. (2021). Electrospinning of flexible poly(vinyl alcohol)/MXene nanofiber-based humidity sensor self-powered by monolayer molybdenum diselenide piezoelectric nanogenerator. *Nano-Micro Lett.* 13 (1): 1–13. https://doi.org/10.1007/s40820-020-00580-5.

57 Nazemi, H., Joseph, A., Park, J., and Emadi, A. (2019). Advanced micro-and nano-gas sensor technology: a review. *Sensors (Switzerland)* 19 (6): https://doi.org/10.3390/s19061285.

58 Sanchez, M. and Rinco, M. (2011). Ammonia sensors based on composites of carbon nanotubes and titanium dioxide. *Carbon Nanotubes: Growth and Applications* 287 (January): 622–625. https://doi.org/10.5772/20466.

59 Wang, C., Yin, L., Zhang, L. et al. (2010). Metal oxide gas sensors: sensitivity and influencing factors. *Sensors* 10 (3): 2088–2106. https://doi.org/10.3390/s100302088.

60 Song, P., Wang, Q., and Yang, Z. (2011). Ammonia gas sensor based on PPy/$ZnSnO_3$ nanocomposites. *Materials Letters* 65 (3): 430–432. https://doi.org/10.1016/j.matlet.2010.10.087.

61 Park, J. and Tabata, H. (2021). Gas sensor array using a hybrid structure based on zeolite and oxide semiconductors for multiple bio-gas detection. *ACS Omega* 6 (33): 21284–21293. https://doi.org/10.1021/acsomega.1c01435.

62 Seekaew, Y., Lokavee, S., Phokharatkul, D. et al. (2014). Low-cost and flexible printed graphene-PEDOT:PSS gas sensor for ammonia detection. *Organic Electronics* 15 (11): 2971–2981. https://doi.org/10.1016/j.orgel.2014.08.044.

63 Korotcenkov, G. and Cho, B.K. (2017). Metal oxide composites in conductometric gas sensors: achievements and challenges. *Sensors and Actuators B: Chemical* 244: 182–210. https://doi.org/10.1016/j.snb.2016.12.117.

64 Dey, A. (2018). Semiconductor metal oxide gas sensors: a review. *Materials Science & Engineering, B: Solid-State Materials for Advanced Technology* 229 (December 2017): 206–217. https://doi.org/10.1016/j.mseb.2017.12.036.

65 Kou, X. et al. (2020). High-performance acetone gas sensor based on Ru-doped SnO_2 nanofibers. *Sensors and Actuators B: Chemical* 320 (May): https://doi.org/10.1016/j.snb.2020.128292.

66 Avossa, J. et al. (2019). Electrospinning of polystyrene/polyhydroxybutyrate nanofibers doped with porphyrin and graphene for chemiresistor gas sensors. *Nanomaterials* 9 (2): https://doi.org/10.3390/nano9020280.

67 Baek, S.H. et al. (2020). Cu-nanoflower decorated gold nanoparticles-graphene oxide nanofiber as electrochemical biosensor for glucose detection. *Materials Science and Engineering: C* 107 (August 2019): 110273. https://doi.org/10.1016/j.msec.2019.110273.

68 Liu, T., Guo, Y., Zhang, Z. et al. (2019). Fabrication of hollow CuO/PANI hybrid nanofibers for non-enzymatic electrochemical detection of H_2O_2 and glucose. *Sensors and Actuators B: Chemical* (December 2018): 286, 370–376. https://doi.org/10.1016/j.snb.2019.02.006.

69 Adabi, M., Esnaashari, S.S., and Adabi, M. (2021). An electrochemical immunosensor based on electrospun carbon nanofiber mat decorated with gold nanoparticles and carbon nanotubes for the detection of breast cancer. *Journal of Porous Materials* 28 (2): 415–421. https://doi.org/10.1007/s10934-020-01004-w.

70 Khan, N., Hakeem, A., Sultana, S. et al. (2022). Journal of environmental chemical engineering effective toxicity assessment of synthetic dye in microbial fuel cell biosensor with spinel nanofiber anode. *Journal of Environmental Chemical Engineering* 10 (2): 107313. https://doi.org/10.1016/j.jece.2022.107313.

71 Jiang, N. et al. (2021). Enhanced bioelectricity output of microbial fuel cells via electrospinning zeolitic imidazolate framework-67/polyacrylonitrile carbon nanofiber cathode. *Bioresource Technology* 337 (April): 125358. https://doi.org/10.1016/j.biortech.2021.125358.

72 Hui, N., Sun, X., Niu, S., and Luo, X. (2017). PEGylated polyaniline nanofibers: antifouling and conducting biomaterial for electrochemical DNA sensing. *ACS Applied Materials & Interfaces* 9 (3): 2914–2923. https://doi.org/10.1021/acsami.6b11682.

73 Liu, Y.L. et al. (2016). Stretchable electrochemical sensor for real-time monitoring of cells and tissues. *Angewandte Chemie, International Edition* 55 (14): 4537–4541. https://doi.org/10.1002/anie.201601276.

15

Preparation and Application of Electrospun Liquid-Metal-Based Stretchable Electronics

Maorong Zheng and Liming Wang

Donghua University, College of Textiles, Key Laboratory of Textile Science & Technology, Ministry of Education, 2999 North Renmin Road, Songjiang District, Shanghai 201620, China

15.1 Introduction

With the rapid development of wearable smart electronic devices [1], electronic skins [2], flexible robots [3, 4], and health-monitoring devices [5–8] are in great demand. Stretchable electronics, with high stretchability while maintaining high electrical conductivity, have proven to be indispensable components of wearable electronics [9], flexible displays [10], soft robotics [11], and deformable supercapacitors/batteries [12]. Therefore, flexible, stretchable electronics attract widespread attention and applications. Stretchable electronics are currently prepared by pre-strain buckling, which primarily results in geometric distortion of conductive paths, or by structuring conductive materials such as metal, carbon nanotubes, graphene, and their composites into transformable geometries such as buckle, helical, and spring [13–15]. Liu et al. [13], for instance, described the production of highly flexible sheath–core conducting fibers by wrapping carbon nanotube sheets with the fiber direction in the cores of stretched rubber fibers. Inspired by the peristaltic behavior of arthropods, by applying the worm-shaped graphene microlayer to polyurethane filaments, Sun et al. [16] were able to create a graphene coating that resembled the caterpillar structure on the polyurethane (PU) fiber surface. These worm-shaped filaments can be stretched up to 1010% with a wide reversible electroresponse range ($0 < \varepsilon < 815\%$), long-term durability (>4000 stretching/releasing cycles), and good initial conductivity ($\sigma_0 = 124\,\mathrm{S\,m^{-1}}$). While these studies have led to tremendous progress in the field of stretchable electronics, there are some obvious drawbacks to this approach. On the one hand, nonstretchable conductive materials are prone to fracture and delamination from elastic polymers, resulting in low conductivity and durability [17]. On the other hand, the fabrication process is mostly complex two-step or multi-step process that does not allow continuous fabrication of long strain-insensitive fiber conductors.

Liquid metals (LMs), with excellent fluidity, low viscosity, high metallic conductivity, and low toxicity, are potential inorganic materials for flexible and deformable

Electrospinning: Fundamentals, Methods, and Applications, First Edition.
Edited by Liming Wang and Xiaohong Qin.

electronics [18]. Recently, gallium-based LMs have attracted widespread attention and provided great help for the development and research of flexible and stretchable electronics [19, 20]. Gallium (Ga), as we are all aware, has a melting temperature of roughly 29 °C and a bulk viscosity that is comparable to that of water. Ga is alloyed with indium (In) and tin (Sn) to produce eutectic gallium indium (EGaIn) and eutectic gallium indium tin (EGaInSn), whose melting points are further lowered below the melting point, allowing LMs to operate and process at a wider range of temperatures. Gallium-based LMs are also less hazardous than typical mercury, making them safer to use [21] and having good biocompatibility [22–25]. Most importantly, due to the high electrical conductivity and excellent fluidity of LMs, the combination of LMs and elastomers is an effective strategy for the fabrication of advanced stretchable electronic devices [26].

Stretchable devices not only have flexibility and can adapt to various nonplanar working environments, but also need to maintain stable device functions under large tensile stress. In recent years, the fabrication of flexible devices by electrospinning has also attracted great interest from researchers, and electrospun fiber materials have a wide range of potential applications in fields such as nanogenerators, flexible strain sensors, and stretchable electrodes. This is mainly related to some characteristics of electrospinning. For making nanofibers, one straightforward and low-cost approach is electrospinning. The generated nanofibers or nanofiber membranes exhibit high specific surface area to volume ratio, high porosity and tiny pore size, good flexibility, light weight, and other properties [27]. Electrospinning is an efficient method for preparing flexible composite fibers, and LMs has good fluidity and conductivity. The combination of the two has opened up broad development prospects for stretchable electronic products. Therefore, researchers have done a lot of research and efforts on how to combine electrospinning and LMs to fabricate high-performance stretchable devices.

15.2 Combination Method of Electrospinning and LMs

Combination of LMs and electrospinning technologies holds great promise for stretchable electronics to simultaneously achieve electrical stability and cyclic durability when stretched. Although the LMs have high surface tension and the poor interfacial connection between the LM and the elastic matrix makes it impossible to maintain the stability of the resistance under large deformation, there are many methods to solve these problems, and great progress has been made.

15.2.1 Direct Spinning

15.2.1.1 In Situ Assembly of Electrostatic Spraying and Electrospinning

Cao et al. [28] designed a LM nanoparticle-based highly robust stretchable electrode (NHSE) with a self-adaptable interface, which can mimic water-to-net interaction to solve the stretchability–conductivity dilemma. The specific preparation process is mainly through the in situ assembly of electrospun elastic nanofiber scaffolds and electrosprayed LM nanoparticles (Figure 15.1a).

The main properties of the material prepared by the above mentioned method are superior to the state-of-the-art reports. Firstly, the sheet resistance of NHSE is extremely low, about 52 $m\Omega\,sq^{-1}$ (Figure 15.1b). Moreover, NHSE is insensitive to a large degree of mechanical stretching and cyclic deformation, such that the electrical resistance change is 350% upon 570% tensile deformation and 5% electrical resistance increases after 330000 stretching cycles with 100% elongation (Figure 15.1c,f). In addition, the robustness and stability of NHSE under long-term exposure to air (420 days) cyclic submersion (30000 times) has been demonstrated (Figure 15.1g,h).

15.2.1.2 Dope Blending

The method of dope blending is to mix a certain amount of LM into the polymer solution to form the spinning solution required for electrospinning. Sha et al. [29] reported a method for introducing LM (EGaInSn with Ga 68.5 wt%, In 21.5 wt%, and Sn 10 wt%) nanodroplets into electrospun polyvinylidene fluoride-co-hexafluoropropylene (PVDF-HFP) nanofibers to enhance their triboelectric performance. In the pictures of the surface morphologies of LM-modified PVDF-HFP nanofibers, most of the nanofibers were bead-on-string morphology, with diameters ranging from 100 to 200 nm, and often appeared in electrospun fibers from PVDF-based polymers, and only a small portion of the LM droplets were deposited outside the fibers; the majority were nanoscale in size and scattered within the nanofibers. The comprehensive properties of electrospun materials are closely related to the LM content. When the mass ratio of 2% was added to the PVDF-HFP, EGaInSn demonstrated the best multifunctionality in terms of mechanical performance and electrical generation. For instance, when TPU is utilized as the positive friction layer and PVDF-HFP/2% LM nanofiber membrane is employed as the negative friction layer, the resulting triboelectric nanogenerators (TENG) can have a peak open-circuit voltage and power density of 1680 V and 24 $W\,m^{-2}$, respectively.

Ye et al. [30] also adopted a spinning solution blending method to incorporate LM into polyacrylonitrile (PAN) polymer, which was then electrospun into a nanofibrous membrane to prepare TENG. The outcomes show that enhancing the PAN nanofiber friction layer with a modest quantity of LM concentration can greatly enhance TENG performance. With 1.5 wt% LM concentration, the current density rises by roughly 40%, and the output voltage and charge density both rise by over 70%.

15.2.2 Post Finishing

15.2.2.1 Coating

The coating method mentioned here is mainly to combine LM with electrospun nanofiber materials by scraping, painting, and brushing. Ma et al. [31] designed a permeable superelastic LM fiber mat that enables biocompatible and monolithic stretchable electronics. Three steps make up its preparation process: electrospinning of the superelastic poly(styrene-block-butadiene-block-styrene) (SBS) fiber mat, coating the stretchable fiber mat with LM, and activating permeability through

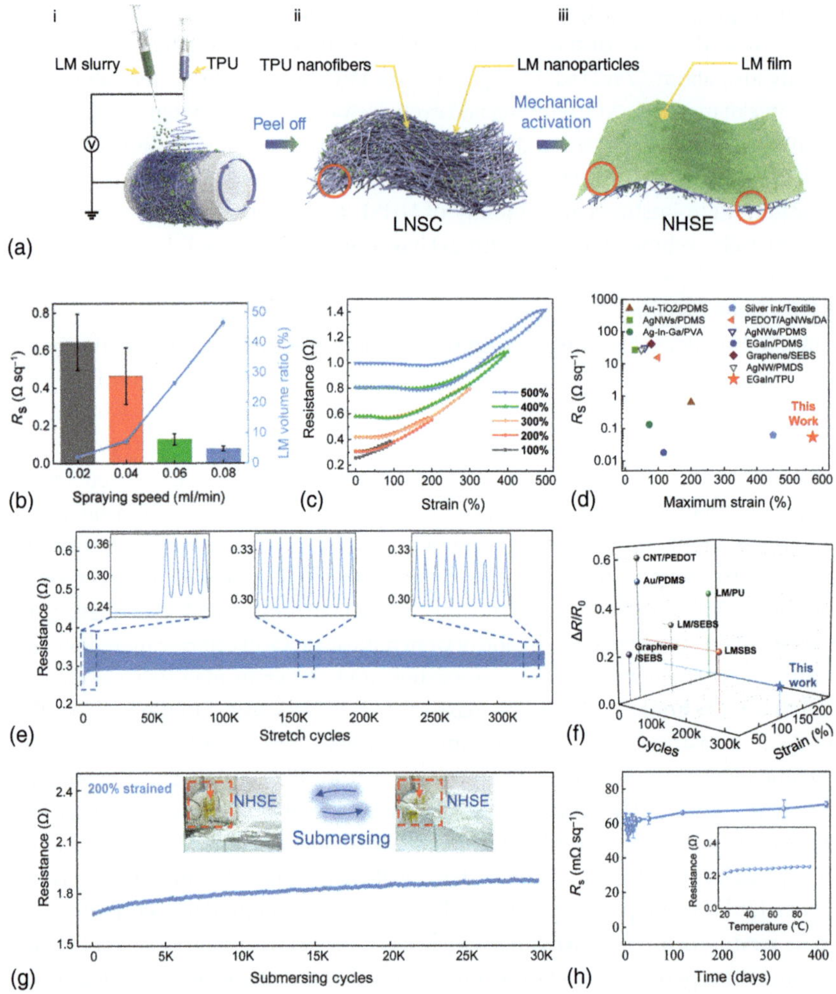

Figure 15.1 (a) The preparation of NHSE. (b) Sheet resistance of the NHSE and volume ratio of LM in the LM nanoparticle/thermoplastic polyurethane (TPU) scaffold composite (LNSC) as a function of EGaIn nanoparticles solution electrospray speed. (c) Resistance variation of the NHSE against stretching cycles (sample size: 1 cm × 1 cm, strain speed: 45% min^{-1}). (d) Comparison of the initial sheet resistance and stretchability with previously published electrodes. (e) Resistance of the NHSE that lasted for ~330 000 stretching cycles under 100% strain (0.2 Hz, sample size: 1 cm × 1 cm). The insets show the detailed resistance values at the beginning, in the middle, and at the end of the cyclic tensile test. (f) Comparison of resistance variation for recently published stretchable electrode under cyclic deformation. (g) The resistance of the 200% stretched NHSE during 30 000 water submersing cycles. (h) Sheet resistance of the NHSE as exposed in the air for 420 days, and the inset shows the resistance variation of the NHSE under an increasing temperature from 20 to 90 °C. Source: Cao et al. [28] / John Wiley & Sons / CC BY 4.0.

pre-stretching. The resulting fiber mat is composed of LM suspended between elastic fibers and self-organized into a laterally meshed and vertically buckled structure, while also providing high permeability, stretchability, electrical conductivity, and electrical stability. Specifically, LM fiber mat maintains superelasticity (over 1800% strain) and ultrahigh conductivity (up to 1 800 000 S m^{-1}) over 10 000 tensile cycles and has exceptional permeability to air and moisture. Furthermore, when applied directly to the skin, LM fiber mats exhibit high biocompatibility according to in vivo and in vitro tests.

Similar to the abovementioned method, Wang et al. [32] fabricated an LM-TPU film-based flexible strain sensor. The flexible three-dimensional (3D) conductive network obtained by coating LM on electrospun PU nanofiber membrane via ordinary mechanical pre-stretching. The flexible devices based on LM-TPU composite films have been proved by tests to have many practically desired properties, such as an extensive stretchable range (0–200%), adequate sensitivity under tensile strain (gauge factor [GF] of 0.2 at 200% strain), ultrahigh cycle stability, and durability (9000 cycles). Furthermore, cytotoxicity tests reveal that the LM-TPU composite film can adhere to the skin directly, demonstrating the LM-TPU composite fiber film's high biocompatibility.

One-dimensional (1D) elastic conductors are essential in the development of high-performance stretchable electronic systems because of their compact size, light weight, and excellent integration abilities [33]. Zhou et al. [34] has fabricated a superstretchable and ultrastable 1D LM–elastomer wire from SBS and EGaIn as building blocks. As shown in, the preparation process is mainly divided into four steps. The first is to obtain a porous SBS fiber mat by electrospinning. The second is to cut the porous SBS fiber mat into wires with a width of 2 mm. The SBS wire is next brush-coated with EGaIn, and after that, post-electrospinning of SBS microfibers was carried out employing the EGaIn-coated SBS wire mentioned above as the collector. The LM-SBS wire exhibits superior stretchability (2300% strain), low electrical resistance (as low as 0.05 Ω cm^{-1}), and robust cycling stability (stretched at 60% strain for 10 000 cycles with no resistance increase). These properties are made possible by the wrinkled structure of the LM and the hierarchical nanoscale buckling structure after repeated stretching. Moreover, this composite wire exhibits extremely high electrical stability (a resistance rise of only 12 times at a 1900% strain)

The advantages of LM itself make it widely used in stretchable and wearable electronics [35, 36]. However, LM also has some challenges, the most obvious of which is that LM has high surface tension (e.g. 718 mN m^{-1} for EGaInSn), which makes it difficult to wet and spread on the substrate surface [37, 38]. To address this challenge, Zhuang et al. [24] fabricated an LM-superlyophilic and stretchable fibrous thin-film scaffold that could be loaded with high-quality LM and exhibit a smart conductivity–strain-enhancing feature. The three main steps in creating the 3D LM-superlyophilic scaffold are electrospinning the SBS porous fiber membrane, chemically growing a silver (Ag) layer on the surface of each SBS fiber to obtain an LM-lyophilic Ag-SBS mat, and the last step is to pre-stretch the LM-lyophilic Ag-SBS mat to 2500% tensile strain to yield the porous LM-lyophilic Ag-SBS mat (Figure 15.2a–e). After 220 000 cycles of the stretch–release test, the LM-based

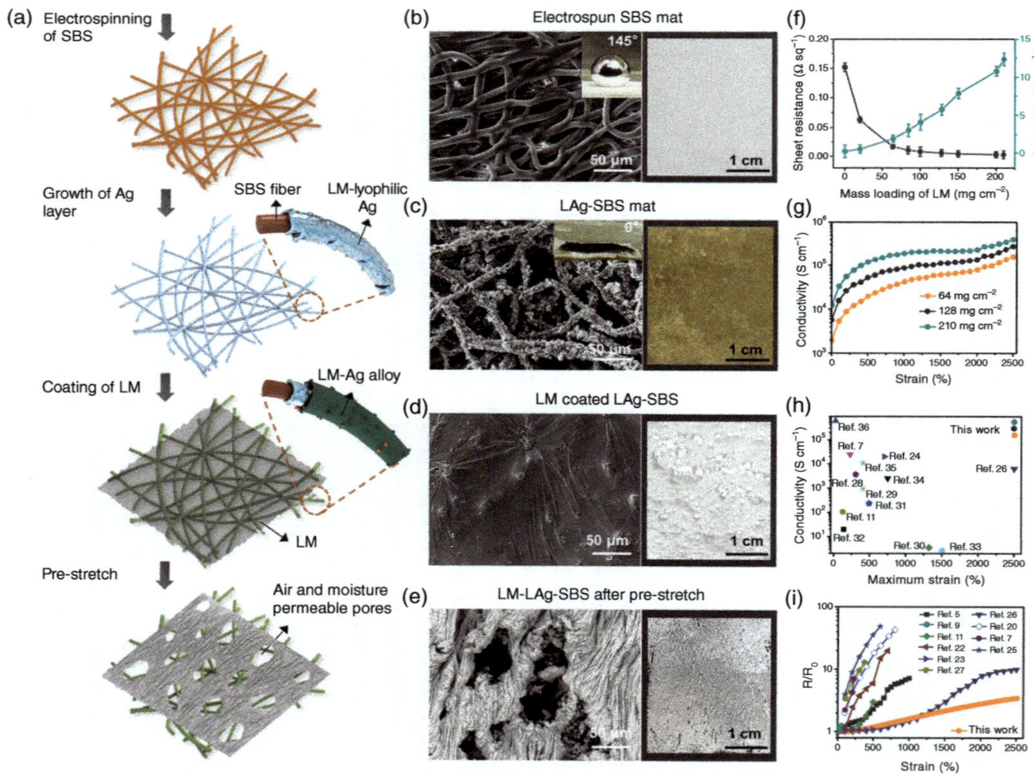

Figure 15.2 (a) Schematic illustration of the fabrication process of the LM-lyophilic Ag-SBS mat. (b–e) SEM of electrospun SBS mat, lyophilic Ag-SBS mat, LM-lyophilic Ag-SBS composite mat, and LM-lyophilic Ag-SBS composite mat after pre-stretching activation, respectively; (f) Sheet resistance and electrical conductivity of LM- lyophilic Ag-SBSs with various mass loading of LM after pre-stretching activation. (g) Electrical conductivity as the function of strains in LM-lyophilic Ag-SBSs. (h) and (i) A schematic diagram of the comparison of the electrical conductivity and resistance change of LM-lyophilic Ag-SBSs and other stretchable conductors. Source: Zhuang et al. [24] / with permission of John Wiley & Sons.

conductor exhibits outstanding conductivity of 155 900 S cm^{-1} at a very high tensile strain of 2500% and a marginal resistance change of less than 25% (Figure 15.2f–i).

Similarly, in order to prepare LM-based stretchable conductors and solve the problem of poor wettability of LM, Ma et al. [39] reported a LM micromesh on electrospun microfiber textile as a highly permeable and ultrastretchable conductor.

Styrene-ethylene-butylene-styrene (SEBS) is collected by electrospinning onto a nonwoven collector. After that, a 10 nm/100 nm chromium (Cr)/ copper (Cu) layer was formed on the surface of SEBS microfiber textile by thermal evaporation, and the textile showed an orange color. A silver-gray LM micromesh was subsequently created on the SEBS microfiber textile in the form of conformal coatings on individual microfibers and thin membranes spanning some microfibers after LM was drop-casted on the metallized microfiber textile and surplus LM was removed by a high-speed rotation. The LM micromesh has many advantages such as low sheet resistance (0.38 Ω sq^{-1}), ultrahigh stretchability (>1000% strain), and mechanical durability In addition, porous morphology of LM micromesh enables permeability and comfort comparable to standard textiles.

15.2.2.2 Stencil Printing

Nowadays, the materials of the device and the manufacturing method are the main factors preventing the development and commercialization of stretchable electronic devices. An inventive solution to the aforementioned issues is offered by the material optimization and procedure simplification in fabrication [40]. Therefore, Wang et al. [41] fabricated high-performance stretchable electronics by printing LM on electrospun TPU fiber membranes by stencil printing. The specific preparation process is to cover the premade template on the electrospun TPU film, then print the LM in situ pattern, and finally encapsulate a layer of electrospun fiber film. The materials are multilayered and reconfigurable, and they offer high stretchability, air permeability, and stability. Moreover, development of stretchable electronic systems and widespread commercial applications are anticipated.

15.2.2.3 Vacuum Filtration

Li et al. [42] fabricated a LM-sliver nanowire (AgNW)-based superelastic permeable membrane (SPM) by vacuum filtration. First, the fibrous membrane prepared by electrospinning is referred to here as SPM. After, due to their conducting channel having fewer connections than the deposited conductive layer, AgNW networks demonstrate greater conductivity. They are also more advanced in the construction of permeable devices. However, AgNWs cannot withstand large tensile deformation, and cannot simultaneously achieve superstretchability and electrical conductivity of flexible electronic products. Therefore, Li et al. added LM microparticles to the AgNW suspension for filtering in order to achieve significant stretchability of the permeable electrode on SPM. These particles might function as dynamic islands to bind the nanowires and accommodate deformation of the AgNW network upon stretching. Finally, the obtained material has excellent electrical conductivity, stretchability, permeability, and cycling stability.

15.3 Application of LM-based Stretchable Electronic System

15.3.1 Stretchable Electronics for Strain Sensing

In recent years, sensors fabricated from LM and electrospun elastic fiber materials have attracted wide attention due to their high sensitivity, excellent stretchability, wide work range, and biocompatibility [43, 44]. They demonstrated excellent potential in areas such as motion detection, human–computer interfaces, and human health [45, 46].

Wang et al. [32] coated EGaInSn onto the surface of the 3D TPU fiber film via electrospinning to fabricate a strain sensor, which can detect the change of human physiological signals, such as finger joints, elbow joints, knee joints, and wrist joints. Install the LM-TPU-based sensor on the joint of the finger. As the finger rotates from 0° to 90°, the sensor's current changes, and the signal recovers as the finger rotates back to its initial position. Additionally, taping the sensor to the knee joint allows for flexible detection of various knee joint movements. When the knee was bent 90°, the electrical signal was stronger than when it was bent 45°. Studies including five straight knee bends of 45° and 90° demonstrate the sensor's excellent stability in detecting various knee bending angles. More importantly, the LM-TPU-based sensor can recognize various knee motion states in real time, including squatting, walking, running, and jumping. In addition, the sensor can also detect large-scale human motion by simply attaching to the wrist and clothing.

15.3.2 Stretchable Strain-Insensitive Electrode

With the in-depth study of strain sensors, sensitivity has always been the focus of current research, but the stability of the electrical conductivity of the material itself is easily overlooked [46]. Stretchable electrodes with stable electrical signal transmission have great potential in wearable electronics [47], such as long-term electrocardiography (ECG) monitoring, electrothermal heater, and stretchable devices.

Cao et al. [28] fabricated a stretchable electrode via in situ assembly of electrospun TPU fibrous elastic scaffolds and electrosprayed LM nanoparticles. This flexible electrode is applied to the epidermis of a human being in order to record an ECG. In the static state, the signal from the stretchable electrode and the commercial gel electrode were nearly identical. (Figure 15.3a,b). Additionally, the stretchable electrodes produce a flatter T-peak signal than commercial electrodes due to their increased impedance in the 10–1000 Hz region. But using the concept of ECG electrode sensitivity, it was discovered that the stretchable electrode's sensitivity (0.43) was greater than that of the commercial electrode (0.36). The main benefit of this application is specifically stable signal identification in spite of outside disturbances. The signal alterations that occur during wrist movement are depicted in Figure 15.3c,d. The findings demonstrate that the signal from the gel electrode fluctuates while the signal from the stretchable electrode does not. If water is sprayed on the gel electrode while recording the signal, noise may obfuscate the signal and prevent it from

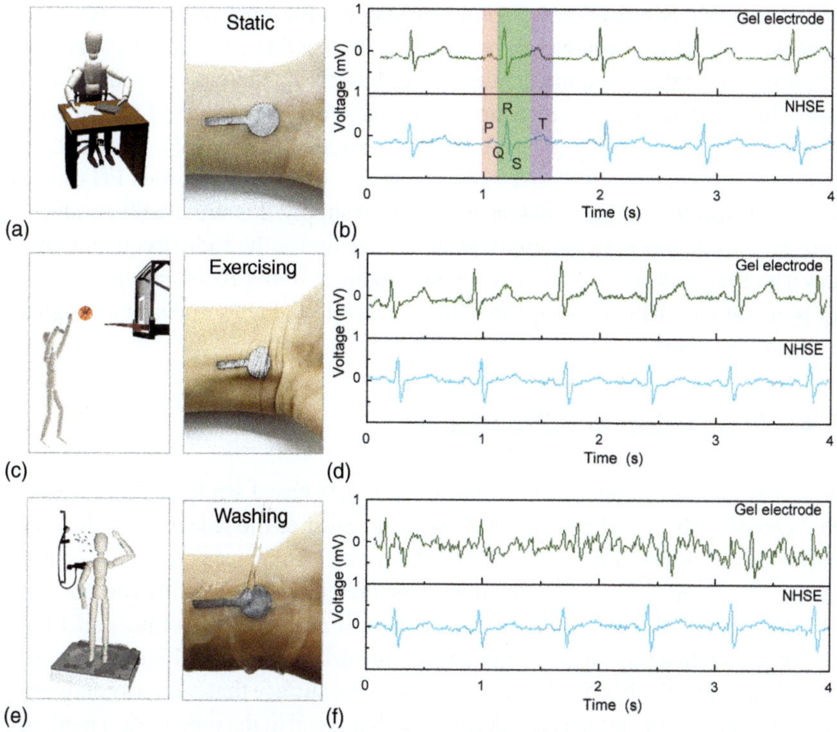

Figure 15.3 (a)–(f) Schematics photographs of the stretchable electrode for on-body healthcare monitoring and corresponding signals of commercial gel electrode (upper) and the stretchable electrode (lower). Source: Cao et al. [28] / John Wiley & Sons / CC BY 4.0.

being precisely identified (Figure 15.3e,f). In contrast, this stretchable electrode is still preserved and remains diagnosable.

By repeatedly electrospinning SBS and printing EGaIn, Ma et al. [31] created vertically stacked multilayer circuits that were then incorporated into a monolithic elastic pad with three layers of printed EGaIn electrodes that each served as an ECG sensor (top layer), sweat sensor (middle layer), and electric heater (bottom layer). The superelasticity of EGaIn-SBS allows for the collection of trustworthy low-noise ECG signals in both the compressive and tensile states. The intermediate layer's interdigital electrode pair functions as a capacitor to measure the volume change of sweat that permeates from the epidermis of human skin to the ECG's top layer. The capacitance increased with the amount of sweat received. The underlying electrothermal heater may react to the input voltage instantly and precisely. For instance, by gradually raising the applied voltage, the surface temperature can be changed from 30 to 95 °C. This shows that it can precisely control the temperature of the output at very low power. In addition, the heater also showed stable cycle performance.

Zhou et al. [34] obtained ultrahigh electrical stability EGaIn-SBS wires by electrospinning SBS nanofiber mats and coating EGaIn, and encapsulating them through SBS nanofiber mats, which are suitable for the transmission circuit of devices that require stable power supply or signal transmission. A piece of EGaIn-SBS wire is used in the circuit to power the light emitting diode (LED) chip. The LED chip could still be steadily lighted even when the wire was stretched to 500% strain with only a slight brightness fall. When the EGaIn-SBS wire is stretched from 0% strain to 100% strain using a spectrometer to measure the emission intensities of the LEDs, it is discovered that the LED's spectral peak intensity reduces by only 16%. The emission intensity was reduced by 95% at 1000% tensile strain when the porous EGaIn-SBS wire was substituted. Along with consistent power supply, this stretchable conductor can also offer stable electrical signal transfer. For example, the audio frequency emitted by a section of EGaIn-SBS wire in a relaxed state or under 1000% strain when the wire is utilized to carry audio signals. The waveforms and amplitudes of the signals are almost the same.

15.4 Conclusion and Prospects

The preparation and application of electrospun-LM-based stretchable electronics have been described in this chapter. At present, most of the research on the combination of electrospinning and LM is in the post-finishing aspect, mainly coating or printing LM on the electrospun nanofiber membrane [48], while the one-step method to realize the combination of electrospun nanofiber materials and LM is rarely reported. In addition, although LM has the disadvantages of high surface tension and exposure to air, an oxide layer forms on the surface that reduces conductivity [49, 50], but the advantages of LM can be well utilized in the preparation methods mentioned in this chapter. Moreover, this electrospun-LM-based stretchable electronic device has played a pivotal role in smart wearable electronic products. The demand for electrospun-LM-based stretchable electronics will continue to increase in the future, and new ways of combining electrospinning with LM are needed for further research.

References

1 Kim, D.C., Shim, H.J., Lee, W. et al. (2020). Material-based approaches for the fabrication of stretchable electronics. *Advanced Materials* 32 (15): 1902743.

2 Vatankhah-Varnosfaderani, M. et al. (2017). Mimcking biological stress–strain behaviour with synthetic elastomers. *Nature* 549: 497–501.

3 Wehner, M., Truby, R.L., Fitzgerald, D.J. et al. (2016). An integrated design and fabrication strategy for entirely soft, autonomous robots. *Nature* 536: 451–455.

4 Park, S.J., Gazzola, M., Park, K.S. et al. (2016). Phototactic guidance of a tissue-engineered soft-robotic ray. *Science* 353 (6295): 158.

5 Ouyang, H., Tian, J., Sun, G. et al. (2017). Self-powered pulse sensor for antidiastole of cardiovascular disease. *Advanced Materials* 29 (40): 1703456.

6 Gao, W., Emaminejad, S., Nyein, H.Y.Y. et al. (2016). Fully integrated wearable sensor arrays for multiplexed in situ perspiration analysis. *Nature* 529: 509.

7 Mimee, M., Nadeau, P., Hayward, A. et al. (2018). An ingestible bacterial-electronic system to monitor gastrointestinal health. *Science* 360 (6391): 915–918.

8 Boutry, C.M., Kaizawa, Y., Schroeder, B.C. et al. (2018). Skin-inspired electronics: emerging semiconductor devices and systems. *Nature Electronics* 1: 314.

9 Rogers, J.A., Someya, T., and Huang, Y.G. (2010). Materials and mechanics for stretchable electronics. *Science* 327: 1603–1607.

10 Tan, Y.J., Godaba, H., Chen, G. et al. (2020). A transparent, self-healing and high-κ dielectric for low-field-emission stretchable optoelectronics. *Nature Materials* 19: 182–188.

11 Hartmann, F., Baumgartner, M., and Kaltenbrunner, M. (2020). Becoming sustainable, the new frontier in soft robotics. *Advanced Materials* 33 (19): 2004413.

12 Mackanic, D.G., Kao, M., and Bao, Z. (2020). Enabling deformable and stretchable batteries. *Advanced Energy Materials* 10 (29): 2001424.

13 Liu, Z.F., Fang, S., Moura, F.A. et al. (2015). Hierarchically buckled sheath-core fibers for superelastic electronics, sensors, and muscles. *Science* 349 (6246): 400–404.

14 Choi, C., Lee, J.M., Kim, S.H. et al. (2016). Twistable and stretchable sandwich structured fiber for wearable sensors and supercapacitors. *Nano Letters* 16 (12): 7677–7684.

15 Woo, J., Lee, H., Yi, C. et al. (2020). Ultrastretchable helical conductive fibers using percolated Ag nanoparticle networks encapsulated by elastic polymers with high durability in omnidirectional deformations for wearable electronics. *Advanced Functional Materials* 30 (29): 1910026.

16 Sun, F., Tian, M., Sun, X. et al. (2019). Stretchable conductive fibers of ultrahigh tensile strain and stable conductance enabled by a worm-shaped graphene microlayer. *Nano Letters* 19 (9): 6592–6599.

17 Gong, S. and Cheng, W. (2017). One-dimensional nanomaterials for soft electronics. *Advanced Electronic Materials* 3 (3): 1600314.

18 Daeneke, T., Khoshmanesh, K., Mahmood, N. et al. (2018). Liquid metals: fundamentals and applications in chemistry. *Chemical Society Reviews* 47 (11): 4073–4111.

19 Teng, L., Ye, S., Handschuh-Wang, S. et al. (2019). Liquid metal-based transient circuits for flexible and recyclable electronics. *Advanced Functional Materials* 29 (11): 1808739.

20 Zhang, J., Liu, M., Pearce, G. et al. (2021). Strain stiffening and positive piezoconductive effect of liquid metal/elastomer soft composites. *Composites Science and Technology* 201: 108497.

21 Kim, J.-H., Kim, S., So, J.-H. et al. (2018). Cytotoxicity of gallium–indium liquid metal in an aqueous environment. *ACS Applied Materials & Interfaces* 10 (20): 17448–17454.

22 Dickey, M.D. (2017). Stretchable and soft electronics using liquid metals. *Advanced Materials* 29 (27): 1606425.

23 Wang, H., Yao, Y., He, Z. et al. (2019). A highly stretchable liquid metal polymer as reversible transitional insulator and conductor. *Advanced Materials* 31 (23): 1901337.

24 Zhuang, Q., Ma, Z., Gao, Y. et al. (2021). Liquid–metal-superlyophilic and conductivity–strain-enhancing scaffold for permeable superelastic conductors. *Advanced Functional Materials* 31 (47): 2105587.

25 Lu, Y., Hu, Q., Lin, Y. et al. (2015). Transformable liquid-metal nanomedicine. *Nature Communications* 6: 10066.

26 Cheng, J., Shang, J., Yang, S. et al. (2022). Wet-adhesive elastomer for liquid metal-based conformal epidermal electronics. *Advanced Functional Materials* 32 (25): 2200444.

27 Yan, X., You, M.H., Lou, T. et al. (2016). Colorful hydrophobic poly(vinyl butyral)/cationic dye fibrous membranes via a colored solution electrospinning process. *Nanoscale Research Letters* 11: 540.

28 Cao, J., Liang, F., Li, H. et al. (2022). Ultra-robust stretchable electrode for e-skin: in situ assembly using a nanofiber scaffold and liquid metal to mimic water-to-net interaction. *Info Mat* 4 (4): e12302.

29 Sha, Z., Boyer, C., Li, G. et al. (2022). Electrospun liquid metal/PVDF-HFP nanofiber membranes with exceptional triboelectric performance. *Nano Energy* 92: 106713.

30 Ye, Q., Wu, Y., Qi, Y. et al. (2019). Effects of liquid metal particles on performance of triboelectric nanogenerator with electrospun polyacrylonitrile fiber films. *Nano Energy* 61: 381–388.

31 Ma, Z., Huang, Q., Xu, Q. et al. (2021). Highly permeable and ultrastretchable liquid metal micromesh for skin-attachable electronics. *Nature Materials* 20 (6): 859–868.

32 Wang, X., Liu, J., Zheng, Y. et al. (2022). Biocompatible liquid metal coated stretchable electrospinning film for strain sensors monitoring system. *Science China Materials* 65 (8): 2235–2243.

33 Lee, J., Shin, S., Lee, S. et al. (2018). Highly sensitive multifilament fiber strain sensors with ultrabroad sensing range for textile electronics. *ACS Nano* 12 (5): 4259–4268.

34 Zhou, N., Jiang, B., He, X. et al. (2021). A superstretchable and ultrastable liquid metal-elastomer wire for soft electronic devices. *ACS Applied Materials & Interfaces* 13 (16): 19254–19262.

35 Liu, S., Sweatman, K., McDonald, S., and Nogita, K. (2018). Ga-based alloys in microelectronic interconnects: a review. *Materials* 11 (8): 1384.

36 Wang, J., Cai, G., Li, S. et al. (2018). Printable superelastic conductors with extreme stretchability and robust cycling endurance enabled by liquid-metal particles. *Advanced Materials* 30 (16): 1706157.

37 Park, J., Wang, S., Li, M. et al. (2012). Three-dimensional nanonetworks for giant stretchability in dielectrics and conductors. *Nature Communications* 3: 916.

38 Zhu, S., So, J.H., Mays, R. et al. (2013). Ultrastretchable fibers with metallic conductivity using a liquid metal alloy core. *Advanced Functional Materials* 23 (18): 2308–2314.

39 Ma, X., Zhang, M., Zhang, J. et al. (2022). Highly permeable and ultrastretchable liquid metal micromesh for skin-attachable electronics. *ACS Materials Letters* 4 (4): 634–641.

40 Luo, Y.F., Wang, M., Wan, C.J. et al. (2020). Devising materials manufacturing toward Lab-to-Fab translation of flexible electronics. *Advanced Materials* 32 (37): 2001903.

41 Wang, M., Ma, C., Uzabakiriho, P.C. et al. (2021). Stencil printing of liquid metal upon electrospun nanofibers enables high-performance flexible electronics. *ACS Nano* 15 (12): 19364–19376.

42 Li, Y., Xiao, S., Zhang, X. et al. (2022). Silk inspired in-situ interlocked superelastic microfibers for permeable stretchable triboelectric nanogenerator. *Nano Energy* 98: 107347.

43 Wang, D., Wang, L., and Shen, G. (2020). Nanofiber/nanowires-based flexible and stretchable sensors. *Journal of Semiconductors* 41 (4): 041605.

44 Ding, Q., Wu, Z., Tao, K. et al. (2022). Environment tolerant, adaptable and stretchable organohydrogels: preparation, optimization, and applications. *Materials Horizons* 9 (5): 1356–1386.

45 Shintake, J., Cacucciolo, V., Floreano, D. et al. (2018). Soft robotic grippers. *Advanced Materials* 30 (29): 1707035.

46 Xiong, J., Chen, J., and Lee, P.S. (2020). Functional fibers and fabrics for soft robotics, wearables, and human-robot interface. *Advanced Materials* 33 (19): 2002640.

47 Hao, Y., He, X., Wang, L. et al. (2021). Stretchable thermoelectrics: strategies, performances, and applications. *Advanced Functional Materials* 32 (13): 2109790.

48 Ma, Z., Huang, Q., Xu, Q. et al. (2021). Permeable superelastic liquid-metal fibre mat enables biocompatible and monolithic stretchable electronics. *Nature Materials* 20: 859.

49 Zheng, R.-M., Wu, Y.-H., Xu, Y.-H. et al. (2019). Advanced stretchable characteristic of liquid metal for fabricating extremely stable electronics. *Materials Letters* 235: 133–136.

50 Zhang, Q., Roach, D.J., Geng, L. et al. (2018). Highly stretchable and conductive fibers enabled by liquid metal dip-coating. *Smart Materials and Structures* 27 (3): 035019.

16

Preparation and Application of Electrospun Photocatalysts

Wendi Liu and Hongnan Zhang

Donghua University, Key Laboratory of Textile Science & Technology, Ministry of Education, College of Textiles, 2999 Renmin North Road, Songjiang District, Shanghai 201620, China

16.1 Introduction

With the completion of the industrial revolution, humankind took a significant step in the process of history. However, the contradiction between the growing energy demand and the growing energy shortages is intensifying. In addition, while consuming energy, human beings also discharge a large number of pollutants into the environment, resulting in serious environmental pollution problems such as the greenhouse effect, air pollution, water pollution, and acid rain. Consequently, precisely because of these reasons, sustainable social development is facing challenges, such as ecological imbalances caused by environmental pollution and accumulatively increasing energy shortages. In this context, developing and utilizing cleaner, more economical, and sustainable new energy sources is of great significance to promoting environmental protection and alleviating the energy crisis [1]. Solar energy has always been an essential component of clean energy due to its easy access, which can effectively solve the energy and environmental problems caused by the excessive use of traditional fossil fuels [2].

Photocatalysis technology, which can directly use solar energy, has attracted wide attention in the field of semiconductor applications due to its unique advantages such as low cost, simple operation, material stability, and environmental protection, and is considered one of the most attractive and promising technologies to deal with ecological problems. Photocatalysis technology can not only utilize solar energy but also alleviate energy shortages. It can also mineralize entirely organic pollutants into small inorganic molecules, thus solving part of environmental problems [3]. The current research and applied photocatalysts are mainly nanocatalyst particles. However, despite the relatively high photocatalytic efficiency of nanocatalyst particles, there are still many problems in practical applications. For example, due to the increased surface energy of nanoscale particles, the agglomeration phenomenon is more serious, leading to the reduction

Electrospinning: Fundamentals, Methods, and Applications, First Edition.
Edited by Liming Wang and Xiaohong Qin.

of specific surface area and the weakening of the catalytic effect, which induces the incomplete oxidation reaction and thus produces other harmful substances. In addition, it is challenging to prepare photocatalysts with high dispersion, stability, and easy recycling, which are steep to runoff and form potential secondary pollution.

Nanofibers can well solve the problem of nanoparticle agglomeration and recycling compared with nanoparticles because of their high aspect ratio and large specific surface area. Over decades, researchers have evolved diverse technical methods to produce nanofibers. Furthermore, electrospinning is recognized as an advanced technology for industrialized production of nanofiber materials based on its characteristics of low manufacturing cost, convenient preparation process, various high polymers, controllable fiber scale, and large yield.

16.2 Photocatalysis

In 1972, Honda-Fujishima reported that a photochemical system consisting of TiO_2 photoelectrodes and platinum electrodes was used to decompose water into hydrogen and oxygen, which was known as a milestone in the new field of semiconductor photocatalysis [4]. Photocatalysts are a general term for chemicals that can catalyze under the excitation of photons. In particular, they are not involved in the reaction, but only the chemicals that accelerate the chemical reaction.

16.2.1 Principle of Photocatalysis

Semiconductor photocatalytic performance is determined by its band structure. The semiconductor energy band contains a valence band (VB) and a conduction band (CB). Electrons are filled with the VB. The CB is an empty energy band. In addition, the VB is separated by a forbidden band. The size of the forbidden band is called the energy gap (E_g).

The photocatalytic reaction can be divided into three steps (Figure 16.1).

In the first stage, the photocatalyst absorbs the photon energy greater than or equal to the E_g; the electrons in the VB absorb the photon energy and then transition to the CB to form the photogenerated electrons. The photogenerated holes will be generated in the VB to form the photogenerated electron–hole pairs. This stage is called photoexcitation.

In the second stage, the electron–hole pair resolves into a free electron and a free hole and then migrates to the photocatalyst surface. The hole on the VB has a vigorous ability to capture electrons. And it can have an oxidation reaction to produce hydroxyl radical ($^{\bullet}OH$). Electrons on the CB will reduce the trapped oxygen molecules into the superoxide radical superoxide radicals ($^{\bullet}O^{2-}$) and eventually continue to react to generate the $^{\bullet}OH$. In the third stage, the pollutants are mineralized into nontoxic and harmless small molecules under the action of $^{\bullet}OH$ [5].

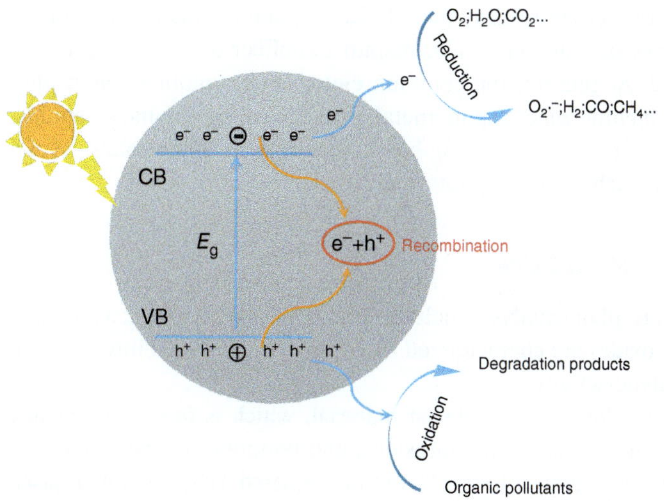

Figure 16.1 The photocatalytic reaction.

16.2.2 Current Challenges of Photocatalysis

(1) Although the energy of sunlight is enormous, semiconductor materials are weak in the use of the sun. The reason is that the Eg of the photocatalysis is broad, but the absorption range of the visible (Vis) light is very narrow. Since most photocatalysis cannot absorb and use Vis light, they can only be excited by ultraviolet (UV) light sources.
(2) In general, there are many defects in semiconductor materials. These defects deteriorate the stability of the electron–hole pair in the reaction process. The electrons and holes are easy to recombine and have a short life. So only a few electrons and holes can participate in the reaction. As a result, the photoelectric conversion efficiency of photocatalysts is significantly reduced.
(3) Many semiconductor photocatalysts are nanoparticles. The agglomeration phenomenon of nanoparticles is relatively severe, which leads to the reduction of specific surface area and the weakening of the catalytic effect, thus inducing incomplete oxidation reactions and producing other harmful substances.

16.3 Electrospun Photocatalyst

The nanoparticles always incline to agglomeration and cannot adequately contact the contaminants to be degraded during the photocatalytic reaction. Moreover, it is difficult to recycle them scattered in the liquid phase because the nanoparticles are small, which is easier to cause secondary pollution to the environment. Compared with traditional nanoparticles, the electrospun nanofiber has excellent dispersion and a larger specific surface area, which is conducive to complete contact between catalyst and pollutant for photocatalytic reaction. In addition, the electrospun nanofiber is easier to carry out efficient separation and recovery of the photocatalyst. A growing number of electrospun nanofiber-based photocatalysts have been produced. At present, the common electrospun nanofiber-based photocatalysts mainly include electrospun metal oxide, electrospun metal sulfide, Bi-based electrospun photocatalyst, Ag-based electrospun photocatalyst, and electrospun graphitic carbon nitride photocatalyst.

16.3.1 Electrospun Metal Oxide

Common metal oxide photocatalysts include TiO_2, ZnO, MgO, CeO_2, WO_3, and Fe_3O_4. These metal oxides are characterized by nontoxicity, high stability in water, and good photocatalytic activity.

TiO_2 is the first developed photocatalyst material, which is favored by people because of its high catalytic activity, nontoxicity, and pollution-free advantages. It is the most studied photocatalyst today. Lee et al. prepared TiO_2 nanofiber-based photocatalysts with electrospinning technology. Experimental studies have shown that the TiO_2 nanofiber-based photocatalysts have higher photocatalytic activity in the degradation of acetaldehyde and ammonia gas compared with commercial nanopowders, which was due to the larger specific surface area and higher porosity of nanofibers. Moreover, it helped to improve the efficiency of electron–hole separation and thus improve the catalytic efficiency [6].

In addition, ZnO is also one of the earliest discovered photocatalysts with good photocatalytic performance. At the same time, it was found that ZnO exhibited excellent photodegradation efficiency for Azocarmine G dye in polystyrene (PS)/ZnO nanofibers prepared by electrospinning. This was because that as the photon energy generated by UV light interacts with the ZnO nanomaterial, an electron–hole pair was generated. This separation can be found in the CB where the electron resides and the VB where the hole was present on the surface of the catalyst. The holes created in the VB and the electrons excited to the CB react with the OH and O_2 species in the environment. The free radicals produced by this reaction were very active and convert the dye into degradation products CO_2 and H_2O [7].

Generally speaking, metal oxide photocatalysts are semiconductors. But MgO is an insulator with high E_g. Mantilaka et al. synthesized MgO nanofibers using electrospinning and using polyvinyl alcohol as the polymer carrier and vibrant yellow as the pollutant model. The experimental results showed that, compared with MgO nanospheres, the MgO nanofiber-based photocatalysts could completely degrade reactive dyes under UV light [8].

16.3.2 Electrospun Metal Sulfide

Compared with metal oxides, metal sulfides have narrower Eg and are easy to use Vis light. Common electrospun metal sulfide photocatalysts mainly include CdS, MoS_2, CuS, SnS_2, and ZnS.

Among them, the CdS and MoS_2 are common photocatalytic sulfide materials, which are similar to graphene in the two-dimensional layered structure. The properties of layered structure materials are determined by the number of layers. The CdS and MoS_2 of a single layer have unique properties, and the change of band gap can change its optical and electrical properties. Liu et al. prepared CdS nanoparticle-functionalized natural cotton cellulose electrospun nanofibers by combining electrospun technology with the chemical deposition method. Compared with the pure CdS nanoparticles, the prepared electrospun nanofibers have good photocatalytic degradation activity for Rhodamine B (RhB) under Vis light irradiation [9].

The MoS_2 belongs to layered transition metal sulfide, which has a special three-layer sandwich structure. At the same time, the MoS_2 has a narrow band gap. So the MoS_2 has good band gap characteristics and will become a promising photocatalyst.

Pant et al. prepared MoS_2/CdS/TiO_2 nanocomposites by electrospinning and calcination. The experimental results showed that compared with pure TiO_2 nanofiber photocatalyst, the composite fiber photocatalyst had a higher degree of inhibition on the photoluminescence intensity, indicating that the prepared composite nanofiber had a lower recombination rate of excited electron–hole and fewer defects [10].

16.3.3 Bi-Based Electrospun Photocatalyst

The price of Bi is relatively low, and it is widely used in photocatalysis. The common Bi-based electrospun photocatalysts include Bi_2O_3, Bi_2S_3, BiOX (X = F, Cl, Br, I), Bi_2MO_6 (M = W, Mo), and $BiVO_4$.

Among the Bi-based oxides, Bi_2O_3 is an important semiconductor material. It is very popular because of its rich and diverse structure, unique electronic structure, and Vis light response characteristics. Wang et al. successfully prepared Bi_2O_3 nanofibers by electrospinning and calcination. It was found that Bi_2O_3 nanofibers calcined at different high temperatures had different photocatalytic activities. This was because electrospun nanofibers photocatalysts have different band gaps at different temperatures. As the temperature increases, the band gap of the electrospun nanofibers photocatalyst will be decreased, and the ability to absorb light will be increased. It will produce several electrons and holes so that the photocatalytic activity will be more vigorous. Moreover, the higher surface area also increased the photocatalytic activity of sites [11].

BiOX (X = F, Cl, Br, I) is a semiconductor material with an anisotropic structure belonging to the tetragonal system. BiOCl, BiOBr, and BiOI are relatively common indirect transition semiconductors. When the indirect transition semiconductor is excited, the electrons need to pass through the electron layers to reach the CB, thus preventing electron–hole pair recombination. Because of its efficient photocatalytic effect, it is widely used in the degradation of organic pollutants [12]. Babu et al. prepared BiOCl nanofibers employing electrospinning. With alizarin red S dye as the target pollutant, the alizarin red S dye could be degraded entirely under UV irradiation for 70 minutes [13].

16.3.4 Ag-Based Electrospun Photocatalyst

Typical photocatalysts in Ag-based electrospun photocatalysts include Ag_3PO_4, Ag_2CrO_4, and AgX (X = F, Cl, Br, I). These materials can produce photocatalytic reactions under low Vis light due to their narrow band gap.

Jin et al. prepared Ag_3PO_4@polylactic acid composite nanofibrous membranes and tested their photocatalytic properties. The results showed that with the increase in the doping amount of Ag_3PO_4, the adsorption and degradation rate of methylene blue (MB) increased significantly, and thus in the dark environment, the adsorption capacity of MB reached about 8%. This also showed that Ag-based electrospun photocatalysts could produce photocatalytic reactions under low Vis light [14].

Compared with AgCl and AgI, AgBr is more stable. AgI is difficult to exist under sunlight stably, and it is easy to decompose even if loaded on other carriers. Fu et al. prepared a new photocatalyst Ag/AgBr/TiO_2 nanofiber based on electrospinning technology. The experimental results showed that compared with TiO_2 nanofibers and Ag/AgBr/TiO_2 nanofibers, the degradation degree of RhB is the highest, and the Ag/AgBr/TiO_2 composite can decompose 98% of the MB aqueous solution within 90 minutes. In addition, the material is recyclable and has good chemical stability [15].

16.3.5 Electrospun Graphitic Carbon Nitride Photocatalyst

Graphitic carbon nitride (g-C_3N_4) is a nonmetallic semiconductor photocatalyst. It has the characteristics of Vis light response, chemical stability, and thermal stability. Although g-C_3N_4 has a semiconductor band gap structure, it differs from traditional organic semiconductors and metal-containing semiconductors. The difference is that it only contains elements (C and N) that are abundant on Earth. Moreover, due to the strong covalent bond between carbon and high chemical and thermal stability, it forms a conjugate layer with a nitrogen atom in the structure.

Saha et al. prepared g-C_3N_4/polyvinylidene fluoride nanofiber felt by electrospinning technology, and the nanofiber had good photocatalytic self-cleaning properties for MB and RhB. It was also found that the photocatalytic reduction rate of pure g-C_3N_4 was slower, while the photocatalytic reduction rate of low-concentration formic acid was significantly improved. This is due to the fact that formic acid may increase electrons in the CB of the g-C_3N_4 and produce hydrogen and carbon monoxide, which also contribute to the reduction of hexavalent chromium. Another study confirmed that g-C_3N_4 retained its performance after repeated reduction cycles [16]. The g-C_3N_4 can be used for photo-splitting water to generate hydrogen. Liu et al. successfully constructed a close contact heterojunction in the interface between g-C_3N_4 nanosheets and hollow porous TiO_2 nanofibers through coaxial electrospinning, which significantly improved the separation efficiency of photo-induced electron–hole pairs in the Vis light region. The hollow porous structure with abundant surface active sites helps to induce multiple light reflections to achieve more efficient light collection. In addition, the hollow and porous nanofibers exhibited the highest hydrogen evolution rate [17]. This work is helpful to understand further the preparation of hollow-structure heterogeneous nanofiber photocatalysts with hydrogen production capacity.

There are many research foundations for the combination of electrospinning and photocatalysis technology. The nanofiber-based photocatalysts prepared by electrospinning can significantly improve the utilization efficiency of sunlight. Electrospun nanofiber-based photocatalysts can also effectively solve the problem of separation and recovery of nanoparticles, which is of great significance for promoting the application of photocatalytic technology.

16.4 Composite Electrospun Photocatalyst

The composite electrospun photocatalyst is obtained by modifying the original monocomponent electrospun photocatalyst. At present, the main modification methods of electrospun photocatalysts include element doping, surface noble metal loading, semiconductor recombination, dye photosensitization, and graft-conjugated polymer.

16.4.1 Element Doping

At present, the element doping methods of electrospun photocatalysts include metal doping, nonmetal doping, and co-doping. The element-doped electrospun photocatalyst can be obtained by introducing the doped-element precursor and the photocatalyst precursor into the spinning solution for blending and spinning, and then heat treatment. The principle of element doping is to change the chemical composition of a semiconductor by introducing ions into the crystal structure of the semiconductor and finally achieve the purpose of regulating the semiconductor energy band.

16.4.1.1 Metal Doping

The metal ions used for doping are mainly transition metal ions and rare-earth metal ions. There are many studies on the doping of transition metal ions, including Fe^{3+}, Mn^{2+}, Cu^{2+}, Co^{2+}, Mo^{5+}, Ni^{3+}, Zn^{2+}, and Cr^{3+}. Liu et al. prepared Fe-doped ZnO nanofibers by electrospinning. The analysis of photocatalytic performance showed that the addition of iron ions can effectively reduce the recombination of electron–holes. Transition metal ions significantly influenced the photocatalytic performance of photocatalysts [18]. The rare-earth metal ions for doping mainly include La^{3+}, Er^{3+}, Ce^{3+}, Sm^{3+}, Pr^{3+} and so on. In order to investigate the effect of rare-earth metal ions on the performance of photocatalysts. Pascariu et al. added La element to the precursor of the ZnO electrospinning solution. The results showed that La^{3+} promoted the degradation of organic dyes and improved the catalytic activity of the photocatalyst [19].

The electrons in the CB were captured by metal ions, which were electron acceptors, thus reducing the recombination of electrons and holes. The photocatalytic activity of the photocatalyst can be improved. In addition, impurity energy was formed in the band gap of metal ions so that a minor incident energy could excite electrons, which helped to widen the optical response range. Moreover, metal ion doping can cause lattice defects, which can not only introduce new energy levels into the semiconductor and enhance light absorption capacity but also facilitate the formation of more active centers and improve photocatalyst activity.

16.4.1.2 Nonmetal Doping

Currently, nonmetal elements commonly used for nonmetal doping of photocatalysts include C, N, B, S, and F.

Song et al. doped element C into TiO_2 nanofibers using electrospinning. Electrons at CB of TiO_2 migrated to C, which resulted in the efficient separation of electrons and holes. In addition, experimental tests showed that the composite could be recycled many times [20]. The element C is an element with similar chemical properties to the element O, which can eliminate some oxygen vacancies and promote the effective separation of electrons and holes of the composite photocatalyst. The characteristic of doping with nonmetallic elements is that doping with nonmetallic elements can directly change the band gap of semiconductors and improve the absorption range of light. In addition, nonmetallic element doping will cause more lattice defects on the semiconductor surface, which can better capture photogenerated electrons and photo-generated holes, thus inhibiting the recombination of photogenerated charges.

16.4.1.3 Co-Doping

Co-doping is doping of two or more metal elements, nonmetal elements, or metal elements into electrospun photocatalysts, mainly based on the synergistic effect between various doping elements to improve their photocatalytic performance. Co-doping proved to be an efficient method to synthesize Vis-light photocatalysts. Studies have shown that co-doping with different ions can expand the photocatalysts' photoresponse range and significantly improve their photocatalytic activity [21].

Yousef et al. investigated the photocatalytic properties of Cu and S co-doped TiO_2 nanofibers. This nanocomposite had a good removal effect on organic pollutants. The reason was that Cu element and S element could reduce the band gap and inhibit the recombination of electrons and holes. [22].

16.4.2 Modified with Noble Metals

Noble metals (Au, Ag, Pt, Pd) and their alloys (AuAg, AuPt, AuPd) modified electrospun photocatalysts can not only promote the effective separation of photogenerated electrons and holes but also expand the light absorption range.

Tang et al. doped noble metal Au into TiO_2 nanofibers by electrospinning, and the composite nanofibers obtained after doping Au showed good photodegradation performance against RhB and MB. Under both Vis light and UV irradiation, they can lead to the separation of electrons and holes [23].

The noble-metal-doped photocatalysts can effectively inhibit the recombination of electrons and holes, improve the separation efficiency of photogenerated electron–hole pairs, and improve photocatalytic activity. In addition, the noble-metal-doped photocatalysts can expand the light absorption range of the photocatalyst. They can absorb not only UV light but also Vis light.

16.4.3 Semiconductor Composite

Semiconductor composite can be divided into two types. One type is a heterojunction formed by combining two semiconductors with different band gap widths, and the other type is a phase junction formed by combining different surface phase structures in the same semiconductor. The heterojunction can be prepared by adding different semiconductor precursors to the electrospinning solution followed by electrospinning and post-processing, while phase junctions can be obtained by adjusting the post-processing process of the same semiconductor precursors.

16.4.3.1 Heterojunction

In the heterojunction photocatalytic system, the electrons can migrate from the semiconductor with a higher energy band to the semiconductor with a lower energy band and accumulate in the CB to realize the effective separation of electron–hole in the two kinds of semiconductors and inhibit their recombination. Moreover, the band structure of the semiconductor composite system with different band gap widths can be adjusted to expand the range of optical response.

Ge et al. prepared TiO_2/CdS nanofiber photocatalysts by electrospinning and produced S-type heterojunctions (Figure 16.2). The existence of CdS effectively hinders the electron–hole recombination in TiO_2. In addition, the experimental results showed that the nanofiber photocatalyst had a remarkable effect on photocatalytic hydrogen production [24].

The construction of heterojunction by electrospinning has the characteristics of simple operation, controllable structure, and high activity photocatalyst preparation. However, the shortcoming of this method is that the oxidation–reduction ability of electron–hole pairs after migration of most heterojunctions is reduced, and it

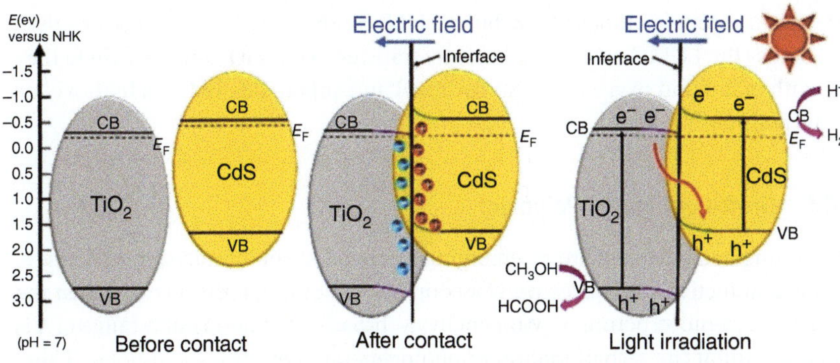

Figure 16.2 Schematic of the S-type heterojunction. Source: Ge et al. [24] / John Wiley & Sons.

is challenging to combine the effective separation and strong oxidation–reduction power of electron–hole pairs, resulting in reduced pollutant degradation ability.

16.4.3.2 Phase Junction

Phase junction is the establishment of a phase junction structure between two different crystal phases in the same semiconductor, so as to separate electrons and holes between different phases on the surface area, inhibit the recombination of electrons and holes, and thus improve the photocatalytic performance.

Lv et al. constructed α–β phase junctions on $Bi_4V_2O_{11}$ via electrospinning. The α–β junction determined the separation and transfer of electrons and holes. At the same time, the α–β junction showed excellent photocatalytic activity for Cr(VI) reduction, MB degradation, and nitrogen fixation [25]. Similarly, Lim et al. prepared Bi_2O_3 nanofiber photocatalysts with α–β junctions by electrospinning and heat treatment, considering adequately the fact that α–β junctions can limit the separation and transfer of electrons and holes. The experimental results showed that the degradation rate of RhB is towering under Vis light [26].

16.4.4 Dye Photosensitization

Using electrospun nanofiber photocatalysts as a matrix to adsorb photosensitizers on the fiber surface can improve the activity of photocatalysts. The photosensitizers are generally organic dyes. Compared with other research methods, dye photosensitization is one of the most promising methods to extend the photoresponse range of photocatalysts to the Vis light region.

Ghafoor et al. used Ag_2S nanoparticles with narrow band gaps to photosensitize electrospun TiO_2 nanofiber to prepare Ag_2S/TiO_2 nanofiber photocatalyst, which enabled TiO_2 nanofibers to have a more comprehensive Vis light absorption range and lower-energy band gap. Besides, they showed excellent photodegradation activity of MB in water under simulated sunlight [27].

Hou et al. prepared polydopamine (PDA)-coated TiO_2 composite nanofibers. Regarding photocatalytic properties, PDA coatings excite electrons by absorbing Vis light. Electrons jump from the highest occupied molecular orbital (HOMO) to the lowest unoccupied molecular orbital (LUMO) (Figure 16.3). The electrons then travel across the LUMO orbitals and rapidly transfer to the TiO_2 surface, where they react with water and oxygen to form hydroxyl and superoxide radicals that oxidize organic pollutants [28].

16.4.5 Graft-Conjugated Polymer

Graft-conjugated polymers are an emerging method for photocatalyst modification. Grafting conductive conjugated polymers on electrospun nanofibers can adjust the internal electronic structure of photocatalysts, broaden the absorption range of Vis light, and inhibit the recombination of photogenerated electrons and holes. There are two specific synthesis methods: the electrospun nanofibers are directly added to the liquid monomer as a matrix for polymerization, and the monomers polymerize

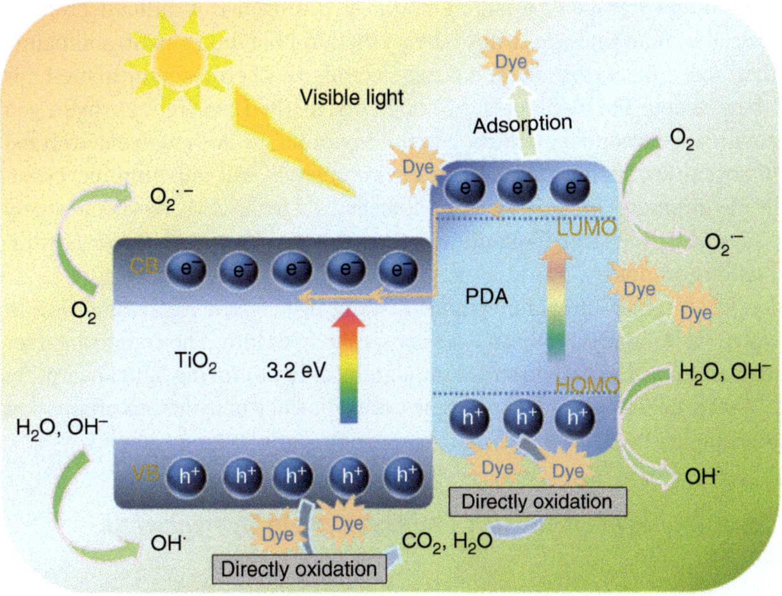

Figure 16.3 Schematic illustration for the mechanism. Source: Hou et al. [28] / John Wiley & Sons.

on the surface of the nanofibers to form polymer-coated functional composites. Instantly dissolving monomers and photocatalyst precursors in organic solvents for electrospinning can also realize the direct loading of catalysts on polymers.

Xie et al. adopted grafting modification followed by spinning. They first grafted polymethyl methacrylate (PMMA) onto xylan and then prepared xylan-g-PMMA/TiO_2 nanofiber photocatalyst by electrospinning. The degradation rate of MB by the photocatalyst was more than 80% [29].

16.5 Application

With the expansion of the research area and the continuous increase of the depth, photocatalysis research has been extended to many fields, such as energy, environmental protection, disinfection, and carbon dioxide reduction.

16.5.1 Applications of Electrospun Photocatalysts in Energy

Due to the continuous consumption of nonrenewable energy in traditional industries, the search for renewable green energy that can replace nonrenewable energy has become a current research focus. Hydrogen is a clean energy with high combustion calorific value and renewable cycle. Hydrogen can generate huge amounts of energy when it is burned, and its combustion product is pure water. Therefore, hydrogen energy is known as a new green, clean, and renewable energy. However, the acquisition and storage of a large amount of hydrogen is a difficult problem currently faced by mankind. Up to now, the methods for obtaining hydrogen mainly include thermochemical methods and water electrolysis. Thermochemical methods require a large amount of fossil fuels. Although this method can generate hydrogen energy, it will cause secondary pollution to the environment. Although electrolyzed water will not cause direct pollution to the environment, it will cause indirect pollution to the environment. In recent years, photocatalytic technology has been widely studied by domestic and foreign scientific and technological workers as a new hydrogen production technology.

In order to efficiently prepare hydrogen, wang et al. used electrospinning to prepare poly(L-lactic acid) (PLA)/TiO_2/Pt composite fiber film. The composite fiber membrane had a porous channel structure, significantly improving light absorption capacity. Compared with pure TiO_2 particles, the efficiency of hydrogen production was 30 times higher [30].

16.5.2 Applications of Electrospun Photocatalysts in Environmental Protection

In the field of environmental treatment, researchers have used electrospinning photocatalysis technology to conduct a lot of research on wastewater treatment and air purification and have made a series of progress. In this field, the oxidative power of photocatalysts is mainly used, and finally the pollutants are completely decomposed.

16.5.2.1 Wastewater Treatment

Due to the discharge of industrial wastewater, many toxic and harmful refractory chemical substances are produced, seriously affecting the ecological environment and human health. Therefore, wastewater treatment has always been the focus of attention in the field of environmental governance. Industrial wastewater mainly comes from textile printing and dyeing and petrochemical industries, and the composition of industrial wastewater is complex and diverse [31].

The discharge of organic pollutants into the water has caused damage to the ecosystem. Therefore, the degradation of organic pollutants has become the focus of attention. Pascariu et al. prepared ZnO–SnO_2 ceramic nanofibers by electrospinning for the degradation of RhB dyes. The experimental results showed that the concentration of RhB dye decreased significantly after only 240 minutes (four hours) UV irradiation (Figure 16.4a). In addition, the degradation rates of RhB dyes by the nanofibers under UV light irradiation and Vis light irradiation were 99.35% and 94.20%, respectively, showed excellent removal ability (Figure 16.4b) [33].

Cr(VI) is the main heavy metal in industrial wastewater, and the toxicity of Cr(VI) is more than 100 times that of Cr(III). Cr(VI) will cause digestive tract and lung cancer, upper abdominal pain, and nausea in humans. Cr(III) is an essential nutrient at trace level. Therefore, the best treatment is to reduce Cr(VI) to Cr(III). In addition, the experiment also showed that it could promote the removal of harmful Cr(VI) in an acidic environment [32].

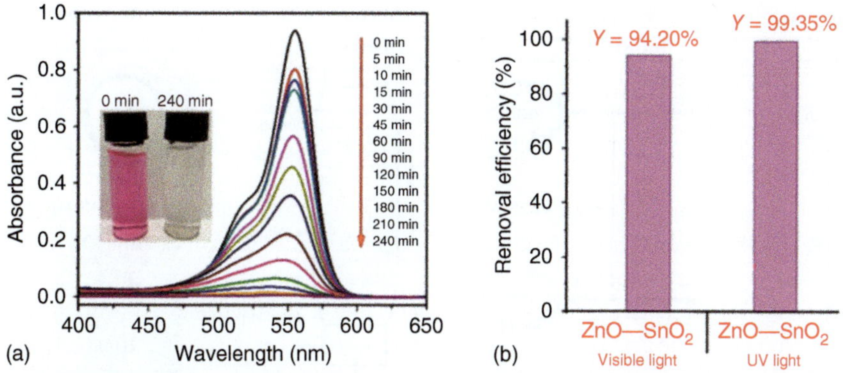

Figure 16.4 Photocatalytic data of Zno–SnO_2 nanofibers. (a) Profiles of UV–Vis spectra for the photodegradation of RhB dye. (b) Removal efficiency (%) obtained under Vis and UV light in the presence of ZnO–SnO_2 nanofibers [32]. Source: Pascariu et al. [33] / John Wiley & Sons.

16.5.2.2 Air Purification

Air pollutants are mainly volatile organic compounds. These substances have complex components and low concentrations, but they are highly toxic and uncomfortable. In addition to being used in water treatment, electrospun photocatalysts can also be used to degrade volatile organic compounds and achieve air purification.

In order to achieve effective degradation of volatile gases, Boaretti et al. prepared a photocatalytic system in which graphene-based materials were coupled with TiO_2. Under UV irradiation, the material can degrade acetaldehyde and methanol very well, which is suitable for indoor air pollution [34].

16.5.3 Applications of Electrospun Photocatalysts in Disinfection

The electrons and holes generated in the process of electrospun photocatalytic reaction react with O_2 and H_2O adsorbed on the surface of electrospun photocatalyst to generate free radicals and superoxide, which can effectively destroy the structure of bacteria, viruses, and other microorganisms, so as to achieve the purpose of sterilization.

Cheng et al. used electrospinning technology to prepare a Cu–TiO_2 nanofiber photocatalyst to study antiviral properties. Under acidic conditions, the electrostatic force between the photocatalyst and the virus was enhanced, thereby improving the virus removal efficiency [35].

Chen et al. prepared a nanofibrous membrane that can be used as a photocatalytic sterilization material. The experiments showed that a large number of hydroxyl radicals were generated on the surface of the nanofibrous membrane, thereby producing a large number of active oxidizing substances, which can enter the interior of pathogens, destroy protein functions and DNA, and disruption of bacterial cell membranes, resulting in pathogen death and antibacterial and antiviral effects [36].

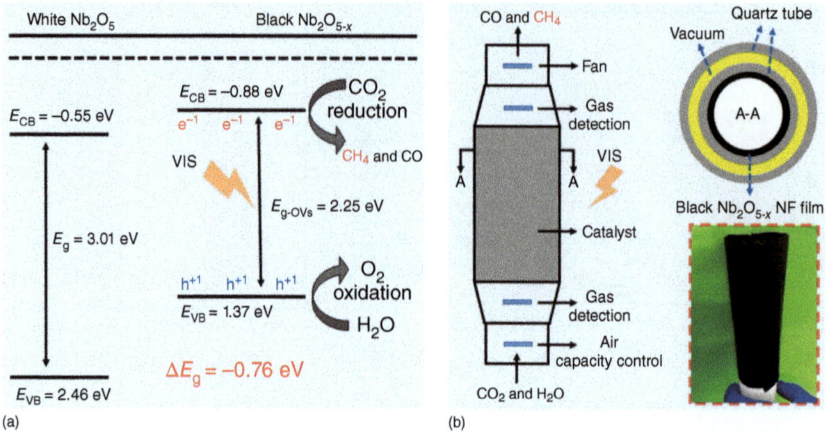

Figure 16.5 Illustrations of using the black Nb_2O_5 nanofibers in photocatalysis devices. (a) Compared with other Nb_2O_5 nanofibers in the photocatalytic ability. (b) The final product. Source: Lin et al. [37] / John Wiley & Sons.

16.5.4 Applications of Electrospun Photocatalysts in CO_2 Reduction

The development and utilization of fossil fuels, while promoting the development of the world economy, will also release a lot of CO_2 gas. CO_2 is also considered the main greenhouse gas, and it is believed that the cause of climate change is the increasing greenhouse gases. Artificial photosynthesis is an ideal way to reduce the growing amount of CO_2 in the atmosphere. So far, catalysts based on various semiconductors have been extensively studied for the photocatalytic reduction of CO_2. Under the optical drive, CO_2 can be converted into valuable energy sources such as carbon monoxide, methane, and methanol.

Lin et al. prepared black Nb_2O_5 nanofibers by electrospinning. Compared with Nb_2O_5-based photocatalysts reported so far, black Nb_2O_{5-x} nanofibers have the smallest band gap, and their absorption range extends from UV to Vis light (Figure 16.5a). Black Nb_2O_5 nanofibers can photocatalyze CO_2 into CH_4. In addition, the flexible catalyst film prepared by this research group can be directly used in devices (Figure 16.5b) [37].

16.6 Conclusion and Prospect

Compared with the traditional nanoparticle photocatalysts, the electrospun nanofiber photocatalysts have good dispersion and a larger specific surface area, which is conducive to the photocatalytic reaction of the catalyst and pollutants in full contact, and it is easier to separate and recover the photocatalyst efficiently. At present, the research of electrospun nanofiber photocatalysts has made well progress, but there are still many problems to be solved in practical application:

(1) The single-component electrospun nanofiber photocatalyst has a wide band gap and a narrow light absorption range. It can only absorb UV light and has low

utilization of light energy. Electrons and holes are easy to recombine, resulting in lower photocatalytic efficiency.

(2) To establish a heterojunction structure between two semiconductors through the method of composite doping. Driven by the internal electric field, electrons and holes can be effectively separated and transferred, thus inhibiting their recombination. However, it is difficult for the composite photocatalyst to have both effective separation of electrons and holes and strong oxidation–reduction ability, resulting in reduced pollutant degradation ability.

(3) The composite photocatalyst obtained by dye photosensitization and graft-conjugated polymer has poor adhesion between photosensitizer and polymer and is easy to fall off.

Therefore, the development of electrospun nanofiber photocatalysts with multi-functional synergies is a difficult task and a future trend.

References

1 Naya, S.I. and Tada, H. (2020). Au-Ag alloy nanoparticle-incorporated AgBr plasmonic photocatalyst. *Scientific Reports* 10 (1): 19972.

2 Huang, C., Hou, H., Hu, E. et al. (2017). Impact of power station capacities and sizes of solar field on the performance of solar aided power generation. *Energy* 139: 667–679.

3 Wang, F., Feng, Y., Chen, P. et al. (2018). Photocatalytic degradation of fluoroquinolone antibiotics using ordered mesoporous g-C_3N_4 under simulated sunlight irradiation: kinetics, mechanism, and antibacterial activity elimination. *Applied Catalysis B: Environmental* 227: 114–122.

4 Fujishima, A. and Honda, K. (1972). Electrochemical photolysis of water at a semiconductor electrode. *Nature* 238 (5358): 37–38.

5 Fujishima, A., Zhang, X., and Tryk, D. (2008). TiO_2 photocatalysis and related surface phenomena. *Surface Science Reports* 63 (12): 515–582.

6 Lee, Y.I., Lee, J.S., Park, E.S. et al. (2014). Effect of calcination temperature on the photocatalytic properties of electrospun TiO_2 nanofibers. *Journal of Nanoscience and Nanotechnology* 14 (10): 8005–8009.

7 El-Samak Ali, A., Ponnamma, D., Hassan, K.M., and AlMaadeed Mariam Al, A. (2022). A stable porous vessel for photocatalytic degradation of Azocarmine G dye. *Microporous and Mesoporous Materials* 341: 111994.

8 Mantilaka, M.P.G., De Silva, R.T., Ratnayake, S.P. et al. (2018). Photocatalytic activity of electrospun MgO nanofibres: synthesis, characterization and applications. *Materials Research Bulletin* 99: 204–210.

9 Liu, Q., Li, J., Zhao, Y. et al. (2015). CdS nanoparticle-functionalized natural cotton cellulose electrospun nanofibers for Vis light photocatalysis. *Materials Letters* 138: 89–91.

10 Pant, B., Park, M., and Park, S.-J. (2019). MoS_2/CdS/TiO_2 ternary composite incorporated into carbon nanofibers for the removal of organic pollutants from water. *Inorganic Chemistry Communications* 102: 113–119.

11 Wang, C., Shao, C., Wang, L. et al. (2009). Electrospinning preparation, characterization and photocatalytic properties of Bi_2O_3 nanofibers. *Journal of Colloid and Interface Science* 333 (1): 242–248.

12 Yang, X., Sun, S., Cui, J. et al. (2021). Synthesis, functional modifications, and diversified applications of hybrid BiOCl-based heterogeneous photocatalysts: a review. *Crystal Growth & Design* 21 (11): 6576–6618.

13 Babu, V.J., Bhavatharini, R.S.R., and Ramakrishna, S. (2014). Bi_2O_3 and BiOCl electrospun nanosheets and morphology-dependent photocatalytic properties. *RSC Advances* 4 (57).

14 Jin, X., Lou, Y., Zhang, X. et al. (2022). Broccoli-liked silver phosphate nanoparticles supported on green nanofiber membrane for Vis-light driven photodegradation towards water pollutants. *Nanotechnology* 33 (18): 185703.

15 Fu, F., Zhang, Y., Yan, L. et al. (2016). Preparation of efficient Ag/AgBr/TiO_2 Vis light photocatalyst for destruction of MB. *Journal of Materials Science: Materials in Electronics* 28 (1): 691–696.

16 Saha, D., Hoinkis, T.J., and Van Bramer, S.E. (2020). Electrospun, flexible and reusable nanofiber mat of graphitic carbon nitride: photocatalytic reduction of hexavalent chromium. *Journal of Colloid and Interface Science* 575: 433–442.

17 Liu, J., Li, D., Liu, X. et al. (2021). TiO_2/g-C_3N_4 heterojunction hollow porous nanofibers as superior vis-light photocatalysts for H_2 evolution and dye degradation. *New Journal of Chemistry* 45 (47): 22123–22132.

18 Liu, L., Liu, Z., Yang, Y. et al. (2018). Photocatalytic properties of Fe-doped ZnO electrospun nanofibers. *Ceramics International* 44 (16): 19998–20005.

19 Pascariu, P., Homocianu, M., Cojocaru, C. et al. (2019). Preparation of La doped ZnO ceramic nanostructures by electrospinning–calcination method: effect of La^{3+} doping on optical and photocatalytic properties. *Applied Surface Science* 476: 16–27.

20 Song, M., Cao, H., Zhu, Y. et al. (2020). Electrochemical and photocatalytic properties of electrospun C/TiO_2 nanofibers. *Chemical Physics Letters* 747: 137355.

21 Zhou, Y.Y., Zheng, Y.M., Wu, X.Q., and Shao, Z.D. (2021). Research progress of performance enhancement methods for electrospun nanofiber-based photocatalyst. *Journal of Textile Research* 42 (11): 179–186.

22 Yousef, A., El-Halwany, M.M., El-Newehy, M.H., and Al-Deyab, S.S. (2017). Synthesis of Cu–S-Co doped TiO_2 nanoparticles supported on carbon nanofibers for simultaneous adsorption and photocatalytic decomposition of reactive black 5. *Journal of Nanoscience and Nanotechnology* 17 (6): 3998–4004.

23 Tang, K.Y., Chen, J.X., Legaspi, E.D.R. et al. (2021). Gold-decorated TiO_2 nanofibrous hybrid for improved solar-driven photocatalytic pollutant degradation. *Chemosphere* 265: 129114.

24 Ge, H., Xu, F., Cheng, B. et al. (2019). S-scheme heterojunction TiO_2/CdS nanocomposite nanofiber as H_2-production photocatalys. *ChemCatChem* 11 (24): 6301–6309.

25 Lv, C., Chen, G., Sun, J., and Zhou, Y. (2016). Construction of alpha-beta phase junction on $Bi_4V_2O_{11}$ via electrospinning retardation effect and its promoted photocatalytic performance. *Inorganic Chemistry* 55 (10): 4782–4789.

26 Lim, G.-D., Yoo, J.-H., Ji, M., and Lee, Y.-I. (2019). Vis light driven photocatalytic degradation enhanced by α/β phase heterojunctions on electrospun Bi_2O_3 nanofibers. *Journal of Alloys and Compounds* 806: 1060–1067.

27 Ghafoor, S., Ata, S., Mahmood, N., and Arshad, S.N. (2017). Photosensitization of TiO_2 nanofibers by Ag_2S with the synergistic effect of excess surface $Ti^{(3+)}$ states for enhanced photocatalytic activity under simulated sunlight. *Scientific Reports* 7 (1): 255.

28 Hou, X., Cai, Y., Song, X. et al. (2019). Electrospun TiO_2 nanofibers coated with polydopamine for enhanced sunlight-driven photocatalytic degradation of cationic dyes. *Surface and Interface Analysis* 51 (2): 169–176.

29 Xie, Y., Sun, X.F., Li, W. et al. (2022). Fabrication of electrospun Xylan-g-PMMA/TiO_2 nanofibers and photocatalytic degradation of methylene blue. *Polymers (Basel)* 14 (12): 2489.

30 Wang, Q., Zhang, Z., Liu, L. et al. (2021). Degradable ultrathin high-performance photocatalytic hydrogen generator from porous electrospun composite fiber membrane with enhanced light absorption ability. *Journal of Materials Chemistry A* 9 (16): 10277–10288.

31 Uddin, F. (2021). Environmental hazard in textile dyeing wastewater from local textile industry. *Cellulose* 28 (17): 10715–10739.

32 Piao, H., Zhao, J., Liu, M. et al. (2022). Ultra-low power light driven lycopodium-like nanofiber membrane reinforced by PET braid tube with robust pollutants removal and regeneration capacity based on photo-Fenton catalysis. *Chemical Engineering Journal* 450: 138204.

33 Pascariu Petronela, C.C., Niculae, O., and Anton, A. (2019). Photocatalytic Activity of ZnO–SnO_2 Ceramic Nanofibersfor RhB Dye: degradation experimental design, modeling, and process. *Optimization* 256: 1800474.

34 Boaretti, C., Vitiello, G., Luciani, G. et al. (2020). Electrospun active media based on Polyvinylidene Fluoride (PVDF)-Graphene-TiO_2 nanocomposite materials for methanol and acetaldehyde gas-phase abatement. *Catalysts* 10 (9): 1017.

35 Cheng, R., Kang, M., Shen, Z.P. et al. (2019). Vis-light-driven photocatalytic inactivation of bacteriophage f2 by Cu-TiO_2 nanofibers in the presence of humic acid. *Journal of Environmental Sciences (China)* 77: 383–391.

36 Chen, P., Yang, Z., Mai, Z. et al. (2022). Electrospun nanofibrous membrane with antibacterial and antiviral properties decorated with Myoporum bontioides extract and silver doped carbon nitride nanoparticles for medical masks application. *Separation and Purification Technology* 298: 121565.

37 Lin, X., Xia, S., Zhang, L. et al. (2022). Fabrication of flexible mesoporous black Nb_2O_5 nanofiber films for vis-light-driven photocatalytic CO_2 reduction into CH_4. *Advanced Materials* 34: 2200756.

17

Smart Electrospun Actuators

Li Liu and Lei Zhang

Donghua University, Key Laboratory of Textile Science & Technology, Ministry of Education, College of Textiles, 2999 Renmin North Road, Songjiang District, Shanghai 201620, China

17.1 Introduction

Industrial revolution is a transition process to machine-manufacturing economy from one dominated by handicrafts and agriculture. The foundation to this transition was machines, which introduced novel ways of living and working by greatly increasing the productivity [1]. Machines are the engine behind the development of modern society and are capable of converting energy into mechanical motions [2]. To a certain extent, unlike machines, which are a kind of actuator made from rigid metals, soft actuators enable applicability in some areas that are not suitable for employing traditional rigid machines [3]. Flexibility and compliance, as two of the greatest advances, make them exceptionally adaptable to complex applications, such as bionic fingers [4], smart fabrics [5], and artificial blood vessel fabrication [6]. Flexibility generally comes from materials, known as polymers, which are composed of macromolecular chains that are flexible enough to entangle together [7]. Therefore, soft actuators are always called polymeric actuators, or origami because of the preprogrammed actuation [8]. The nature or preference of macromolecular chains can provide soft actuators with some unique properties through interactions with the stimulus from the surrounding environment [9]. Those polymeric actuators made from these smart polymers are given the name of thermos-, pH-, light-, and alternating magnetic-field-responsive actuators that are able to be triggered and controlled by temperature [10], pH [11], light [12], and alternating magnetic field [13]. The utilization of smart polymers, indeed, creates a possibility that those actuators made from smart materials can sense environment and show feedback. In other words, the interface between actuator and environment can affect the actuation, the more interfaces possessed, the more possibility of actuation will occur. Electrospun fibrous mat, comprised of thousands of nanofibers, can supply numerous interfaces or sites for the actuator to interact with the surrounding environment due to the large specific area of nanofibers. When diameter of fibers drops from 500 to 5 nm, the corresponding surface area changes greatly, from 10 000 to 1 000 000 m^2kg^{-1} [14].

Electrospinning: Fundamentals, Methods, and Applications, First Edition.
Edited by Liming Wang and Xiaohong Qin.

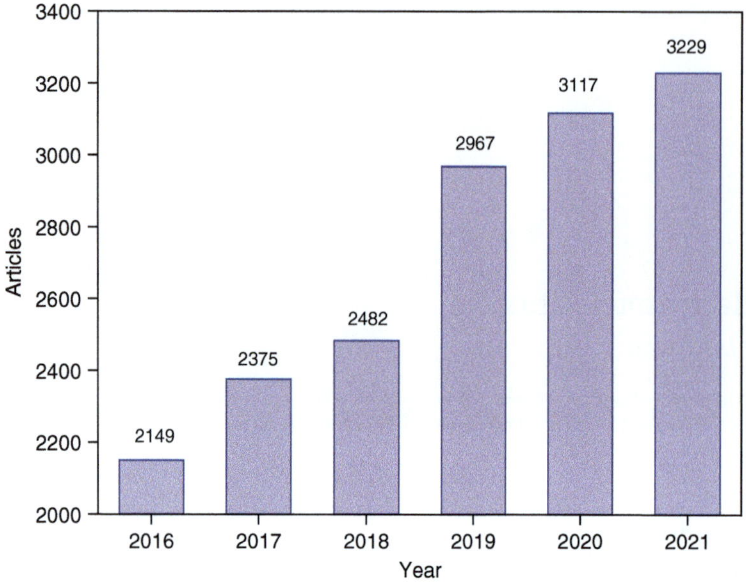

Figure 17.1 The recent six-year article publication trend graph on electrospinning actuators. Source: Lei Zhang.

Together with another characteristic that the large aspect ratio of nanofibers can create anisotropy, electrospun actuators have been a special candidate among soft actuators, and attracted a lot of scientists' interest in recent years, as illustrated in Figure 17.1.

17.2 Mechanism of Soft Actuators

Soft actuators can be inspired by the very common phenomenon that a piece of paper rolls from the edge after it is put on the surface of water, then returns to its original peaceful state. Cellulose, the raw material of paper, swells gradually from the bottom to the top side, which forces the paper to roll as water wets the paper from the bottom. After the whole piece of paper is wet, this swollen gradient state disappears, and the piece of paper turns back to being flat on the water [15, 16]. Soft actuators are inspired not only by piece of paper but also by plants [17] or animals [18]. Pine cone, known as forecast weather, is so sensitive to humidity in the air around it that their scales swell shut in humid air, otherwise they keep their open state [19–21]. The secret behind them is anisotropy as a result from a compromise effect in the form of actuation. To create such anisotropy, a bilayer structure is accepted by scientists, in which the active layer is sensitive to the stimulus from the environment but the other passive layer is inert [22], as illustrated in Figure 17.2 (prototype). Upon being stimulated, only the active layer swells or shrinks, showing size change, and then a compromise actuation is triggered by such an anisotropy across the section. This process is always accompanied with energy transfer between the actuator and

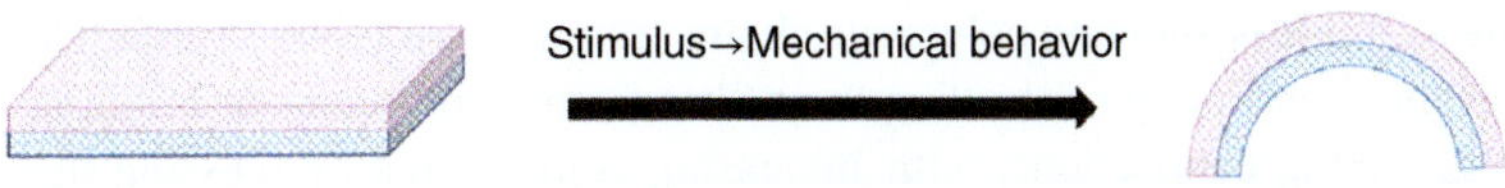

Figure 17.2 A bilayer structure actuator inspired by anisotropy, where the active layer (pink) will change in size under stimulation, while the inert layer is unaffected, thus driving the structure. Source: Zhou et al. [22] /Taylor & Francis / CC BY 4.0.

the environment. High energy transfer efficiency can facilitate this kind of actuation [23]. Electrospun fibrous mats with a large specific surface area and high porosity can have an effective energy transformation by providing numerous sites and thousands of tunnels. The selection of smart materials determines the stimulus, which the actuator will respond.

17.3 Fabrication of Electrospun Actuators

In the formation of polymeric film bilayer structure, the polymer chains, either from the active layer or from the passive layer, entangle together randomly, generating a strong interface without any delamination. However, nanofibers fabricated by electrospinning are highly relied on the electric field, in which the fibers are stretched constantly along with the solidification before arriving at the collector. This would cause a problem of delamination when constructing bilayer electrospun actuator, since the fibrous structure narrows the contact area at the interface and the solid state hinders the motion of polymeric chains between the two layers. Then, there have been several ways to avoid delamination when fabricating electrospun

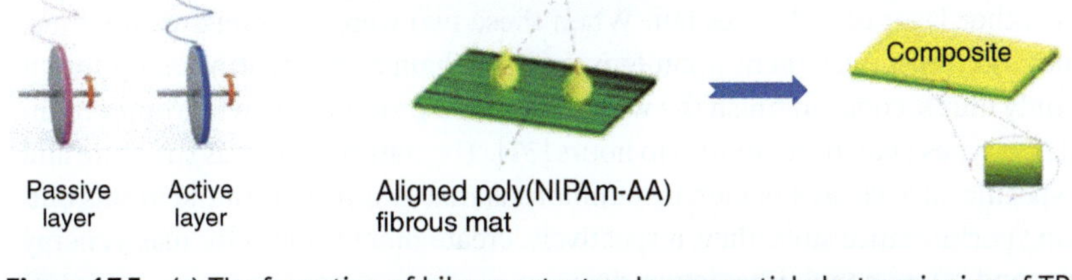

Figure 17.3 (a) The formation of bilayer actuators by sequential electrospinning of TPU and P(NIPAM-ABP) followed by UV cross-linking (steps 1 and 2) with increasing contact area by pressing the sample. And the actuator has no significant delamination. Source: Liu et al. [24] / John Wiley & Sons. (b) Schematic fabrication of the composite membrane. The aligned fibers were embedded in a TPU matrix to generate a composite membrane with gradient TPU content along the thickness. The composite membrane was generated by dropping a TPU solution on one side of poly(NIPAm-AA) mat. Source: Adapted content from Liu et al. [26].

actuator. One example is allowing polymeric chains from the two layers to entangle adequately at the interface by controlling their fiber-formation process or building covalent cross-linking connections, with increasing contact area by pressing the sample. Following this principle, our two previous reports [24, 25] showed a tenacious interface with no delamination in the bilayer electrospun fibrous actuator, as illustrated in Figure 17.3a. Another one is the combination of electrospun fibrous mat and film forming a fiber–film type of bilayer actuator. The key relies on the combination moment that the electrospun fibrous mat was solid but the film was a solution. This approach has substantially increased the contact area, compared to fiber–fiber type, since the flowability of the solution turned the way of contact from point-to-point to area-to-area. Additionally, during the solidification process, the entanglement state between the fiber and film initiated by the solution was finally frozen, as depicted in Figure 17.3b.

17.4 Evaluation of Electrospun Actuators

Polymeric actuators made by electrospinning offer advantages of high sensitivity over film-type actuator due to the large specific area, which provides more sites for the actuators to interact with stimulus from the surrounding environment. In effect, this interaction is a kind of stimulus–feedback procedure, which is highly dependent on energy transfer, mostly accompanied by mass exchange. Efficient energy transfer can facilitate actuation in the form of actuation speed. Jiang et al. innovatively fabricated a thermoresponsive actuator by electrospinning thermoresponsive polymers of poly(*N*-isopropylacrylamide) P(NIPAM) and thermoplastic polyurethane (TPU) into the active fibrous layer and the passive fibrous layer, respectively. P(NIPAM), as one of the most typical thermoresponsive polymers, switches its solubility in water depending on the temperature. It becomes soluble below the lower critical solution temperature (LCST) in water and vice versa as a result of the polymer chains converting between coil and globule states, which are sensitive to temperature change [27], as presented in Figure 17.4a. This P(NIPAM) electrospun fibrous layer still retains the thermoresponsive ability, showing temperature-dependent size change in water, but the other layer of TPU does not. When these two were combined without any delamination, they were bent upon temperature change. It is worth noting that it takes only four seconds to finish the whole actuation process [30], while other traditional film types take from minutes to hours [31]. The nature of porous structure and large specific surface area of electrospun fibrous mat mainly contribute to this outstanding performance since they, respectively, create more tunnels for mass/energy transfer and offer enough interaction sites.

Upon being triggered, polymeric actuators would perform their mechanical motions, mostly in the form of bending. Curvature, the reciprocal of the radius, was used to evaluate the bending degree. The larger the curvature is, the more the actuator bends (Figure 17.4b). Based on electrothermal effect, Yao et al., selected polydimethylsiloxane (PDMS) film as the heat-induced active layer that expands by the joule heat from Ag nanowires upon being powered on. Combining with

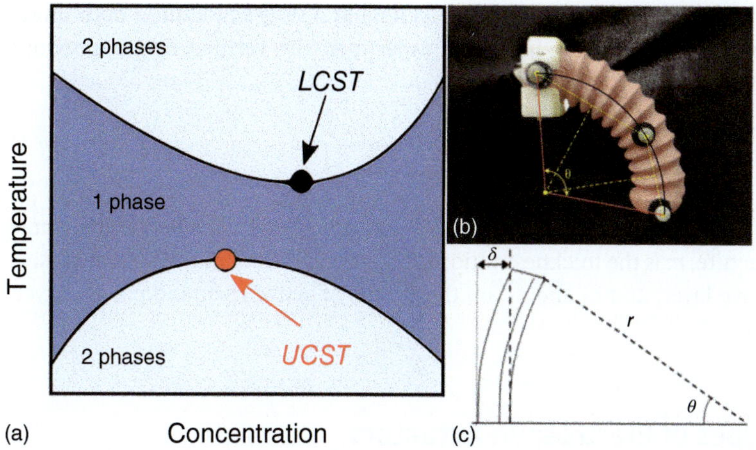

Figure 17.4 (a) The solubility of P(NIPAM) in water as a function of temperature, soluble in water at temperatures below the LCST or above the UCST. Source: Niskanen et al. [27] / Royal Society of chemistry. (b) The actuator will bend when triggered to achieve mechanical movement, the smaller the θ angle, the greater the degree of bending. Source: Chen et al. [28] / Mary Ann Liebert, Inc. / CC BY 4.0. (c) The deflection δ and the angles θ was used to describe the actuation behavior as well. Source: Shivashankar et al. [29] / IOP Publishing.

the other passive layer of polyimide layer, the bilayer structure curls, which is quantitatively described by the curvature [32].

Not only curvature, sometimes deflection, the degree to which a part of a bilayeractuator is displaced when actuating, or the angles between the bending end and the previous axis ahead, was used to describe the actuation behavior as well (Figure 17.4c).

The actuation is caused by the anisotropy across the section, which is analogous to the bending behavior of beam structure, in which the thickness ratio and modulus ratio mainly affect the final performance [33–35]. According to the Timoshenko equation (17.1), the roles of thickness ratio (m) and Young's modulus ratio (n) are depicted in Figure 17.5. When other variables are constant, the curvature

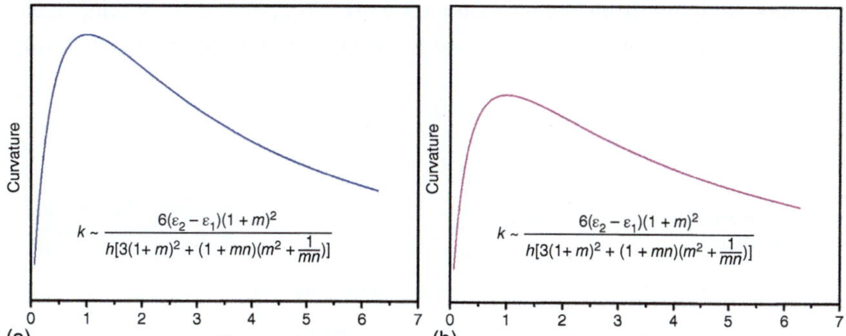

Figure 17.5 Quantitative analysis of effect of (a) thickness ratio (m) and (b) Young's moduli ratio (n) on curvature. Source: Li Liu.

can reach the maximum as either the thickness or Young's moduli of both layers are close to each other. This has been experimentally verified in our previous research [25].

$$k = \frac{1}{r} = \frac{6(\varepsilon_2 - \varepsilon_1)(1+m)^2}{h\left[3(1+m)^2 + (1+mn)\left(m^2 + \frac{1}{mn}\right)\right]} \tag{17.1}$$

where k is the curvature; r is the radius of the circle; h is the thickness of the whole bilayer structure; m is the thickness ratio; n is the Young's moduli ratio of the passive layer to active layer; and ε_2 and ε_1 are the strains of active layer and passive layer, respectively.

17.5 Types of Electrospun Actuators

The classification of electrospun actuators is consistent with the types of smart materials used, which is determined by the stimulus, such as temperature, pH, light, electricity or electric field, magnetic field, or multiresponsive stimulus. These electrospun actuators will be triggered and present mechanical motions upon being stimulated by the corresponding stimulus from the environment (Figure 17.6). And their characteristics and mechanism are discussed with details.

17.5.1 Thermoresponsive Electrospun Actuator

Thermoresponsive smart materials can show size/solubility change, either swell or shrinkage, as they are suffering from temperature change in the presence of water molecules. Owing to their nature of macromolecular chains, totally converse performance can be found even under the same conditions. P(NIPAM), a typical LCST thermoresponsive polymer with a cloud point of 32 °C, is soluble in water below 32 °C but precipitated above 32 °C because the interaction with water molecules is profoundly affected by the temperature. Cross-linking agents can be introduced during polymerization to turn the solubility into dimensional size change. According to the description from Seema Agarwal's work [30], photo-cross-linking agent 4-acryloylbenzophenone (ABP) was added to the P(NIPAM) polymer chains during

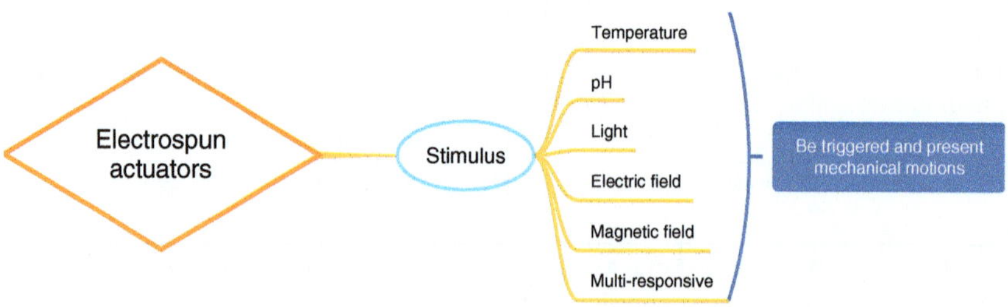

Figure 17.6 The classification of electrospun actuators according to different stimulus factors. Source: Lei Zhang.

the radical polymerization, forming a photo-cross-linkable thermoresponsive polymer of poly(*N*-isopropylacrylamide-*co*-4-acryloylbenzophenone) (P(NIPAM-ABP)) that only shows size change instead after cross-linked by UV lamp (Figure 17.7a). They have demonstrated that the sensitivity of thermoresponsive actuator can be highly improved when the P(NIPAM-ABP) was electrospun into nanofibers because of the high porosity and large specific surface area. Compared with the traditional film-like actuators, another advantage of electrospun fiber-like actuators is the anisotropy created in-plane by the arrangement of fibers, which can diversify the actuation behavior. In our previous report [24], aligned P(NIPAM-ABP) fibrous mat was fabricated by employing a fast-rotating wheel as the collector, by which the flying P(NIPAM-ABP) fibers were stretched and forced to deposit parallelly on the surface of rotating wheel. Thus, an anisotropy was created in-plane that the dimensional size change, featuring swell or shrinkage, becomes different along fiber alignment and in the direction perpendicular to the fiber alignment. Therefore, such an actuator gets a directionally controlled actuation ability (Figure 17.7b). Later, it was demonstrated that the directionally controlled actuation was not only the result of anisotropic dimensional size change, but also a synergistic effect of anisotropy in dimensional size change and e-modulus in different directions (Figure 17.7c) [25]. In addition to LCST, upper critical solution temperature (UCST) polymers are another type of thermoresponsive polymers that have a converse phase transition in that they are soluble at high temperatures and precipitate at low temperatures in water.

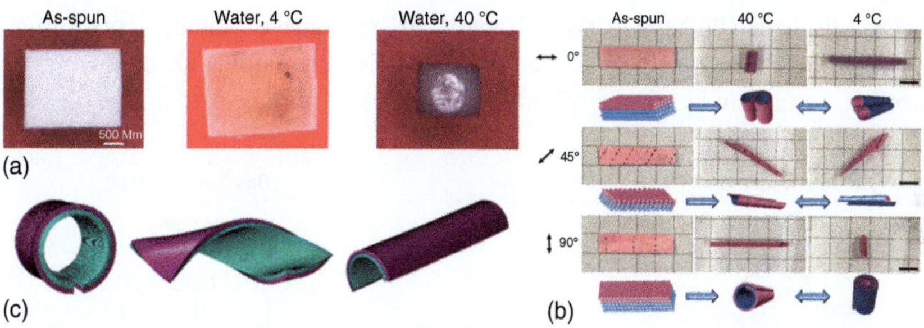

Figure 17.7 (a) Dimensional size changes shown in the thermoresponsive polymer P(NIPAM-ABP) formed by the addition of the photo-cross-linker 4-acryloylbenzophenone (ABP) to the P(NIPAM) polymer chain when it is exposed to cold (4 °C) and warm (40 °C) water. Source: Jiang et al. [30] / With permission from John Wiley & Sons. (b) Fiber orientation-dependent actuation behavior of bilayer TPU (pink)/P(NIPAM-ABP) blue) fibrous membranes (length: 2.0 cm, width: 0.5 cm) in water at different temperatures. Black arrows show the fiber orientation direction. 0°, 45°, and 90° are angles between the fiber direction and the long axis of the sample, as indicated by a black dotted line on the sample. Scale bar = 0.5 cm. Actuator gets highest directionally controlled actuation abilities when the fiber orientation is 90° to the long axis of the sample. Source: Li et al. [24] / John Wiley & Sons. (c) Equilibrium shapes of bilayer mats obtained from finite-element simulations in 40 °C water. From left to right are 0°, 45°, and 90° fiber alignments in one of the two layers. By using anisotropic expansion coefficients in combination with an anisotropic elastic modulus, good agreement with the experimental shapes in (c) is obtained. It is demonstrated that the directionally controlled actuation is the synergistic effect of anisotropic dimensional changes and different orientations of e-modulus. Source: Li et al. [25] / John Wiley & Sons.

Liu et al. [36] fabricated an electrospun fibrous mat from a UCST polymer synthesized from acrylamide (AAm) and acrylonitrile (AN), and this UCST electrospun fibrous mat presents a significant dimensional size change in water with different temperatures, featuring swell and shrinkage at 70 and 5 °C, respectively. It has demonstrated a great potential in the application of electrospun UCST actuator.

17.5.2 pH-Responsive Electrospun Actuator

pH-responsive polymers are a series of materials which are capable of detecting pH change of the surrounding environment and then give some response by varying their dimensional change [37]. pH polymers can be divided into either acidic or basic type depending on the mechanism, which lies in, respectively, the protonation or the deprotonation process of the functional groups of macromolecular chains [38]. Polyacids have acidic functional groups that tend to accept protons at a low pH environment lower than its acid dissociation constant (pK_a) such as carboxylic acids (–COOH) and sulfonic acids. The polymer chains are negatively charged due to the deprotonation of these acidic groups at high pH values and then repel each other, leading to dissolving or swelling [39]. Conversely, it can be found that the polybases are dissolving or swelling at low pH values since the basic group, such as $-NH_2$, in the polymer chains accepts protons, resulting in the positive charge of the chains (Figure 17.8).

Polyacrylic acid (PAA) is a typical acidic pH-responsive polymer containing –COOH group. Cho et al. [41] fabricated a reversible actuator with directionally controlled actuation by electrospinning PAA into a fibrous mat as the active

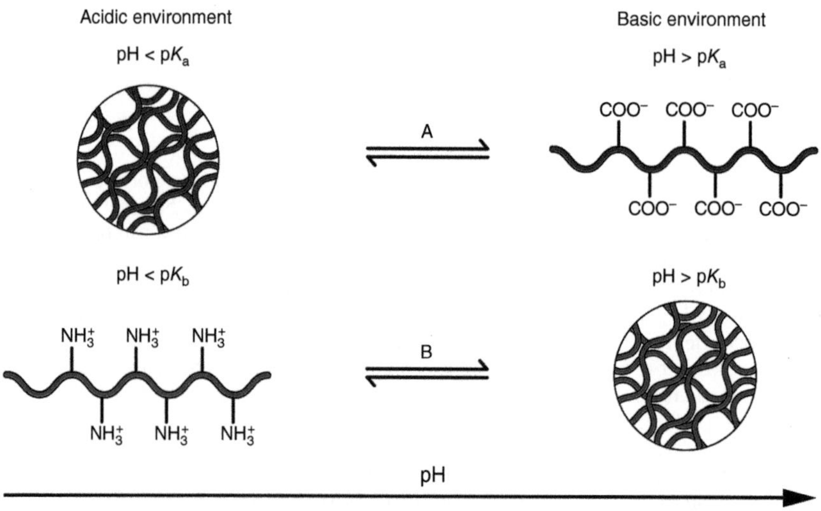

Figure 17.8 Polyacids (a) and polybases (b) polymeric chain state depending on the ionization degree. Source: Reproduced with permission from Ofridam et al. [40]. Copyright 2021 John/Wiley & Sons Ltd.

layer, together with photolithography. This bilayer structure was covered by cross-linkable polyethylene glycol (PEG) hydrogel, but with different cross-linked area in the form of aligned stripes by photomask under UV lamp. When stimulated by pH change, the created stripes with different cross-linked degrees would cause anisotropic dimensional change, leading to directionally controlled actuation. It opens and closes in the width direction of the sample upon being triggered by the pH value change. Additionally, it has to be mentioned here, Sarikaya et al. [42] presented a different pH actuator, a yarn, which performs its mechanical motions in one-dimension rather than bending in three-dimension. By hydrolyzing the stabilized aligned electrospun PAN, the pH responsive ability was afforded to the aligned PAN electrospun fibrous mat due to the formation of acid group of –COOH. Then the aligned fibrous mat was twisted to form a pH-responsive yarn. And a counterintuitive result comes out that the twisted yarn made from coiled pH-responsive fibers showed a contrary expansion and contraction to bundles, with the bundles shrinking but the yarns elongating in acid. This lies in the synergistic effect of the change of fiber diameter, modulus, and torsional stiffness.

17.5.3 Light-Responsive Actuator

Incorporating photothermal components into thermal-induced size-change polymers and employing photochemical polymers are nowadays two main approaches to fabricate light-responsive actuators [43]. The first one, in effect, is still relying on the thermal-induced mechanism with turning the thermal source outside into a built-in heater source. This allows the actuator to be manipulated remotely, the photothermal effect in the actuator. Reduced graphene oxide (rGO) [44], golden nanocages [45], carbon black [46], Fe_3O_4 nanoparticles [47], and other black materials [48] are typical photothermal components that are able to effectively transfer light, generally near infrared (NIR), into heat. Duan et al., constructed a NIR-sensitive actuator by electrospinning Fe_3O_4 nanoparticles into polyacrylonitrile (PAN) fibrous mat acting as the built-in heater, which causes the shrinkage of the active layer, electrospun fibrous P(NIPAM) mat [49]. This actuator can not only show simple bending behavior that can mimic grappling but also move forward as a result of the directionally controlled bending and the remote manipulation by NIR.

Besides the incorporation of photothermal components, the application of photochemical polymers, in which configuration change occurs when triggered by light, provides a new strategy to fabricate light-sensitive actuator [50]. This kind of polymer contains light-induced photoreaction segments, of which two states at molecular level are conversed depending on the light wavelength. For instance, azobenzene group is transformed between trans state and cis isomer state, upon being triggered by light, specifically trans–cis conversion, at 300–400 nm UV wavelength and cis–trans conversion at >400 nm visible light. Nohadani et al. [51] introduced azobenzene moiety into polysaccharide-based materials, then electrospun them into photo-responsive cellulose fibrous mat that present light-based regulating ability in hydrophobicity and actuation deformation (Figure 17.9).

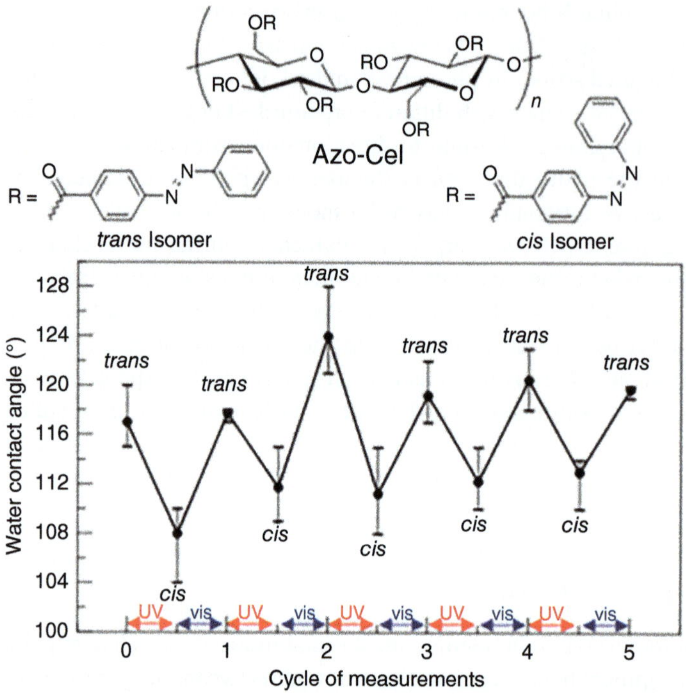

Figure 17.9 Light-responsive fiber mats made by introducing azobenzene into polysaccharide-based materials. The WCA changes of the Azo-Cel fabric over five cycles of UV/visible light irradiations. The plotted WCA values are the mean values of four sets of the measurements using four different pieces of the Azo-Cel fabric, and the error bars represent their maximum and minimum values. Source: Reproduced with permission from Mohd Ghazali et al. [52]. Copyright 2022 John Wiley & Sons – Books.

17.5.4 Electric-Field-Responsive Actuator

Compared with the thermos- and pH-responsive types, manipulation through electric field or electricity seems a more modern way, for these electric types can be operated under ambient condition in air without aqueous environment, which makes them more attractive to be applied in a broad field. It still follows the mechanism discussed previously that gradient or anisotropy should be generated upon being triggered. Actuation is closely related to the occurrence of anisotropy in the soft actuator. Drawing on this knowledge, it mainly concludes two types: electroactive type and electricity type.

Electroactive polymers (EAP) are a class of polymers that present size changes under electric field [53]. Dielectric elastomers, belonging to the EAP group, respond to electric field with large deformation. Typically, the dielectric elastomer is sandwiched between two electrodes, which are compressed in thickness direction and expanded in-plane due to the generated Maxwell's compressive electrostatic force when electric field is applied. It is worth noticing that both the electrodes should be made of soft materials so that it is possible for these kinds of actuators to be deformed. One of the biggest challenges for the application of dielectric elastomer actuators is the high voltage applied, ranging from several hundred up to several thousand voltages [52, 54, 55].

In contrast, polyelectrolyte is more sensitive to voltages as low as several voltages. The charged groups of polyelectrolytes in salts tend to transport to the oppositely charged electrodes, generating anisotropy in osmotic pressure. Nafion is a hydrophobic polyelectrolyte in which the hydrated cations move toward the counter electrode of cathode, forming osmotic gradient and causing expansion there, upon being stimulated by electric field [56]. Sodium alginate (SA), PAA, and gelatin have the similar performance in electric field, showing great potential for fabricating biomedical soft actuators.

Electricity actuators, in which direct current is generated, include electromagnetic and the derivation of thermoresponsive actuators. The former is the utilization of Lorentz force, and the latter is electrothermal type, with the contribution of Joule heat. The Lorentz force generated in a well-arranged current-carrying conductive route in soft polymer in a magnetic field is the reason to drive the actuator. Mao et al. fabricated an innovative actuator by engraving organized channels in a silicone elastomer shell for conductive liquid metal Galinstan. In magnetic field, the Lorentz force exerted on the liquid metal in the channel allows the silicone elastomer shell for an ultrafast and powerful bending upon being charged with direct current [57]. (PDMS, a soft polymer, swells at high temperature. Aouraghe et al. used carbon nanotube film as the Joule-heat-generating source to combine the PDMS, forming an electrothermal actuator that bends when the power source is on [58].

Electric/electricity-stimulus-responsive actuator has unique advantages; however, until now, it has not induced scientists' interest to take advantages of electrospinning technology to fabric these types of actuators. It still holds great potential for the future.

17.5.5 Magnetic-Field-Responsive Actuator

Magnetic field, capable of obstacle-ignoring transmission, highly efficient energy transfers, and no need for specific pathway, can be considered as a promising driving force to inform future untethered soft robot design. Besides the electromagnetic effect driven by Lorentz force discussed in the last section [57], magnetic force and magnetothermal heat are another two main direct stimuli to design and fabricate magnetic-field-responsive actuators. A thin composite was fabricated by spin coating of mixture of magnetic neodymium–iron–boron (NdFeB) microparticles and PDMS precursor, followed by magnetization. This soft actuator showed an ultrafast magnetic field response in milliseconds by adjusting the strength and frequency of the magnetic field [59]. Described in one of our last papers [60], magnetothermal particles of Fe_3O_4 were embedded into thermoresponsive polymer of P(NIPAM) fibers by electrospinning. It has demonstrated that the loading amounts of particles and the high porosity contributed to the ultrafast actuation, in several seconds. And the actuation process can be manipulated remotely even if the actuator is covered by a piece of polytetrafluoroethylene (PTFE) film. Furthermore, in our other paper

[61], the aligned fibers applied allow the electrospun fibrous mat to actuate in a desired direction.

Unlike the actuator operated by light, the magnetic-field-responsive actuator does not need the specific pathway to guarantee the transmission of magnetism. It provides a novel strategy to design wireless end actuator.

17.6 Conclusions and Perspectives

This chapter has presented the basic concept of a soft actuator, its actuation mechanism, ways to evaluate soft actuators, with an emphasis on the combination of smart materials and electrospinning technology. The driving force for the soft actuator essentially originates from reconstruction of smart materials, which are a kind of responsive material capable of responding to the environment by giving feedback with occurrence of chemical or physical configuration upon receiving the stimulus. Electrospinning is more of a technology to shape the smart materials in physical, constructing the smart materials by stacking numerous continuous nanofibers, forming smart fibrous mat. The large specific surface area and porous structure with high porosity provide interacting sites for the stimulus and the pathways for energy or mass transfer. For thermo- and pH-responsive actuators or other actuators that rely on energy/mass transfer to trigger the actuation, electrospun fibrous mat can facilitate the actuation dramatically. Conversely, for the actuators that somewhat independent of mass transfer from surrounding environment, such as light-, electric-field-, and magnetic-field-responsive actuators, electrospun mats cannot contribute too much to facilitate the actuation. However, there is an attractive point that anisotropy in-plane can be created just by changing the arrangement of electrospun fibers, then the soft actuator can gain a directionally controlled ability. Currently, electrospun actuators are layered structures that restrict their degree of freedom. In future, the improvement of degree of freedom and miniaturization of electrospun actuators could be promising points. And more work should be going further into the real applications, not just grabbing a polystyrene ball or bending somehow. Only by doing so can it be to possible bring the electrospun soft actuators and the other soft actuators to a new world.

References

1 Xu, X. (2017). Machine Tool 4.0 for the new era of manufacturing. *The International Journal of Advanced Manufacturing Technology* 92 (5): 1893–1900.

2 Sandakalum, T. and Ang, M.H. (2022). Motion planning for mobile manipulators—a systematic review. *Machines* 10 (2): 2075–1702.

3 Ahmed, F., Waqas, M., Jawed, B. et al. (2022). Decade of bio-inspired soft robots: a review. *Smart Materials and Structures* 31 (7): 073002.

4 Zhang, Z., Chen, H., and Zhang, Z. (2018). Configuration synthesis and performance analysis of finger soft actuator. *Applied Bionics and Biomechanics* 2018: 1–14.

5 Yin, R., Yang, B., Ding, X. et al. (2020). Wireless multistimulus-responsive fabric-based actuators for soft robotic, human–machine interactive, and wearable applications. *Advanced Materials Technologies* 5 (8): 2000341.

6 Zhao, Z., Fang, R., Rong, Q. et al. (2017). Bioinspired nanocomposite hydrogels with highly ordered structures. *Advanced Materials* 29 (45): 1703045.

7 Rajan, L., Sidheekha, M.P., Shabeeba, A. et al. (2022). Conducting polymers as bio-mimetic multistep macromolecular sensors of working conditions: polyindole/polyvinyl alcohol hybrid film senses electrical and chemical working ambient. *Materials Chemistry Frontiers* 6 (12): 1706–1718.

8 Farhan, M., Behl, M., Kratz, K. et al. (2021). Origami hand for soft robotics driven by thermally controlled polymeric fiber actuators. *MRS Communications* 11 (4): 476–482.

9 Mohammad, S.A., Shingdilwar, S., Banerjee, S. et al. (2020). Macromolecular engineering approach for the preparation of new architectures from fluorinated olefins and their applications. *Progress in Polymer Science* 106: 101255.

10 Khanh, T., Keskūla, A., Zondaka, Z. et al. (2019). Role of polymerization temperature on the performance of polypyrrole/dodecylbenzenesulphonate linear actuators. *Synthetic Metals* 247: 53–58.

11 Wang, W., Xu, X., Zhang, C. et al. (2022). Skeletal muscle fibers inspired polymeric actuator by assembly of triblock polymers. *Advanced Science* 9 (13): 2105764.

12 Zheng, C., Jin, F., Zhao, Y. et al. (2020). Light-driven micron-scale 3D hydrogel actuator produced by two-photon polymerization microfabrication. *Sensors and Actuators, B: Chemical* 304: 127345.

13 Seo, Y., Ghazanfari, L., Master, A. et al. (2022). Poly(2-oxazoline)-magnetite NanoFerrogels: magnetic field responsive theranostic platform for cancer drug delivery and imaging. *Nanomedicine: Nanotechnology, Biology, and Medicine* 39: 1549–9634.

14 Gibson, P., Schreuder-Gibson, H., and Rivin, D. (2001). Transport properties of porous membranes based on electrospun nanofibers. *Colloids and Surfaces A: Physicochemical and Engineering Aspects* 187: 469–481.

15 Coley, A. and Mcmanus, D.J. (1994). On spacetimes admitting shear-free, irrotational, geodesic time-like congruences. *Classical and Quantum Gravity* 11 (5): 1261–1282.

16 Douezan, S., Wyart, M., Brochard-Wyart, F. et al. (2011). Curling instability induced by swelling. *Soft Matter* 7 (4): 1506–1511.

17 Xu, Z., Zhou, Y., Zhang, B. et al. (2021). Recent progress on plant-inspired soft robotics with hydrogel building blocks: fabrication, actuation and application. *Micromachines* 12 (6): 608.

18 Cao, X., Zhang, M., Zhang, Z. et al. (2019). Review of soft linear actuator and the design of a dielectric elastomer linear actuator. *Acta Mechanica Solida Sinica* 32 (5): 566–579.

19 Dawson, C., Vincent, J.F.V., and Rocca, A.M. (1997). How pine cones open. *Nature* 390 (6661): 668.

20 Paris, O., Burgert, I., and Fratzl, P. (2010). Biomimetics and biotemplating of natural materials. *MRS Bulletin* 35 (3): 219–225.

21 Burgert, I. and Fratzl, P. (1893). Actuation systems in plants as prototypes for bioinspired devices. *Philosophical Transactions of the Royal Society A: Mathematical, Physical and Engineering Sciences* 2009 (367): 1541–1557.

22 Zhou, S., Zhou, Q., Wang, M. et al. (2020). Preparation of bilayer shape memory intelligent hydrogel actuators and their structural characteristics. *International Journal of Food Properties* 23 (1): 470–480.

23 Liu, Y., Cao, Y., Feng, X.-Q. et al. (2019). Phase transition and optimal actuation of active bilayer structures. *Extreme Mechanics Letters* 29: 100467.

24 Liu, L., Jiang, S., Sun, Y. et al. (2016). Giving direction to motion and surface with ultra-fast speed using oriented hydrogel fibers. *Advanced Functional Materials* 26 (7): 1021–1027.

25 Liu, L., Ghaemi, A., Gekle, S. et al. (2016). One-component dual actuation: poly(NIPAM) can actuate to stable 3D forms with reversible size change. *Advanced Materials* 28 (44): 9792–9796.

26 Liu, L., Bakhshi, H., Jiang, S. et al. (2018). Composite polymeric membranes with directionally embedded fibers for controlled dual actuation. *Macromolecular Rapid Communications* 39 (10): 1800082.

27 Niskanen, J. and Tenhu, H. (2017). How to manipulate the upper critical solution temperature (UCST)? *Polymer Chemistry* 8 (1): 220–232.

28 Chen, W., Xiong, C., Liu, C. et al. (2019). Fabrication and dynamic modeling of bidirectional bending soft actuator integrated with optical waveguide curvature sensor. *Soft Robotics* 6 (4): 495–506.

29 Shivashankar, H., Sangamesh, R., and Kulkarni, S.M. (2019). Optimization of bilayer actuator based on carbon black/polymer composites. *IOP Conference Series: Materials Science and Engineering* 522 (1): 012007.

30 Jiang, S., Liu, F., Lerch, A. et al. (2015). Unusual and superfast temperature-triggered actuators. *Advanced Materials* 27 (33): 4865–4870.

31 Belmonte, A., Lama, G.C., Cerruti, P. et al. (2018). Motion control in free-standing shape-memory actuators. *Smart Materials and Structures* 27 (7): 075013.

32 Yao, S., Cui, J., Cui, Z. et al. (2022). Correction: soft electrothermal actuators using silver nanowire heaters. *Nanoscale* 14 (17): 6671–6671.

33 Timoshenko, S. (1925). Analysis of Bi-metal thermostats. *Journal of the Optical Society of America* 11 (3): 233–255.

34 Arranz, M.A., Sánchez, E.H., Moreno, Y. et al. (2022). Thickness gradient related magnetic anisotropy of wedged Co nanocluster assemblies deposited on glass plates. *Journal of Magnetism and Magnetic Materials* 547: 0304–8853.

35 Jiménez, S.M.A. and Mcmeeking, R.M. (2016). A constitutive law for dielectric elastomers subject to high levels of stretch during combined electrostatic and mechanical loading: elastomer stiffening and deformation dependent dielectric permittivity. *International Journal of Non-Linear Mechanics* 87: 125–136.

36 Liu, F., Jiang, S., Ionov, L. et al. (2015). Thermophilic films and fibers from photo cross-linkable UCST-type polymers. *Polymer Chemistry* 6 (14): 2769–2776.

37 Deirram, N., Zhang, C., Kermaniyan, S. et al. (2019). pH-responsive polymer nanoparticles for drug delivery. *Macromolecular Rapid Communications* 40 (10): e1800917.

38 Yu, C., Wang, L., Xu, Z. et al. (2020). Smart micelles self-assembled from four-arm star polymers as potential drug carriers for pH-triggered DOX release. *Journal of Polymer Research* 27 (5): 111.

39 Jannat, N.E., Alam, M.A., Rahman, M.A. et al. (2019). Carboxylic acid modified pH-responsive composite polymer particles. *Journal of Polymer Engineering* 39 (7): 671–678.

40 Ofridam, F., Tarhini, M., Lebaz, N. et al. (2021). pH-sensitive polymers: classification and some fine potential applications. *Polymers for Advanced Technologies* 32 (4): 1455–1484.

41 Cho, K., Kang, D., Lee, H. et al. (2022). Multi-stimuli responsive and reversible soft actuator engineered by layered fibrous matrix and hydrogel micropatterns. *Chemical Engineering Journal* 427: 130879.

42 Sarikaya, S., Gardea, F., Auletta, J.T. et al. (2021). Athermal artificial muscles with drastically improved work capacity from pH-responsive coiled polymer fibers. *Sensors and Actuators B: Chemical* 335: 129703.

43 Bhatti, M.R.A., Kernin, A., Tausif, M. et al. (2022). Light-driven actuation in synthetic polymers: a review from fundamental concepts to applications. *Advanced Optical Materials* 10 (10): 2102186.

44 Punetha, V.D., Ha, Y.M., Kim, Y.O. et al. (2020). Rapid remote actuation in shape memory hyperbranched polyurethane composites using cross-linked photothermal reduced graphene oxide networks. *Sensors and Actuators, B: Chemical* 321: 128468.

45 Feng, B., Xu, Z., Zhou, F. et al. (2015). Near infrared light-actuated gold nanorods with cisplatin-polypeptide wrapping for targeted therapy of triple negative breast cancer. *Nanoscale* 7 (36): 14854–14864.

46 Jin, X., Shi, Y., Yuan, Z. et al. (2022). Bio-inspired soft actuator with contact feedback based on photothermal effect and triboelectric nanogenerator. *Nano Energy* 99: 107366.

47 Cao, Y., Li, W., Quan, F. et al. (2022). Green–light–driven poly(*N*-isopropylacrylamide-acrylamide)/Fe_3O_4 nanocomposite hydrogel actuators. *Frontiers in Materials* 9: 2296–8016.

48 Liu, G., Cai, M., Feng, Y. et al. (2016). Photothermally actuated interfacial hydration for fast friction switch on hydrophilic polymer brush modified PDMS sheet incorporated with Fe_3O_4 nanoparticles. *Chemical Communications* 52 (18): 3681–3683.

49 Wei, X., Chen, L., Wang, Y. et al. (2022). An electrospinning anisotropic hydrogel with remotely-controlled photo-responsive deformation and long-range navigation for synergist actuation. *Chemical Engineering Journal* 433 (3): 134258.

50 Shen, C., Lan, R., Huang, R. et al. (2021). Photochemically and photothermally controllable liquid crystalline network and soft walkers. *ACS Applied Materials & Interfaces* 13 (2): 3221–3227.

51 Ahmadi-Nohadani, H., Nono-Tagne, S., Barrett, C.J. et al. (2022). Electrospun Azo-cellulose fabric: a smart polysaccharidic photo-actuator. *Macromolecular Rapid Communications* 43 (9): 2200063.

52 Mohd Ghazali, F.A., Mah, C.K., Abuzaiter, A. et al. (2017). Soft dielectric elastomer actuator micropump. *Sensors and Actuators A: Physical* 263: 276–284.

53 Maksimkin, A.V., Dayyoub, T., Telyshev, D.V. et al. (2022). Electroactive polymer-based composites for artificial muscle-like actuators: a review. *Nanomaterials* 12 (13): 2272.

54 Ji, X., Liu, X., Cacucciolo, V. et al. (2021). Untethered feel-through haptics using 18-μm thick dielectric elastomer actuators. *Advanced Functional Materials* 31 (39): 2006639.

55 Jiang, L., Wang, Y., Wang, X. et al. (2021). Electrohydrodynamic printing of a dielectric elastomer actuator and its application in tunable lenses. *Composites Part A: Applied Science and Manufacturing* 147: 106461.

56 Yang, L., Zhang, D., Wang, H. et al. (2020). Actuation modeling of ionic–polymer metal composite actuators using micromechanics approach. *Advanced Engineering Materials* 22 (12): 2000537.

57 Mao, G., Drack, M., Karami-Mosammam, M. et al. (2022). Soft electromagnetic actuators. *Science Advances* 6 (26): eabc0251.

58 Aouraghe, M.A., Mengjie, Z., Qiu, Y. et al. (2021). Low-voltage activating, fast responding electro-thermal actuator based on carbon nanotube film/PDMS composites. *Advanced Fiber Materials* 3 (1): 38–46.

59 Wang, X., Mao, G., Ge, J. et al. (2020). Untethered and ultrafast soft-bodied robots. *Communications Materials* 1 (1): 67.

60 Liu, L., Song, W., Jiang, S. et al. (2022). Giving penetrable remote-control ability to thermoresponsive fibrous composite actuator with fast response induced by alternative magnetic field. *Nanomaterials* 12 (1): 53.

61 Shao, D., Song, W., Qin, X. et al. (2022). A precisely designed composite actuator with directionally fast actuation, non-contact operation, and obstacle-penetrating triggering ability using aligned nanofibers and alternating magnetic field. *Macromolecular Materials and Engineering* 2200219.

18

Electrospun Nanofibers for Biomedical Applications

Zhaoxuan Ding and Xinxin Li

Donghua University, College of Textiles, Key Laboratory of Textile Science & Technology, Ministry of Education, 2999 Renmin North Road, Songjiang District, Shanghai 201620, China

18.1 Introduction

With the rapid development of biomedical materials, it is increasingly common to use electrospinning technology to prepare biomedical materials. Electrospun nanofibers have many excellent properties, such as large specific surface area [1, 2], high porosity [3], and outstanding flexibility [4, 5]. The aforementioned advantages make electrospinning a good application prospect in wound dressing.

Another essential advantage of electrospinning technology is that hydrophobic or hydrophilic compounds or biomacromolecules can be directly encapsulated into fibers (e.g. proteins) [6]. It is because electrospinning can be carried out under relatively mild environmental conditions, and the activity of the loaded substances can be maintained during the fiber-forming process [7, 8]. Compared with other traditional methods, it is more suitable for encapsulating active compounds that are unstable under heat [9]. In addition, compared with conventional films produced by solution-casting technology, two-dimensional (2D) films made of electrospun fibers can promote the dispersion of the loaded compounds into surrounding media to improve the efficiency of releasing active substances such as drugs [10]. Due to the abovementioned advantages, electrospun fibers are highly sensitive to the environment and more flexibly regulate the release rate of active substances [11, 12]. In addition, electrospun fibers can also be used to imitate the microstructure of the human extracellular matrix (ECM) [13]. They can significantly improve the biocompatibility of related materials, make the drug loaded on them more stable, and promote the growth of adjacent tissues [14] (such as cartilage tissue, and skin tissue).

Because of the abovementioned advantages, electrospinning technology has a high application prospect in medical materials. In this chapter, the research status of electrospinning in tissue engineering scaffolds, wound dressings, drug delivery, and other fields are mainly introduced (Figure 18.1). Though electrospinning has progressed in many areas, there are still many difficulties in further exploration,

Electrospinning: Fundamentals, Methods, and Applications, First Edition.
Edited by Liming Wang and Xiaohong Qin.

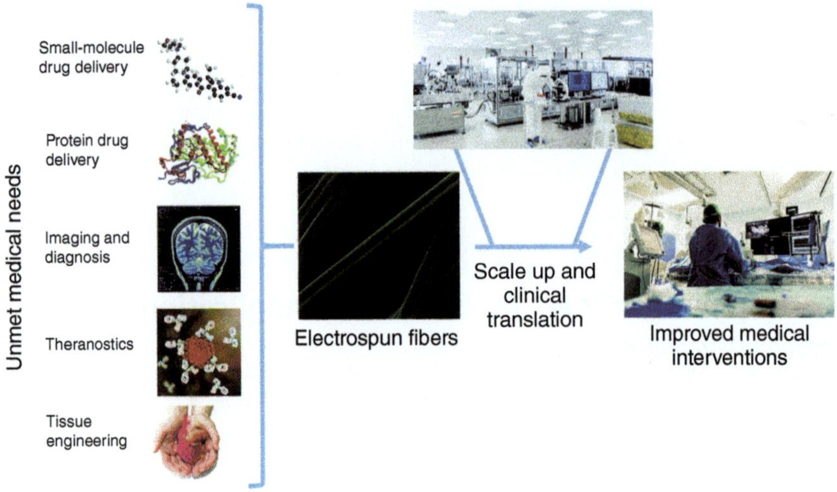

Figure 18.1 A schematic illustration of the significant applications of electrospinning in biomedicine. Source: Dziemidowicz et.al. [10]/Royal Society of Chemistry/CC BY 3.0.

which are discussed in the following sections. By overcoming these questions, medical materials are expected to enter the new world of the nano era as soon as possible.

18.2 Wound Dressing

The medical dressing is an essential application of electrospun nanofiber membranes in the biomedical field. The high specific surface area of electrospun dressings is conducive to the absorption of wound exudate and the release of drugs [15, 16]. The fibers are interlaced to form micropores that effectively block the invasion of external liquids and bacteria while ensuring gas exchange; other advantages are shown in Figure 18.2. These properties provide an ideal microenvironment for promoting cell respiration and skin regeneration [17]. In addition, due to the

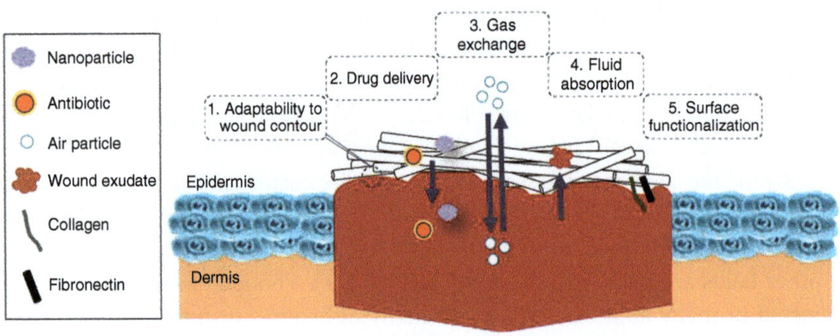

Figure 18.2 Some advantages of the electrospun membrane in the wound dressing field. Source: Juncos Bombin et al. [15]/with permission of Elsevier.

flexible and convenient electrospinning technology, antibacterial agents [18], and vitamins [19], growth factors [20] can be added according to the wound situation to promote further wound healing. The drug release rate loaded on the nanofiber membrane can be adjusted by selecting different types of polymers, controlling the position of drugs in the nanofiber, and changing the drug content to meet the actual needs [21, 22].

18.2.1 Double-Component/Multicomponent Electrospun Medical Dressing

Natural polymers are widely distributed in nature, with rich sources, good biocompatibility, low cytotoxicity, and biodegradability. Meanwhile, their degradation products are friendly to the human body and environment [23]. However, the composition and structure of polysaccharides and proteins are complex, and their physical and chemical properties are easily affected by external factors [24, 25]. Some of them also show high viscosity due to their high molecular weight. At the same time, their poor mechanical properties also limit their applications as raw materials for preparing wound dressings separately. Due to the abovementioned defects, it is challenging to prepare smooth and uniform electrospun fibers using natural polymers alone.

It is found that the natural/synthetic composite electrospun membrane prepared by combining natural polymer with synthetic polymer combines mutual advantages. It can simultaneously have excellent mechanical properties, thermal properties, biocompatibility, and biological activity. It is a critical way to improve the performance of electrospun dressings effectively. Trinca et al. [26] used polycaprolactone/cellulose acetate (PCL/CA) blends as skin dressings. By adding PCL, the mechanical strength of the dressing was improved, and the elongation at break was more than 430%. PCL/CA dressing has good water vapor permeability ($730\,g\,(m^2 \cdot d)^{-1}$), phosphate buffered saline (PBS) solution absorption capacity of up to 369%, and high porosity (about 80%), which meet the requirements of wound dressing application. Cell experiments showed that PCL/CA dressing had low cytotoxicity and could promote the proliferation of mouse fibroblast cells. Gomaa et al. [27] developed electrospun dressings of polylactic acid (PLA) and CA and carried out research and characterization of wound treatment. It was found that 3D nanofiber structure, CA hydrophilicity, and biological activity can promote cell proliferation. In vivo experiments showed that PLA : CA (7 : 3) dressing could control the granulation tissue formation, encourage epithelium formation, and significantly promote wound healing. It was an ideal choice for wound dressing. Spasova et al. [28] prepared electrospun nanofiber membranes with a mass fraction of 7% of poly(levo-polylactic acid) (PLLA) and (PLLA/PEG), respectively, in dichloromethane as a solvent, and immersed the two nanofiber membranes in a mass fraction of 0.05% CS(chitosan) solution. The results of the *Staphylococcus aureus* adhesion experiment showed that compared with the PLLA nanofiber membrane, the addition of CS endows PLLA and PLLA/PEG nanofiber membranes with excellent hemostatic and antibacterial activities, making them more conducive to wound healing.

The natural/synthetic composite electrospun dressing has low biological toxicity and specific mechanical properties. It is an excellent medical dressing that can provide a suitable environment for wound healing. However, there are still some problems with natural/synthetic composite electrospinning flavorings. It is difficult to adapt to different wounds' needs by relying only on hybrid polymers. Various functional factors are added to the hybrid system to meet different use scenarios. These problems limit further research on composite dressings in the biomedical field.

18.2.2 Functional Multicomponent Electrospun Dressing

In various diseases, there are additional requirements for the functionality of wound dressing [16, 29]. For example, in a deep II burn wound, there will be a large amount of exudate on the wound surface [30]. Hence the dressing needs to have a strong permeability and reduce its adhesion. In the electrospinning system, abundant functional substances can be introduced to meet the needs of other diseases.

Wound infection is a problem that most wounds have to face. A conventional solution is to add antibiotics to the electrospinning system. Sang et al. [31] prepared poly(caprolactone lactate) (PLCL) nanofiber membrane containing ciprofloxacin (CPF) by electrospinning. The addition of CPF reduces the nanofibers' diameter and enhances the nanofibers' hydrophilicity and ability to adsorb cells. The results showed that the electrospun membrane exhibited explosive release in the early stage, then slowly and continuously released. The entire CPF release of the nanofiber membrane reached 88% at 35 h. The results of the bacteriostasis experiment showed that the drug-loaded nanofiber membrane had an excellent antibacterial effect on *Escherichia coli* and golden grape balls. Yang et al. [32] developed a polyethylene pyrrolidone (PVP) and ethyl cellulose (EC) electrospun dressing loaded with silver nanoparticles (AgNPs). AgNPs are distributed on one side and can be fully exposed on the fiber membrane surface. AgNPs direct contact with bacterial cell walls to destroy cell wall and plasma membrane components and inactivate proteins, and they play a significant role in wound antibacterial treatment. In vitro experiments also showed that the dressing had good bactericidal activity for the growth of Gram-positive *S. aureus* and Gram-negative *E. coli* and was a very effective wound dressing.

Due to explosive release, synthetic antibiotics/antibacterial nanoions often mean potential side effects [33, 34]. Some researchers introduced natural active substances into the electrospinning system to improve this defect. Since honey has antibacterial and antioxidant properties, it has been recorded since ancient times that honey is applied to wounds to accelerate healing. Tang et al. [35] added different proportions of honey into sodium alginate (SA)/polyvinyl alcohol (PVA) spinning solution to prepare a nanofiber membrane with a smooth surface and uniform 3D structure. With the increase in honey concentration, the degradability of the nanofiber membrane was improved, and the water absorption capacity decreased. The results showed that the prepared nanofiber membrane could not only eliminate free radicals and reduce the oxidative stress reaction of the wound but also inhibit the growth of bacteria, which had a specific inhibitory effect on Gram-positive bacteria.

When trauma occurs, many growth factors are released from platelets and other cells to mediate cell reaction, thus promoting the hemostasis and repair of wounds [36]. Therefore, when preparing electrospun wound dressings, growth factors can be added to the spinning system as active substances to enhance the function of wound dressings to promote wound healing. Schneider et al. [37] used electrospinning technology to add epidermal growth factor (EGF) to regenerated silk fibroin and polyethylene oxide to prepare the spinning solution and regenerated silk fibroin fiber containing EGF. The structure of the electrospun fiber is relatively stable, and the release process of EGF experienced a process from sudden release to continuous release, which is very effective for the repair and treatment of chronic skin injuries. The research shows that the constant release of EGF can effectively promote wound repair and accelerate wound healing by using the regenerated silk fibroin fiber containing EGF in medical dressings. Functional dressings have been used in many treatment processes. Its functionality plays a very important role in the specified wound types. However, with the continuous development of the medical level, researchers have realized that wound recovery is a dynamic process, requiring dressings to give feedback according to the changing wound environment. Therefore, the concept of intelligent dressing is gradually derived.

18.2.3 Intelligent Wound Dressing

During the treatment, we need the dressing to adjust itself in time according to the condition of the wound or have intelligent interaction with the external environment or the wound.

When the wound is infected or inflamed, the body fluid around the wound is acidic. Zhao et al. [38] prepared polyvinylidene fluoride (PVDF) nanofiber yarn by conjugated electrospinning process and formed PANI/PVDF yarn by in situ polymerization of polyaniline (PANI), as shown in Figure 18.3. By using the electrospinning yarn machine, electrospun nanofibers are twisted into electrosun yarn. This sentence means that in the pH range of 4.0–8.0, PANI/PVDF yarn is −48.53 mV per pH for yarn and PANI/PVDF fabric is −38.4 mV per pH. The results show that the prepared PANI-modified PVDF yarn and fabric may have potential applications in intelligent oral and maxillofacial surgical dressings for monitoring wound healing. Yuan et al. [39] took advantage of this feature to add anti-inflammatory

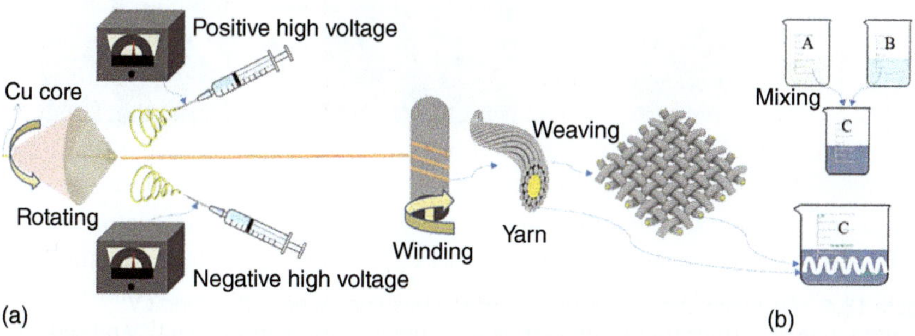

Figure 18.3 (a) Diagram of the preparation of PVDF yarn and fabric and (b) the preparation of PANI solution and PANI/PVDF yarn. Source: Zhao et al. [38]/de Gruyter/CC By 4.0.

drugs ibuprofen and $NaHCO_3$ into the spinning solution prepared by biodegradable PLLA and used the drug-loaded pH-responsive nanofiber membrane prepared by electrospinning as a wound dressing. When the wound is inflamed, the pH value is about 5.0. The CO_2 gas produced by $NaHCO_3$ in the fiber in contact with acid causes the rapid release of ibuprofen, which plays a role in treating wound inflammation. Suppose biosensors are implanted into pH-responsive nanofiber membranes used as wound dressings and combined with imaging technology. In that case, they can play a therapeutic role and monitor the wound in real time, understand the infection, and make predictions before it deteriorates.

With the in-depth research on stimulus-responsive polymers in the field of polymer science in recent years, they can cause their own physical or chemical changes after being stimulated by the outside. More and more researchers combine stimulus-responsive polymers into electrospinning to prepare stimulus-responsive drug-loaded nanofiber systems. Hu et al. [40] mixed polyacrylamide (PNIPAAm), EC, and ketoprofen (KET) to prepare the spinning solution and drug-loaded composite nanofibers by electrospinning. The drug-loaded composite nanofibers have good stability in water, in which KET is distributed in the fibers in an amorphous state. The water contact angle proved that when the temperature gradually increased beyond the lower critical solution temperature (LCST) value (32 °C), the drug-loaded composite nanofibers could change from hydrophilic to hydrophobic with good temperature stimulus response performance and played a role in regulating the release of KET. Elashnikov et al. [41] prepared PNIPAAm nanospheres doped with antibacterial crystal violet (CV) and then blended them with PLLA spinning solution for electrospinning to prepare drug-loaded composite nanofibers with temperature stimulus response. When the LCST (32 °C) of PNIPAAm is higher or lower, the nanofibers have switchable wettability. They can inhibit the release of CV when the temperature is $T > LCST$ and promote the release of CV when it is $T < LCST$, as shown in Figure 18.4.

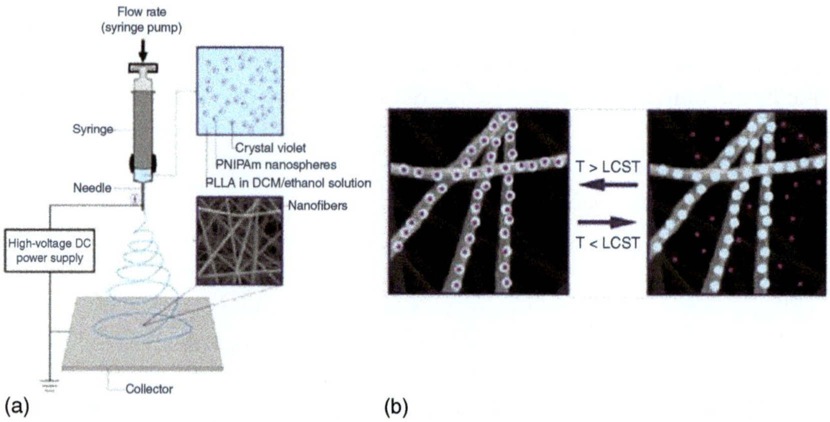

Figure 18.4 (a) Schematic representation of electrospinning procedure and CV encapsulation. (b) Schematic representation of material phase transition and CV release from PLA/PNIPAM nanofibers. Source: Elashnikov et al. [41]/with permission from Elsevier.

Although significant progress has been made in the application research of stimulus-responsive electrospun nanofibers, they are still in the laboratory stage. It is necessary to expand their preparation scale, reduce the preparation cost, and create conditions for their industrialization and practical application. In addition, expanding the application scope of stimulus-responsive electrospun nanofibers is also a future research direction worthy of the effort.

18.3 Tissue Engineering Scaffold

The scaffold material with suitable bionic properties can be made by electrospinning technology. The critical difference between electrospun and traditional materials is that they are composed of ultrafine fibers. Their bionic performance is mainly reflected in fiber size and stent pore. Electrospun fiber size can be controlled from micrometer to nanometer, adapting to the size in the natural ECM [42, 43]. In addition, scaffold materials' porosity and pore diameter can also be controlled by a series of processing parameters during the preparation process. The open pore structure and high porosity are conducive to cell implantation and culture, tissue growth, and the flow of nutrients and metabolites [44, 45]. Based on the aforementioned advantages, the electrospun scaffold can provide good adhesion and growth conditions for cells in vivo. It can be applied to repair and reconstruct blood vessels, bones, muscles, nerves, ligaments, the liver, and other tissues. Some tissue engineering studies are shown in Figure 18.5.

18.3.1 Vascular Tissue Engineering Scaffold

The number of deaths caused by cardiovascular diseases has been high every year. Vascular transplantation is a standard method to treat vascular conditions, and the supply of vascular grafts often exceeds the demand in clinical practice [47, 48]. By improving the electrospinning method, multilayer/double-layer vascular scaffolds with unique nano- to macrostructures can be prepared, effectively simulating the 3D tissue structure composed of collagen and polysaccharide composite nanofibers and mimicking the human natural vascular ECM.

Frerich et al. [49] used gelatin as the vascular scaffold, inoculated endothelial cells, and cultured them for 16 days under dynamic perfusion pulse stimulation. They found that the adhesion of endothelial cells was good, and there were many entangled capillary-like network structures in the scaffold. The results showed that the staging achieved its tissue engineering vascularization to a certain extent. Some research groups have made remarkable achievements using heparin and vascular endothelial growth factor (VEGF) to reduce thrombosis and improve endothelialization. By coaxial electrospinning, Huang et al. [50] prepared a heparin-containing PLA caprolactone scaffold. When the stent is implanted into the femoral artery of dogs, heparin acts as an anticoagulant to improve vascular patency. Endothelial progenitor cells (EPCs), known as angioblasts, can repair damaged blood vessels from bone marrow to peripheral blood vessels under the

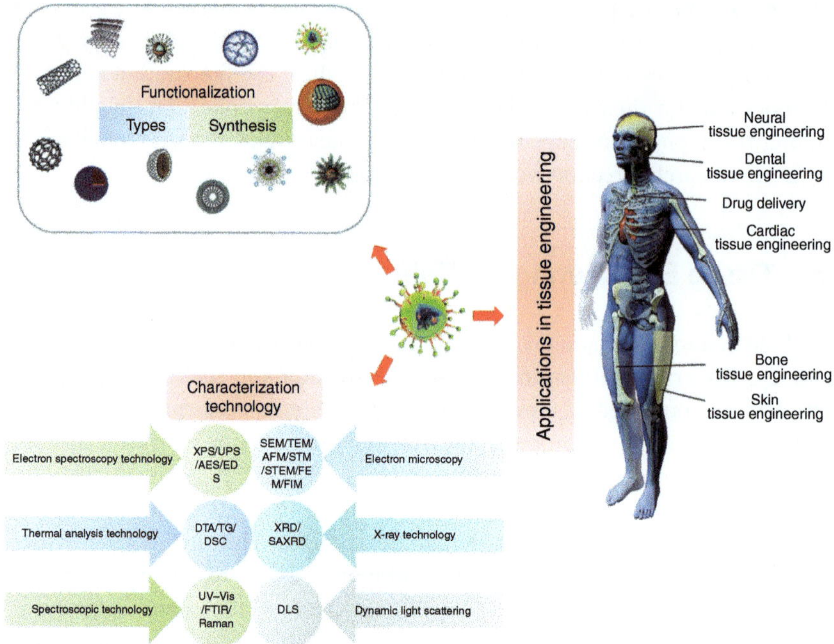

Figure 18.5 Summary of tissue engineering. XPS, X-ray photoelectron spectrometer; UPS, ultraviolet photoelectron spectroscopy; AES, auger electron spectroscopy; EDS, energy dispersive spectrometer; DTA, differential thermal analysis; TG, thermogravimetry; DSC, differential scanning calovimltry; UV, ultraviolet and visible spectrum; FTIR, Fourier transform infra-red; Raman, Raman spectrum; SEM, scanning electron microscope; TEM, transmission electron microscope; AFM, atomic force microscope; STM, scanning tunnel microscope; STEM, scanning transmission electron microscope; FEM, finite element method; FIM, field intensity mete; XRD, X-Ray diffraction; SAXRD, X-ray small angle diffraction; DLS, dynamic light scattering method. Source: Xinmin et al. [46]/Royal Society of Chemistry.

stimulation of physiological or pathological factors. Li et al. [51] coated VEGF into heparin-containing scaffolds through lotion electrospinning to promote the proliferation of EPCs. Heparin and VEGF were controlled to release. The platform had adequate anticoagulation and promoted the proliferation of EPCs.

As the coagulation performance of small-caliber vessels is the most challenging bottleneck, more research focuses on antithrombotic research. Dominik et al. [52] created a double-layer tubular stent with enhanced anticoagulation performance for tissue engineering of small blood vessels, shown in Figure 18.6. The scaffold comprises a PCL-based porous outer layer and a heparin-modified PLA-based electrospun inner layer. The thermally induced phase separation and electrospinning combination produce an asymmetric scaffold with improved mechanical properties. The release test confirmed that heparin was released from the stent. In addition, the anticoagulant activity was measured by activated partial thromboplastin time (APTT).

Interestingly, the endothelial cell culture test showed that after 14 days of culture, human aortic endothelial cell line (HAEC) tended to organize in a chain structure,

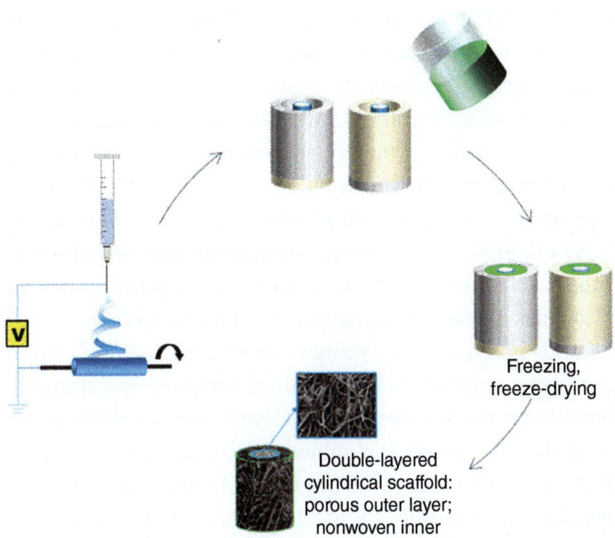

Figure 18.6 The fabrication method of double-layered cylindrical scaffolds using electrospinning and thermally induced phase separation. Source: Domalik-Pyzik et al. [52]/MDPI/CC BY 4.0.

a typical feature of early angiogenesis. The HAEC activity of the heparin-modified scaffold was higher in a longer culture time. The proposed stent design and composition have great application potential in small-vessel tissue engineering. However, up to now, most nanostents still have thrombosis in orthotopic transplantation, which is an essential factor affecting vascular patency. In the future, we will further evaluate the antithrombotic effect of the stent, screen, and optimize the anticoagulation scheme of the stent and find an appropriate and effective solution.

18.3.2 Nerve Tissue Engineering Scaffold

At present, many studies have compared polymer membranes with electrospun fiber scaffolds through in vitro and in vivo experiments to prove the efficacy of these electrospun fiber scaffolds in enhancing nerve regeneration [53, 54]. Many in vivo experiments have demonstrated that electrospun fiber nerve conduits can effectively promote nerve axons' extension and nerve innervation recovery. In addition, the material with appropriate degradation rates and mechanical properties can minimize the inflammatory reaction, prevent nerve compression, and guide axon regeneration.

Xue et al. [45] prepared silk fibroin fiber scaffold by electrospinning as a nerve conduit for surgical transplantation to treat peripheral nerve injury in rats. After 12 months of operation, most of the catheters were degraded and replaced by tissues with nerve-like appearance. The gait analysis system obtained symmetrically and walking gait parameters to evaluate dogs' hind limb motor function comprehensively. The results showed that the transplanted group showed better standing stability and motor function than the nontransplanted group, while there was no significant difference within the transplanted group. This is enough to show that scaffold (SF) stents can promote the behavior recovery of injured hind limbs almost as much as autologous grafts. To further enhance the repair efficiency of nerve tissue, some researchers coated the scaffold with nerve growth factor (NGF). Lundborg et al. [55] further confirmed that the phenomenon of nerve chemotaxis does exist in the process of peripheral nerve regeneration and has tissue, nerve bundle, and functional specificity. Brushart et al. [56] believed that it is precisely because of the role of nerve chemotaxis that regeneration axons can be guided to grow to the distal end of the severed nerve without regeneration to other tissues. In addition, neurotrophic also plays a vital role in the selective process of motor nerve regeneration to the distal end. Based on this, Wang et al. studied the application of NGF in poly(lactic-*co*-glycolic acid) (PLGA) shell fiber, wrapped the spinning support collected directionally on the stainless-steel rod, and sealed it with nylon wire to form neuroglial cyst (NGC). The NGC obtained was transplanted into rats' 13 mm sciatic nerve defect. After 12 weeks of repair, the scaffold promoted nerve regeneration. Electrophysiology and muscle weight tests showed that compared with the control group without NGF growth factor, the functional recovery of the regenerated nerve was significantly improved. These studies indicate that NGF plays a vital role in promoting the repair of damaged peripheral nerves.

18.3.3 Bone Tissue Engineering Scaffold

The incidence of bone defects caused by bone infection is very high globally, so there is substantial clinical demand for bone transplantation. The bone tissue scaffold prepared by electrospinning comprises nanofibers with uniform diameter, structure, and high porosity. During spinning, stable and bio-friendly macromolecular materials can be used as raw materials. At the same time, growth factors or inorganic salts that can promote bone effect can be added, promoting the proliferation and differentiation of bone cells on the scaffold and further depositing into bone. Properties, compositions, synthesis techniques, and outcomes of electrospun nanofibers for bone tissue engineering are shown in Figure 18.7.

Ye et al. [58] cut the fiber scaffold into small segments, formed a 3D bone-like platform through freeze-drying, thermal cross-linking, and other methods, and made a composite scaffold containing the main mineral components of bone (nano-hydroxyapatite), as shown in Figure 18.7. Then, the synthesized bone morphogenetic protein-2 (BMP-2)-derived peptide was fixed on the surface, and the scaffold was evaluated in vitro and in vivo in the rat skull defect model. The experimental results showed that the presence of nano-hydroxyapatite and BMP-2 increased the expression of genes related to osteoblastic differentiation of stem cells, and the release of BMP-2 peptide could last for 21 days. Compared with the control group, the scaffold can better induce bone differentiation. To further strengthen the recovery speed of bone tissue, some researchers added nanoparticles or bone tissue growth factors into the system. Whang et al. [59] developed an electrospun PCL/Gel nanofiber scaffold based on mesoporous silicate nanoparticles (MSN) to achieve dual delivery of alendronate (ALN) and silicate, thus having a synergistic

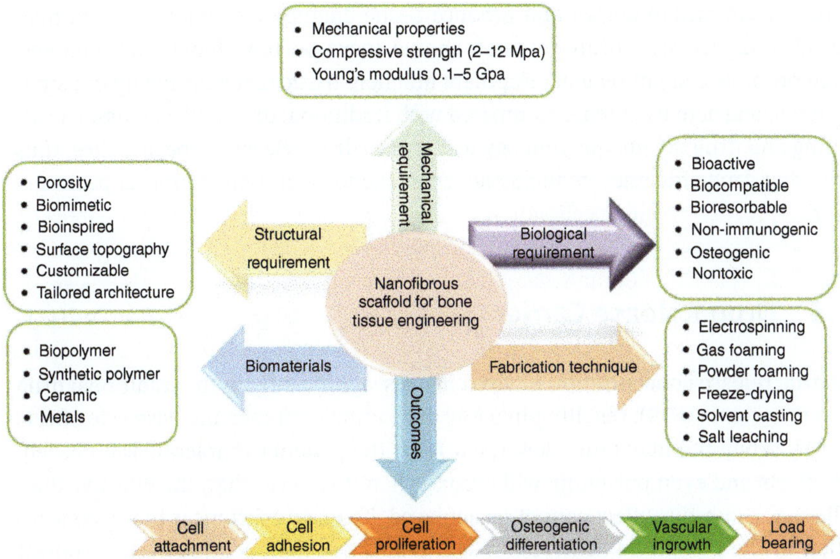

Figure 18.7 Properties, compositions, synthesis techniques, and outcomes of electrospun nanofibers for bone tissue engineering. Source: Anjum et al. [57]/MDPI/CC BY 4.0.

effect in regulating bone remodeling, wherein ALN inhibits the bone absorption process by preventing the expression of guanine triphosphate and guanine-related proteins, and silicate promotes bone formation by improving angiogenesis and bone calcification. The results of bone repair in the critical-size skull defect model of rats show that the electrospun PCL/Gel nanofiber scaffold based on dual delivery of ALN and silicate is promising for clinical bone repair.

The electrospun nanofiber scaffold has recently become a new choice for tissue autograft, the gold standard for tissue regeneration. An essential component of tissue engineering, electrospun nanofiber scaffold, has a broad application prospect, including tendons, blood vessels, neurons, bones, and cartilage. On the one hand, for tissue repair, the electrospun nanofiber scaffold has characteristics like natural tissue and advantages similar to the fiber structure of a biological ECM, such as high surface volume ratio, variable porosity, size, and shape. On the other hand, the electrospun nanofiber scaffold has a variety of tissue regeneration designs. Using innovative nanomaterials combined with more favorable engineering technology has brought the best brightness to the research field of tissue regeneration. However, electrospun nanofibers show promise in tissue engineering applications, but many technical challenges remain to be solved. Most of the published studies were conducted in vitro. Therefore, the content and structure of polymer nanofiber scaffolds still need to be further optimized for use in vivo. Creating 3D porous scaffolds containing cells and growth factors is crucial for future cell infiltration and viability studies. In addition, it is essential to transfer electrospun nanofibers from the laboratory to a commercial scale. Despite many obstacles, electrospinning seems to be a promising technology for manufacturing functional nanofibers, enabling researchers from various disciplines to design and produce tissue engineering innovation matrices with ideal objectives. Future research should focus on multifunctional scaffolds, which promote cell physical development and help tissue regeneration by transmitting biological activity signals. Compared with traditional drug delivery, tissue engineering and drug delivery technology will lead to drug release at specific sites, thus improving drug efficiency, reducing adverse reactions, and protecting unpackaged drugs.

18.4 Drug Release Carrier

The drug release concentration in vivo changes significantly in the traditional drug delivery system (DDS), resulting in a low drug utilization rate and easy side effects [60, 61]. When the maximum dose (peak) that the patient can tolerate is exceeded, side effects and even poisoning will occur. When it is lower than the effective dose (valley), the curative effect cannot be achieved. To ensure a constant drug concentration in the blood and achieve a safe and effective therapeutic effect, sustained drug release has become the primary issue in selecting drug carrier materials [62]. Nanofibers prepared from polymer materials by electrospinning technology can be used as drug carriers and have a specific sustained release performance. Meanwhile, it can promote the absorption of drugs by the human body and significantly improve the utilization rate of drugs by the human body according to their large specific surface area. With the development of research, its extensive solution system and controllable appearance make it a good application prospect in the field of sustained drug release. Sustained Drug release is a relatively complex process strongly related to the coating substrate, drug loading mode, medicine, and fluid body environment. Some related mechanisms of drug release are shown in Figure 18.8.

18.4.1 Diffusion-Driven Electrospun Nanomembranes

The concept of diffusion is that the difference in solute concentration between different regions drives solute migration. Diffusion-driven drug delivery is generally used for hydrophobic drugs. In a diffusion-driven system, the diffusion rate is mainly changed by changing the porosity of the electrospun substrate. Thakur et al. [64] added two drugs (lidocaine hydrochloride and mupirocin) to the solution of poly(levo-polylactic acid) (PLLA) to prepare a spinning solution. They sprayed the two spinning solutions through double-needle electrospinning in an electric field to prepare composite nanofiber membranes with double-drug-loading components. In early wound treatment, lidocaine hydrochloride will be released suddenly to relieve wound pain. In contrast, mupirocin will be released slowly for a long time to protect the wound from bacterial infection during healing. Liao et al. [65] loaded the spinning solution of two components (PCL solution, zein, and calcium lactate [CL] mixed solution) into two nozzles for electrospinning. They prepared a PCL/zein CL nano-spinning membrane. The addition of CL can significantly enhance the composite nano-spinning membrane's wettability, mechanical strength, and

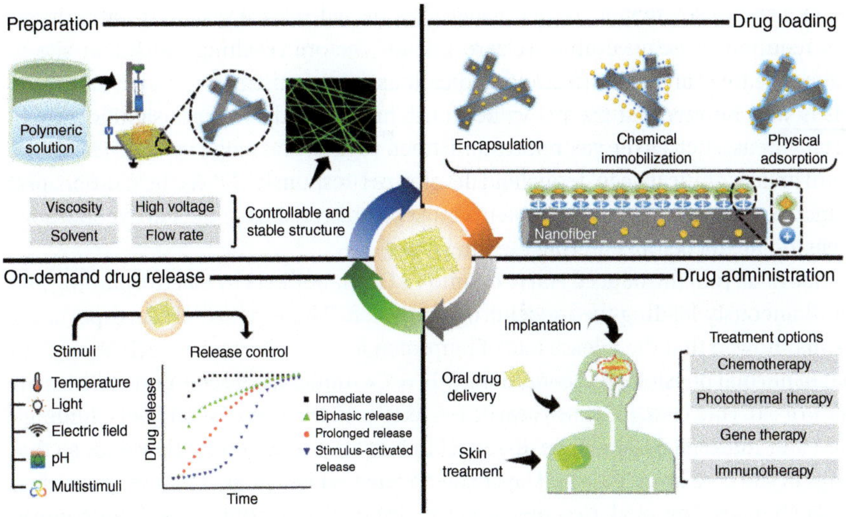

Figure 18.8 Schematic of on-demand DDSs using stimulus-responsive NFs. Source: Singh et al. [63]/MDPI/CC BY 4.0.

biocompatibility. In vitro osteoblast culture test showed that the composite nanospinning membrane can effectively promote calcium phosphate nucleation, which is conducive to cell adhesion, diffusion, and proliferation.

18.4.2 Intelligent Responsive Electrospun Nanomembranes

To reduce the sudden release and explosive release caused by drugs and increase the adaptability of drugs to different environments, researchers have turned their attention to intelligent, responsive drug carriers. The intellectual stimulus-responsive DDS can sense different stimuli inside and outside the organism, respond effectively to pH value, temperature, light, ultrasound, magnetic field, and other environmental triggers, change the structure or performance, and change the physical and chemical properties responsively. It is an innovative system that releases drugs, takes drug release in specific targeted areas, and exerts therapeutic effects. Compared with traditional direct administration, intelligent, responsive drugs have the following advantages, (i) fewer side effects; (ii) more extensive application scenarios; and (iii) stronger drug targeting.

18.4.2.1 pH-Responsive Electrospun Nanomembranes

In recent years, many research reports have focused on developing stimulus-responsive drug carriers. This property is beneficial in many drug delivery applications. Stimulation-responsive drug carriers reach the target in vivo through endocytosis and transportation in the body. Under the effect of environmental stimulation, they release drugs to improve the therapeutic effect and minimize adverse reactions. Usually, the release of internal drugs is controlled by connecting some macromolecular substances or small molecular functional groups at the pore site as a response "switch." When using a controlled-release system to carry drugs, it is required to achieve almost "zero release" before reaching the lesion site to reduce the toxic and side effects of medications on organisms.

It is well known that the pH value in the human body changes significantly in certain areas, such as the gastrointestinal tract, vagina, and blood vessels. Therefore, it is of great significance to design and develop pH-responsive DDSs. In addition, oral drugs containing peptides and proteins must be protected from the harsh conditions of gastric media before entering the intestine.

Zhao et al. [66] introduced $NaHCO_3$ into PLGA nanofibers by electrospinning and simultaneously loading the model drug ibuprofen. The in vitro release experiment results showed that the release rate of ibuprofen in nanofibers was much faster than that in normal physiological conditions (pH = 7.4) under acidic conditions with a pH value of 5.0. Han et al. [67] used acrylic resins to construct core–shell fibers. Eudragit L100 and Eudragit S100 will be dissolved in response to different pH values so that drugs in the core layer and shell layer can be released in a graded manner according to pH changes. Son et al. [68] prepared pH-responsive nanofiber mesh for controlling antibiotic release in response to pH changes. Eudragit EPO (EPO) and Eudragit L100 (L100) are injected through internal and external needles, and electrospinning is carried out through coaxial nozzles composed of internal and external needles. Different amounts of EPO and L100 are co-injected with tetracycline through a needle and electrospun to the fiber mesh simultaneously. At pH 6.0, with the increase of EPO content, the mass erosion rate of the mesh decreases gradually, while at pH 2.0, the difference in the mass erosion rate of the mesh can be ignored. These differences were confirmed by scanning electron microscopy and monitoring dry weight changes. At pH 6.0, compared with acid conditions, the fiber structure of the mesh disappears rapidly because L100 is located in the shell of the nanofiber during electrospinning. The change in pH value and the blending ratio of the two polymers significantly affected the release of tetracycline. Tetracycline is rapidly released from the mesh at pH 6.0, while the release rate is weakened at pH 2.0. For the two pH values, tetracycline releases faster from the mesh, and the mixing ratio of EPO is higher. The electrostatic interaction between EPO and L100 is expected to produce different tetracycline release curves. Therefore, at neutral pH, a higher amount of encapsulated drug is released from the mesh and successfully inhibits bacterial growth.

Moreover, Kaassis et al. [69] prepared a novel and highly adjustable pulsed drug delivery system by electrospinning a mixture of polyethylene oxide (PEO), SA, and ibuprofen sodium (SI). The obtained fibers contain SI microcrystals embedded in the PEO–SA matrix to form a novel 3D structure extending up to the needle. The results showed that at pH 6.8, the fiber dissolved very quickly, releasing all embedded drugs in about 20 minutes. However, an unusual two-stage release mechanism was observed at pH 3, representing feeding status or gastric pH in elderly patients. This includes rapid burst release, followed by no drug release within about 120–150 minutes, followed by the final release stage, releasing the remaining drug into the solution. The amount of release in the initial stage and the length of time between the first and last drug release stages can be controlled by adjusting the SI and SA contents of the fiber, respectively. This results in a highly adjustable pulse release of the material.

18.4.2.2 Temperature-Responsive Electrospun Nanomembranes

Temperature sensitivity is one of the fascinating characteristics of intelligent nano-drug carriers. Compared with other smart nano-drug carriers stimulated by the environment, the temperature is the most convenient and effective environmental factor to control drug release, which has been widely used and studied. Generally, under pathophysiological conditions, such as inflammation and tumor, the temperature of the diseased tissue will be 37 °C higher than that of normal human tissue, and the tumor area can reach 40 to 42 °C. An intelligent nano drug carrier is designed based on the temperature difference between tumor tissue and normal tissue, which can achieve targeted drug release from the tumor under the trigger of temperature. On the other hand, temperature-sensitive nano-drug carriers can use external environmental stimuli, such as ultrasound and magnetic field, to heat the tumor area to reach the temperature required by drug carrier to release drugs. Lin et al. [70] used the single-needle electrospinning technology to co-blend the temperature-sensitive PNIPAm with poly(2-acrylamide-2-methylpropane sulfonic acid) (PAMPS) and the drug model nifedipine (NIF). They controlled the hydrogen bond between the drug and PNIPAm molecules by adjusting the temperature so that the drug release rate changed with the temperature, achieving the goal of drug release control. With the extension of fiber swelling time, the drug release will slowly increase, and the release can be effectively controlled by changing the temperature. Yu et al. [71] used cross-linking agents *N′*,*N′*-methylenebisacrylamide (MBA) to graft polymethylmethacrylate (PMMA) onto PVCL, then dissolved the graft-modified polymer into *N*-dimethylformamide (DMF) to prepare spinning solutions of different concentrations, prepared fiber membranes with different morphologies and wettability through high-voltage electrostatic spinning equipment, and mixed the anti-cancer drug Erlotinib into the spinning solution. They prepared a drug-loaded nanofiber felt, and the drug release rate was measured at different temperatures.

The appearance of polyisopropylacrylamide (PIPAAm) further enhanced the sensitivity of temperature-sensitive materials. Okuzaki et al. [72] prepared poly (isopropylacrylamide [IPAAm]-co-SA) electrospun nanofibers. When the temperature drops from 40 to 25 °C, the poly (IPAAm co SA) fiber membrane has rapid (and reversible) swelling. What is important is that compared with poly(polyacrylamide [PAAM]-*co*-SA) bulk hydrogel, the swelling rate of poly (PAAM-*co*-SA) nanofiber membrane is one order of magnitude higher than that of the bulk hydrogel. This is mainly due to the larger specific surface area and higher porosity. By modifying the functional groups of the material, a composite with temperature and pH responsiveness can be obtained. Aoyagi et al. [73] obtained carboxyisopropylacrylamide (CIPAAM) by connecting a carboxyl functional group to the isopropyl group of PIPAAm through a carefully designed synthesis scheme. Because CIPAAM contains both isomer residues and carboxyl groups, CIPAAM has dual sensitivity to pH and temperature. The instability of temperature sensitivity and strain rate is still a big challenge for temperature-sensitive materials. At the same time, materials' broad temperature response domain is also a problem that needs to be solved in subsequent research.

18.4.2.3 Magnetic-Response Electrospun Nanomembranes

Magnetic-response materials have attracted more and more attention due to their advantages of noncontact response and low preparation cost. They have been widely used in medical imaging and industrial fields. The main characteristics of magnetic capsules are strong drug carrying capacity, good biocompatibility, good magnetization value, and excellent drug release performance. At present, magnetite and magnetic hematite are recognized as the only magnetic materials with good biocompatibility and high safety [74].

Therapeutic drug doxorubicin (DOX) and magnetic nanoparticles are used to prepare magneto-thermal-responsive controlled-release system for skin cancer. Some researchers are committed to changing the drug release performance of materials through the magnetocaloric effect. Kim et al. [75] used chemically cross-linked temperature-responsive polymers as spinning materials to add fiber tumors. Rodrigues et al. [76] synthesized magnetic nanoparticles with copolymer-functionalized hydrophilic graphene to achieve an efficient multifunctional biomedical system for mild hyperthermia and stimulus-responsive drug delivery. In vitro experiments showed that the magnetic nanoparticles had a high loading capacity for DOX, high heating efficiency under a magnetic field, and controlled drug release under the dual response of pH and thermal stimulation.

CALVO-CORREAS T [77] blends bio-based polyurethane with magnetic Fe_3O_4 nanoparticles to prepare magnetically responsive bio-based composites, shown in Figure 18.9. Adding Fe_3O_4 nanoparticles not only endows the material with magnetic responsiveness but also reduces the polymer's crystallinity and increases the material's shape fixation due to the interaction of its polymer hydroxyl groups. In addition, under an external alternating magnetic field, Fe_3O_4 nanoparticles can heat the material, softening the polymer chain segment and restoring its shape, thereby changing the release drug performance of the bio-based material. The results showed that the drug release rate increased significantly under a strong magnetic field. The main problems of magnetic-responsive electrospun

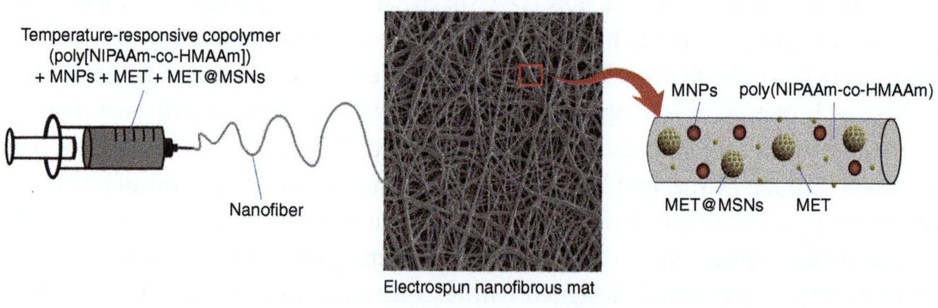

Figure 18.9 Smart implantable NFs were fabricated via electrospinning of MNPs, MET, and MET@MSNs. Source: Samadzadeh et al. [77]/with permission from Elsevier.

membranes are potential toxicity and side effects. Other researchers changed fiber aggregates' porosity and pore size with a solid magnetic effect to change the drug release rate. And in vivo transplantation and magnetic interference resulting in poor responsiveness are also significant problems. Subsequent work will focus on more excellent magnetic-response materials and biocompatible materials.

Samadzadeh et al. [77] developed a kind of intelligent high-temperature nanofiber, which has the dual-stage drug release ability of an "on–off" switch of thermal response and magnetic response (AMF) at the same time and has a good application prospect in hyperthermia chemotherapy. Intelligent thermotherapy nanofiber scaffold is prepared by electrospinning temperature-responsive copolymer, which is mixed with iron oxide (II, III) magnetic nanoparticles (MNPs, 10 nm), metformin (MET), and mesoporous silica nanoparticles (MSN) loaded with MET(MSNs MET). Research shows that all magnetic nanofibers (MNF) have heating characteristics and "on–off" switchable heating ability and have reversible swelling rate and corresponding drug release to respond to the "on–off" switching AMF application.

Electrospinning technology is one of the few methods to prepare polymer nanofibers or micron-long fibers, which has broad application prospects and huge market potential. At the same time, electrospun fiber products have the characteristics of a large specific surface area, flexible shape design, and similar structure to the ECM. As drug carriers and tissue engineering scaffolds, electrospun fiber products have also received widespread attention. Different sustained-release effects can be achieved by constructing drug carriers with other structures or changing factors such as fiber porosity, diameter, and swelling. For some drugs that are difficult to control due to the time of administration and the poor compliance of patients in clinical practice, it is key to preventing drug release to consider how to coat the drug well. At the same time, the solvent of the polymer, drug release mechanism, and drug state should also be considered.

18.5 Conclusion

This chapter mainly introduces electrospinning applications in the medical field, involving wound dressings, tissue engineering scaffolds, drug-release carriers, and other areas. Benefiting from the high porosity, small pore size, and large specific surface area of electrospun membranes, electrospun membranes have excellent drug release ability and can effectively promote cell metabolism. There is hardly space here to introduce enough electrospinning, such as antibacterial materials, medical sensors, and tumor treatments. On the other hand, because many nanoparticles and polymers can be applied to the electrospinning system, electrospun nanofiber membranes' apparent morphology and functionality are infinitely possible. By introducing special nanoparticles, the intelligent interaction of the electrospun membrane with the surrounding environment means that the drug can target the target in a targeted manner, reducing drug side effects while making drug delivery efficient.

We expect more researchers to devote their research enthusiasm to the medical field of electrospinning. However, limited by the fact that most electrospinning solutions are toxic, there is still a long way to go for electrospun membranes to be used in medical scenarios. At the same time, intelligent nanofilms' high sensitivity and narrower response domain are also a direction for future research. Under the dual charm of the nano effect and intelligent interactive system, electrospinning will create a greater miracle in the medical field.

References

1 Buzgo, M., Mickova, A., Rampichova, M., and Doupnik, M. (2018). *Needleless Emulsion Electrospinning for the Regulated Delivery of Susceptible Proteins*, 325–347. Cambridge, UK: Woodhead Publishing.
2 Wu, J., Zhang, Z., Gu, J. et al. (2020). *Journal of Controlled Release* 320: 337–346.
3 Nauman, S., Lubineau, G., and Alharbi, H.F. (2021). *Membranes* 11: 1–38.
4 Ma, H., Burger, C., Chu, B., and Hsiao, B.S. (2020). *Fibrous Materials* 773–806.
5 Wang, Y., Tian, L., Zhu, T. et al. (2021). *Chemical Research in Chinese Universities* 37: 443–449.
6 Zare, M., Dziemidowicz, K., Williams, G.R., and Ramakrishna, S. (1968). *Nanomaterials* 2021: 11.
7 Zhang, Y., Guan, J., Wu, J. et al. (2019). *Carbohydrate Polymers* 219: 405–413.
8 Amiri, N., Ajami, S., Shahroodi, A. et al. (2020). *Macromolecules* 162: 645–656.
9 Dong, W.H., Liu, J.X., Mou, X.J. et al. (2020). *Colloids and Surfaces, B: Biointerfaces* 188: 110766.
10 Dziemidowicz, K., Sang, Q.Q., Wu, J.L. et al. (2021). *Journal of Materials Chemistry B* 9 (4): 939–951.
11 Unalan, I., Enderlein, S.J., Slavik, B. et al. (2019). *Pharmaceutics* 11: 570.
12 Ghiyasi, Y., Salahi, E., and Esfahani, H. (2021). *Materials Today Communications* 26: 102163.
13 Najafiasl, M., Osfouri, S., Azin, R., and Zaeri, S. (2020). *Journal of Drug Delivery Science and Technology* 57: 101708.
14 Peng, Y., Ma, Y., Bao, Y. et al. (2021). *Macromolecules* 183: 68–78.
15 Juncos Bombin, A.D., Dunne, N.J., and McCarthy, H.O. (2020). *Materials Science and Engineering: C* 114: 110994.
16 Naomi, R., Bahari, H., Ridzuan, P.M., and Othman, F. (2021). *Polymers* 13: 1–20.
17 Hadisi, Z., Farokhi, M., Bakhsheshi-Rad, H.R. et al. (2020). *Macromolecular Bioscience* 20: 1–17.
18 Dalwadi, C. and Patel, G. (2015). *Recent Patents on Nanotechnology* 9 (1): 17–25.
19 Khadka, D.B. and Haynie, D.T. (2012). *Nanomedicine and Nanotechnology* 8: 1242–1262.
20 Teixeira, M.A. and Amorim, M.T.P. (2020). *Polymers* 12: 7.
21 Zhu, Z., Liu, Y., Xue, Y. et al. (2019). *ACS Applied Materials & Interfaces* 11: 36141–36153.

22 Khan, A.R., Huang, K., Jinzhong, Z. et al. (2020). *Colloids and Surfaces, B: Biointerfaces* 196: 111352.

23 Kalantari, K., Afifi, A.M., Jahangirian, H., and Webster, T.J. (2019). *Carbohydrate Polymers* 207: 588–600.

24 Xia, J., Zhang, H., Yu, F. et al. (2020). *ACS Applied Materials & Interfaces* 12: 24370–24379.

25 Naomi, R., Ratanavaraporn, J., and Fauzi, M.B. (2020). *Materials* 13: 1–22.

26 Trinca, R.B., Westin, C.B., Fracassi da Silva, J.A., and Moraes, M.A. (2017). *European Polymer Journal* 88 (5): 161–170.

27 Gomaa, S.F., Madkour, T.M., Moghannem, S., and El-sherbiny, I.M. (2017). *International Journal of Biological Macromolecules* 105 (1): 1148–1160.

28 Spasova, M., Paneva, D., Manolova, N. et al. (2008). *Macromolecular Bioscience* 8 (2): 153–162.

29 Wang, M., Hou, J., Yu, D.G. et al. (2020). *Alloys and Compounds* 846: 156471.

30 Wagner, A.D., Özdemir, B.C., and Rüschoff, J. (2019). *Current Opinion in Oncology* 31 (4): 354–361.

31 Qingqing, S., Yao, L.Ü., Heyu, L.I. et al. (2017). *New Chemical Materials* 45 (12): 176–179.

32 Yang, J.K., Wang, K., Yu, D.G., and Williams, G.R. (2020). *Materials Science & Engineering R: Reports* 111: 110805.

33 Pilehvar-Soltanahmadi, Y., Akbarzadeh, A., Moazzez-Lalaklo, N., and Zarghami, N. (2016). *Artificial Cells, Nanomedicine, and Biotechnology* 2020 (446): 1350–1364.

34 Augustine, R., Hasan, A., Patan, N.K. et al. (2020). *ACS Biomaterials Science & Engineering* 6: 58–70.

35 Tang, Y.D., Lan, X.Z., Liang, C.F. et al. (2019). *Carbohydrate Polymers* 219: 113–120.

36 Itxaso, G.O., Garazi, G., Borja, G.F. et al. (2017). *International Journal of Pharmaceutics* 523 (2): 556–566.

37 Schneider, A., Wang, X.Y., Kaplan, D.L. et al. (2009). *Acta Biomaterialia* 5 (7): 2570–2578.

38 Zhao, H., Dai, Z., He, T. et al. (2022). *e-Polymers* 22: 69–74.

39 Guindon, J., Deng, L.T., Fan, B.C. et al. (2014). *Biomaterials Science* 2 (4): 502.

40 Hu, J., Li, H.Y., Williams, G.R. et al. (2016). *Journal of Pharmaceutical Sciences - US* 105 (3): 1104–1112.

41 Elashnikov, R., Slepička, P., Rimpelova, S. et al. (2017). *Materials Science & Engineering C* 72: 293–300.

42 Hongliang, J., Yingqian, H., Yan, L. et al. (2005). *Journal of Controlled Release* 108 (2–3, 243): 237.

43 Saraf, A., Baggett, L.S., Raphael, R.M. et al. (2010). *Journal of Controlled Release* 143 (1): 95–103.

44 Guo, T., Yang, X., Deng, J. et al. (2018). *Journal of Materials Science. Materials in Medicine* 30 (1): 9–17.

45 Xue, C., Zhu, H., Tan, D. et al. (2018). *Journal of Tissue Engineering and Regenerative Medicine* 12 (2): e1143–e1153.

46 Zheng, X., Zhang, P., Zhenxiang, F. et al. (2021). *RSC Advances* 19041–19058.

47 Langer, R. and Vacanti, J. (1993). *Science* 260 (5110): 920–926.

48 Ercolani, E., Gaudio, C.D., and Bianco, A. (2015). *Journal of Tissue Engineering and Regenerative Medicine* 9 (8): 861–888.

49 Frerich, B. and Zuckmantel, K. (2006). Hemprich. *Head & Face Medicine* 2 (1): 26–34.

50 Huang, C., Wang, S., Qiu, L. et al. (2013). *ACS Applied Materials & Interfaces* 5 (6): 2220–2226.

51 Li, H., Williams, G.R., Wu, J. et al. (2017). *Materials Science and Engineering C* 79 (10): 245–254.

52 Domalik-Pyzik, P. and Morawska-Chochol, A. (2022). *Journal of Functional Biomaterials* 13 (11): 1.

53 Yang, F., Xu, C.Y., Kotaki, M. et al. (2004). *Journal of Biomaterials Science, Polymer Edition* 15 (12): 1483–1497.

54 Tian, L., Prabhakaran, M.P., Hu, J. et al. (2016). *Colloids and Surfaces. B, Biointerfaces* 145: 420–429.

55 Lundborg, G., Dahlin, L.B., Danielsen, N. et al. (1986). *Scandinavian Journal of Plastic and Reconstructive Surgery* 20 (3): 279–283.

56 Brushart, T.M. (1987). *Experimental Neurology* 97 (2): 289–300.

57 Anjum, S. and Rahman, F. (2022). *International Journal of Molecular Sciences* 23: 16.

58 Ye, K.Q., Liu, D.H., Kuang, H.Z. et al. (2019). *Journal of Colloid and Interface Science* 534: 625.

59 Wang, Y., Cui, W.G., Zhao, X. et al. (2018). *Nanoscale* 11 (1): 60.

60 Kajdič, S., Planinšek, O., Gašperlin, M. et al. (2019). *Journal of Drug Delivery Science and Technology* 51: 672–681.

61 Akhgari, A., Heshmati, Z., and Makhmalzadeh, B.S. (2013). *Advanced Pharmaceutical Bulletin* 3 (1): 85–90.

62 Aguilar, L.E., Ghavaminejad, A., Park, C.H. et al. (2017). *Nanomedicine and Nanotechnology* 13 (2): 527–538.

63 Singh, B., Kim, K., Park et al. (2022). *Nanomaterials* 11: 12.

64 Thakur, R.A., Florek, C.A., Kohn, J. et al. (2008). *International Journal of Pharmaceutics* 364 (1): 87–93.

65 Liao, N., Joshi, M.K., Tiwari, A.P. et al. (2016). *Journal of the Mechanical Behavior of Biomedical Materials* 60: 312–323.

66 Zhao, J. and Cui, W. (2014). *Nanoscience and Nanotechnology Letters* 6 (4): 339.

67 Han, D. and Steckl, A.J. (2017). *ACS Applied Materials & Interfaces* 9 (49): 42653–42660.

68 Son, Y.J., Kim, Y., Kim, W.J. et al. (2015). *Journal of Pharmaceutics* 104 (8): 2611–2618.

69 Kaassis, A.Y.A., Young, N., Sano, N. et al. (2015). *Journal of Materials Chemistry B.*

70 Lin, X.L., Tang, D.Y., Cui, W.W. et al. (2012). *Journal of Biomedical Materials Research Part A* 100A (7): 1839–1845.

71 Yu, Z., Gu, H., Liu, H. et al. (2015). *RSC Advances* 5 (80): 64944–64950.

72 Okuzaki, H., Kobayashi, K., and Yan, H. (2009). *Macromolecules* 42 (16): 5916–5918.

73 Aoyagi, T., Ebara, M., Sakai, K. et al. (2000). *Journal of Biomaterials Science. Polymer Edition* 11 (1): 101–110.

74 Kun, C., Ming, M., and Ning, G. (2003). *Chinese Journal of Medical Devices* 27 (4): 276–278.

75 Kim, Y.-J., Ebara, M., and Aoyagi, T. (2013). *Advanced Functional Materials* 23 (46): 5753–5576.

76 Ro, R., Baldi, G., Doumeltt, S. et al. (2018). *Materials for Biological Applications* 93: 206–217.

77 Samadzadeh, S., Babazadeh, M., Zarghami, N. et al. (2021). *Materials Science & Engineering C-Materials For Biological Applications* 118: 111384.

Index

Electrospinning: Fundamentals, Methods, and Applications, First Edition.
Edited by Liming Wang and Xiaohong Qin.